MANAGEMENT OF TECHNOLOGY AND OPERATIONS

MANAGEMENT
OF TECHNOLOGY
AND OPERATIONS

R. Ray Gehani, D. Eng., Ph.D.
The University of Akron

JOHN WILEY & SONS, INC.

New York · Chichester · Weinheim · Brisbane · Singapore · Toronto

To

my father for his lifelong pursuit of excellence with honesty
my mother for her selfless love
many teachers for their dedication and help
my children, Rashmi and Gautam, who have provided immense joy and
pride and through whose dreams I live mine
and my wife, Meena, who shared this journey with me

This book is printed on acid-free paper.

Published simultaneously in Canada.

This publication is designed to provide accurate and authoritative information in regard to the subject matter covered. It is sold with the understanding that the publisher is not engaged in rendering professional services. If professional advice or other expert assistance is required, the services of a competent professional person should be sought.

Library of Congress Cataloging-in-Publication Data:

Gehani, R. Ray.
    Management of technology and operations / by R. Ray Gehani.
      p. cm.
    Includes index.
    ISBN 0-471-17906-X (cloth)
    1. Technology—Management—Case studies.   2. Production
management—Case studies.   3. Industrial management—Case studies.
   I. Title.
    T49.5.G44   1998
    658.4—dc21   97-53199

Printed in the United States of America.

10 9 8 7 6 5 4 3 2 1

CONTENTS

INTEGRATION AND PIONEERING OF TECHNOLOGY

This book is an attempt to help the managers, engineers, and the students of technology management to learn about some of the principles and practices technology-driven organizations use. These organizations compete in an intensely competitive and dynamic marketplace. They do so by carefully managing their core and supportive technology-related competencies.

In this book the fast-moving technology-driven organizations are modeled as a high-performance yacht (or a race car) powered by a six-cylinder V-6 turbo-engine of technology. The six cylinders of the engine represent the major technology-related competencies. Managed in an interdependent synchronous manner, these competencies are the alternate sources of sustainable competitive advantage and produce the wealth for the corporation.

HOW IS THIS BOOK DIFFERENT?

Many managers, engineers, and students in the time-compressed work environment of the 1990s are very busy in their day-to-day rush to meet their immediate-term expectations. They can barely pay adequate attention to fighting the fires and managing the latest assignment that they must work on. They hardly have any slack time to sit back, reflect, and learn about all the latest expertise developed for a complex and multidisciplinary subject like the management of technology.

To help a busy learner comprehend the "big picture" of management of technology, the technology-driven organization is modeled here as a yacht. The yacht is powered by a V-6 turbo-engine with six cylinders pumping in unison. The six cylinders of the engine of technology are the six core and supportive technology-related competencies.

The turbo-engine of the yacht is connected to an efficient transmission system and a steering wheel. The transmission integrates and converts the power generated from each cylinder of competencies.

The steering wheel is guided by a pioneering visionary captain. The challenge for the captain of the yacht, or the manager of a technology-driven organization, is to steer the yacht through rough waters and stormy weathers to its destination of a yet-to-be claimed treasure island with big profits. The "captain" managing a technology-driven organization is sort of like Christopher Columbus on his journey to claim the Oriental spice islands that possess an abundant supply of silver and spices. The captain must get to the treasure island while competing with the pirates and captains of other ships trying to get to the same destination first.

BENEFITS TO FIVE GROUPS OF READERS

Based on my accumulated experience of teaching management of technology to a wide variety of students and other professional learners, a number of groups of learners will find this book interesting and useful. The benefits to each group of learners are outlined below.

1. *Working Engineers in Small, Medium, and Big Businesses.* Management of technology has increasingly become a cross-cultural, cross-functional effort, where engineers work side by side with the production managers and marketing supervisors. The professional from different functional areas and diverse training and educational backgrounds must work together from the same "music sheets." To help them manage enterprise-wide technology, this book will provide them a common understanding of the management of technology and operations.

2. *General Business Managers.* The busy decision makers who have to make decisions which affect or are affected by the management of technology wish to know how to deploy their technology-related resources more profitably. They often have neither the time nor the inclination to dig into intricate details. This book will give them some of the best practices of other technology-driven organizations.

3. *Upper-Level Undergraduate and Graduate Students of Science and Engineering.* An increasingly larger number of engineering and science (particularly computer and imaging science) students are taking courses related to management of technology and operations. On graduation, some of them aspire to start their own technology-driven businesses. Others will work for technology-driven organizations. As a chair and executive officer of the Technology Management Section (an international group of professors and professionals interested in management of technology, in the Institute for Operations Research and Management Science), I learned that many universities are offering management of technology programs jointly between their College of Engineering/Science and the College of Business. The students registered in such programs will learn the theoretical principles and the common practices followed by successful technology-driven organizations.

4. *Upper-Level Undergraduate and Graduate Business Students.* They will learn how to integrate principles of management of technology with what they learn in other functional management areas such as marketing, manufacturing, accounting, finance, production/operations, and information management. The material covered here will also facilitate a better understanding of the capstone strategic management course in the business program.

5. *Students Majoring in Management of Technology.* The students majoring in management of technology will find this book a good primer for their expert interest. After a good grasp of some of the basics discussed in this book, they will be well-prepared to tackle the details provided in other sources.

6. *As a Core Course for All Programs.* This book should be particularly useful to the students of a core course in an engineering/science or business program, for either undergraduate or graduate work. This book assumes very little prior knowledge, nor does it contain hard-to-understand quantitative models. Such models are introduced in a narrative form.

7. *A Primer for Foreign Students and Managers.* This book is particularly recommended to the foreign students. In my experience they are often lost because they do not have much background knowledge about America's historical evolution. Many of the milestones in America's growth as a world leader are described here.

DUAL GOALS: TO INFORM AND TO INSPIRE

Managing a complex technology-driven organization is a big challenge. The field of management of technology, as understood today, is inundated with many details. Even among the experts of management of technology there is little consensus about what should be covered and what should be left out in this area.

This book defines and analyzes the multidisciplinary—and in some ways still emerging—field of management of technology. This is done by six major technology-related value-adding competencies. Managing these competencies can provide a sustainable competitive advantage. These six strategic sources for sustainable competitive advantage are like the six cylinders of a V-6 engine that can be used to power a high-performance yacht (or a racing automobile).

The principles and practices of management of technology are presented in this book in an easy-to-understand manner appropriate for managers, engineers, and students with busy schedules. Mainly the major principles and practices of each of the "cylinders" of the technology "turbo-engine" are highlighted. There are many other sources available to gain advanced knowledge on these individual competencies. The reader should be able to capture the essence of managing a technology-driven organization and its operations.

Many real-world examples from technology-driven organizations are used to inform as well as to inspire. From my personal experiences in management of technology-driven organizations and their operations, the need to be inspired is as important as the need to be informed. Most of the time, managing a technology involves making risky decisions under uncertain conditions. The inspiration that can be gained from the successes and failures from the lives of the pioneers of technology is invaluable. The lessons thus learned should help immensely in our journey into wilderness of uncertain future, with a hope for "a pot of gold at the end of the rainbow."

Overview of the Chapters

As stated earlier, we will discuss the management of technology-driven organizations and their operations by discussing their technology-related competencies. For each competency, some theoretical models and principles of management of technology will be illustrated with real-life practices in leading technology-driven organizations. Wherever possible, brief profiles of the pioneers of different technologies, along with the cases of management trends in technology-driven organizations, will be provided to elaborate as well as enlighten the readers.

The stage will be set in Chapter 1, which is an introduction to technology, management of technology, and global competitiveness. These terms will be defined, and the basis of the V-6 engine model of technology will be clarified further. The various different forms of technology which we encounter in our day-to-day life will be discussed. A brief historical review of the role of contextual environment in the development and evolution of technology will help readers understand why certain technologies were deployed at certain times. A general discussion of global competition in technology-driven industries will be provided.

The Core-Transformational Competencies of the Technology Turbo-Engine

Chapters 2, 3, and 4 discuss the primary technology-related competencies in the core-transformational process of a technology. Their scope spans from a concept to the commercialization of products and services. For a technology-driven organization, the primary strategic sources for sustainable competitive advantage are derived from managing three primary competencies. These are: (1) production automation and engineering, (2) proprietary know-how and intellectual property based on research and development, and (3) new product development for customer loyalty in targeted market segments.

Production operations and automation engineering competency will be described and analyzed in Chapter 2. The early mechanization of the textile industry as well as the use of steam power, which started the Industrial Machine Revolution in England, will be reviewed. The learning curve advantages, as well as the economies of scale and scope, will be introduced. Other practices such as Just-In-Time manufacturing, Kanban inventory management, and Lean manufacturing will be discussed.

Chapter 3 discusses the creation of proprietary know-how and intellectual property by scientific research and systematic experimentation. We will review how in the 20th century the American corporations gained world leadership with the Scientific Industrial Revolution. Setting up of "invention factories" and research and development laboratories made America the home of electrical and chemical technologies. Some comments will be made on patenting and ownership of intellectual property. Controversial issues such as the piracy of patents and whether patents are a property or a privilege will be introduced. Management of research and development (R&D) by acquisition and strategic alliances with universities, collaborators, and competitors will be discussed.

Chapter 4 explores the critical role of product development and customer trust. We will also cover the role of positioning of products and technologies for customer satisfaction and long-term loyalty. A clear understanding of the new product development strategies, new product development process, and organizational designs for reducing the cycle time for new product development should help a manager of technology gain competitive advantage over the firm's competitors. Building up relationships with customers rather than bargaining transactions with them should add more to the bottom line. Listening to the voice of customers should help build trust and long-term relationships. Customer satisfaction, dissatisfaction, and delighting should be considered in the context of customer loyalty.

Supportive Competencies of Technology

The above-mentioned primary competencies for converting proprietary know-how or concepts into commercially successful products in the marketplace must be integrated and synergized with three supporting competencies of management of technology. These supportive sources, when deployed effectively and efficiently, can also contribute to gain a sustainable competitive advantage over competitors. These competencies are (1) promise of high quality and reliability, (2) processing of information and cross-functional communication for integration, and (3) the best people, high-performance teams, and human capital for innovation and creativity.

The promise of quality and reliability in products and production processes is the subject of Chapter 5. The shifting focus of total quality management (TQM) from products to processes and profits will be reviewed. The available assessment models and the quality standards, such as the Malcolm Baldrige National Quality Award and ISO 9000/ISO 14000, have played increasingly important roles in the management of technology, particularly in the way customers perceive such achievements of technology-driven organizations when making their purchasing decisions.

The processing of information for cross-functional communication is the subject of Chapter 6. Computer interface and integration is discussed in conjunction with the topics covered in earlier chapters. Advanced manufacturing technologies use a high intensity of information processing and computer technologies. Implications of both the Internet and Intranets will be spotlighted. This is a very fertile field with many fast-changing developments with significant impact on the competitiveness of a technology-driven organization.

Finally, the sixth cylinder of the V-6 engine for a fast-racing technology-driven enterprise involves management of people resource and human capital for innovation, creativity, and change, is discussed in Chapter 7. We will learn from the historical developments in human resource management practices for management of technology and operations. The most recent tides of massive downsizing, as well as the implications of other related issues, need to be considered too.

Integration and Pioneering of Technology

The core and supportive competencies of management of technology must be integrated together and guided carefully, just like the efficient transmission and steering of a yacht. These higher-level competencies are to be managed carefully if the yacht is expected to get to the treasure island before other pirates or competitors do so.

Project management and cross-functional integration is covered in Chapter 8. For success, the six competencies of the technology turbo-engine must be managed to move in a smooth coordinated manner. Project management does that coordination across different competencies for the one-time initiatives, which have clearly defined starting and finishing points. On the other hand, similar ongoing cross-functional integration is needed for the continuous process.

Chapter 9 discusses the role of pioneering vision and leadership in developing sustainable competitive advantages for technology-driven organizations. Analytical tools such as a leader's preferred sources of power to influence others will be briefly discussed. A comparison will be made between the transactional leaders driven by reducing the expense lines and the transformational leaders driven by the top line. In fast-changing technology-driven enterprises, a value-based vision provides a dynamic guidance system.

The projected future of technology and success in society is explored by way of conclusion in Chapter 10. This chapter explores issues such as the macroscopic roles, rights, and responsibilities of different technologies and their management.

ACKNOWLEDGMENTS

I would like to thank many of my fellow journeymen and women trying to decode the distinct patterns for the effective management of a technology-driven enterprise. As the Chairman of the Technology Management Section of The Institute for Operations Research and Management Science (INFORMS), a professional association of about 400 academics and executives involved with management of technology, I learned a lot from my personal exchanges with a wide range of fellow professionals interested in this area. I will particularly like to thank Dave Gibson, Raymond Smilor, and

George Kozmetsky of the University of Texas at Austin; Michael Badawy of the Virginia Polytechnic Institute and State University; Charles Hofer of Georgia State University; Bill Riggs of Georgia Institute of Technology; Dave Tansik of the University of Arizona; Bob Mason and Arnold Reisman of the Case Western Reserve University; Dundar Kocaoglu of Portland State University; R. Balachandra of Boston University; Frederick Betz of NSF; Tarek Khalil of the University of Miami; Glenn Dietrich of the University of Texas at San Antonio; Jeff Liker and John Ettlie of the University of Michigan; Fariborz Damanpour of Rutgers University; Jamaluddin Hussein of Purdue University; William Souder of the University of Alabama; Eli Geisler of University of Wisconsin, and many more.

I have learned a lot from the presentations and papers of a large number of fellow researchers in this field too numerous to list. I have enjoyed learning from the writings of Robert Burgleman of Stanford; David Teece, Richard Nelson, Laura Tyson, and many others at the University of California at Berkeley; Kim B. Clark, Michael Porter, Rosabeth Moss Kanter, John Kenneth Galbraith, and others at Harvard University; Michael Cusumano, Lester Thurow, and others at MIT; C. K. Prahalad of the University of Michigan; Charles Hill at Washington University; Michael Tushman and Kathy Harrigan at Columbia University; and Ed Mansfield at Wharton, to name a few.

I have benefited from my discussions with Dr. Edwards Deming during his visits to Columbia University, with Joseph Juran for a joint executive program I co-organized with the Juran Institute, and with Kaoru Ishikawa for a presentation at a conference in Tokyo sponsored by the Association for Overseas Technological Scholarship of the Ministry of International Trade and Industries of Japan.

At Akron University Dean Stephan Hallam, Dean Frank Kelley, Dean James Strong, Dean James Inman, Professors Ken Dunning, Gary Meek, John Hebert, Jay Patankar, Ken Aupperle, Bruce Simmons, David Myers, Bob Figler, Susan Hanlon, Alan Krigline, and others have provided me a constant source of intellectual stimulation in our collective search for new ways to organize and present knowledge to our students.

In my previous professional life as a manager in the technology-driven enterprises in industrialized and emerging economies, I learned a lot from many caring mentors. Noteworthy among these mentors were Kazunaga Murayama, Michihiko Tanaka, and Shunro Kataoka, who helped me gauge the challenges of developing sophisticated new technologies in Japan for the global markets. Chief executive officers like Sam Gibara of Goodyear, George Fisher of Eastman Kodak, Andy Grove of Intel, David Kearns of Xerox, and others like Arun Bharat Ram, Raunaq Singh, B. B. Mathur, and Manhar Bhagat, from different enterprises, taught me the art of balancing the diverse economic, political, social, and technological forces in running a technology-driven enterprise in the global economy.

I admit that I could not have come to where I am professionally without the help of many of my dedicated teachers, too many to list here. Professor S. Prakash Sethi has been a beacon of inspiration and guidance for me during the past many years. I am thankful to Professor Harris Jack Shapiro for his affectionate nurturing when I needed it most. I will also like to thank Professors Michael Chanin, Donald Vredenburgh, George Sphicas, William McCutchen, T. K. Das, R. Parthasarthy, T. S. Srinivasan, Ramesh Mehta, J. D. Singh, Ichitaro Uematsu, Akihro Abe, Toru Kawaii, Junji Watanabe, Norimasa Okuii, V. B. Gupta, P. Bajaj, D. S. Varma, A. P. Kudchadkar, C. V. Sheshadri, and many more teachers and well-wishers who constantly urged me to give my best.

Beside these mentors, exchanges with many friends, classmates, and acquaintances working worldwide in technology-driven enterprises have helped me mold my ideas about technology, management of technology, and ways to gain a sustained global competitiveness. Friends and colleagues like Julian and B.J. Yudelson, Dean Siewers, Paul Karen, Tom Comte, Paul Bernstein, Richard Rosett, Jeff Lassard, Don Zrebiec, Bill Wiggenhorn, Andrew DuBrin, Janet Barnard, Guy Johnson, Eugene Fram, Bob Pearce, Gary Bonvillian, Michael Prosser, Edward Schilling, Masatoshi Sugiyama, James Chung, Takeshi Utsumi, Paul Petersen, Hideo Hayashi, Ivan Abel, Ken and Patricia Ehrensal, Sharon Badenhop, Ron Hilton, Kumiko and Osamu Shima, Shinya Watanabe, Beat and Franzi Schwarzenbach, and others helped me during the ups and downs of a multidiscipline professional life.

I would also like to thank the team of editors at John Wiley & Sons, particularly Bob Argentieri, Akemi Takada, Bob Hilbert, and many others for copyediting, proofreading, and fine-tuning my ideas and the manuscript.

I salute you all for what I learned from you, and am now able to share with more readers through this book. Thank you very much. I hope that the readers will carry this torch of knowledge further. If you have any suggestions for making improvements, please send them to me.

CHAPTER 1

Introduction to Technology, Management of Technology and Global Competitiveness

PREVIEW: WHAT IS IN THIS CHAPTER FOR YOU?

There is little consensus among professionals and experts about what technology stands for. We will review the different hypothesis professionals have proposed, and we will learn from the experts' opinions. Then we will develop a working understanding of technology: "8P" definition of technology. This is based on the different forms of technology we encounter in our daily lives. We will use this understanding to develop a V-6 engine model of management of technology and operations. Finally, we will use this V-6 engine model to analyze implications for global competitiveness of technology-driven American enterprises.

DEFINING TECHNOLOGY

What Is Technology?: Some Hypotheses

To understand technology and how to manage technology, we could begin by asking some simple questions: What is technology? Or, what images do we see when we hear this term? What can we learn about managing technology from experiences of other technology-driven enterprises? What can technology do for us? Can it hurt us? These simple questions generate a very wide spectrum of responses in managers, engineers, and scientists interested in this subject.

Given below are some of the hypotheses professionals suggest when they are asked what comes to their minds first when they hear the word *technology*. For discussions' sake, let us also consider the counterarguments for each hypothesis.

Hypothesis 1: "Technology is something complex."

Some people feel that technology is something that is complex and difficult to fully understand—like NASA's space shuttle or the Hubble telescope, Internet, virtual reality, and so on. Others counterargue that if there is a complex contraption to do something obvious—like freezing a cube of water, making dough rise, or tying a simple knot—then we do not value that as technology. We quickly recognize that complexity alone does not fully define technology.

Hypothesis 2: "Technology is the state of the art."

Then somebody else could hypothesize that "technology is something which is of the most current state of the art." If one truly believes in the unlimited regenerative power of the human mind demonstrated over the past few centuries, the term "state of the art" would have a very short shelf-life. Consider making beach sandals using state of the art Intel MMX computer chips to flash light as one walks. Most people would not think of such beach sandals as technology. The discussion on this hypothesis starts dwindling down too.

Hypothesis 3: "Technology is something which produces sophisticated high-tech products."

We start getting elaborate and careful. Somebody adds that "technology is a process that produces high technology products." Really? Let us consider 3M's Post-it pads. These were developed as reusable bookmarks for the Bible, but are also handy in jotting down short memos and new ideas on the go. To many users of Post-it pads, they do not seem like a high-tech product. But actually, Post-it paper is produced by using a very sophisticated thin-layer film coating technology to coat uniformly and apply just enough glue so that it will stick, but not so much that the paper will not peel off easily. The development and production of simple-looking Post-it pad papers required a very sophisticated technology. This hypothesis gets into rough waters.

Hypothesis 4: "Technology is unknown."

It seems like a no-win situation. Somebody else covers the risk by adding that "technology is something I don't know." Well, what about Coca-Cola? Very few people in the world know, or are ever likely to know, the recipe of one of the world's most popular drinks, namely, Coca-Cola. The family members of Coca-Cola's owners keep the secret of "the Real Thing" very carefully. Yet, very few people would consider this soft drink a memorable technology. We start losing supporters for this hypothesis too.

By now most people in the discussion are intrigued by the "right" definition of technology, and they start thinking real hard. They are getting tired of the counterarguments. They want to come up with one hypothesis that will prevail.

Hypothesis 5: "Technology is a high-tech process producing high-tech products."

As a last resort, a persistent player suggests that "technology depends on a high-tech process producing high-tech products (versus low-tech products produced by a low-tech process)." We need examples to understand the hypothesis clearly. Somebody quickly mentions computers.

Are computers the symbols of high technology that they are often made out to be? It is not that simple. First of all, in articles or reports on technology we don't often see the terms "high-tech process/high-tech product" industry or "high-tech process/low-tech product" industry. The commonly used terms are either "high-tech" or "low-tech" industries. Secondly, in the computer industry in the Silicon Valley, much of the assembling and making of the electronic circuit boards for the computers is done by some low-tech chemical scrubbing and manual cleaning processes. These could be easily done by using some of the minimum-wage low-tech workers. In other words, not all the process operations in a "high-technology" computer industry are sophisticated and high-tech. Designing a chip may be just one such process. We are still not satisfied by our understanding and definition of technology.

There are two other hypotheses which commonly appear. These are:

Hypothesis 6: "Technology empowers and creates."

To some professionals. technology extends our limited biological powers. The medical, pharmaceutical, and biogenetic technologies have helped extend our lives. And most people would recognize that. Technologies such as artificial insemination have helped families conceive babies when their biological fates would have forced them to reconcile otherwise.

On the other side of the technology coin, people point out technology's ugly underbelly. They suggest:

Hypothesis 7: "Technology is dangerous and a destroyer."

One could rightfully argue that technology has extended the demonic power of mankind and made it easy to destroy thousands of fellow humans, along with everything else around them. Hitler's Holocaust with chemical gas technology is one such example. The technologies that terrorists use in their weapons, namely explosives, are other examples. What about the military uses of nuclear missile technology?

If the oil tanker that transports gasoline for our cars and gas for our stoves hits a rock, destroys overnight the pristine environment, thus affecting generations of sea otters and other marine life. How about the CFC technology that is allegedly making a hole in our ozone layer? Some observers believe that this damage cannot be fully repaired by a few billion dollars donated by a business corporation as compensation.

The above discussion illustrates how complicated the subject of defining technology is. Even the experts cannot agree.

Experts' View of Defining Technology

Even for experts it is hard to define technology—that is, what to include in that definition and what to leave out. Are ideas part of technology, or only the domain of sciences? Do we include people into our definition of technology? And so on.

This author has spent many sessions in national and international conferences where professors and professionals debated these questions in one session after another, year after year. So we will not try to do the same here.

We will instead develop a working understanding of the term *technology*. Later we will consider how we encounter technology in our day-to-day life. This will help us identify the different forms of technology and how to manage technology effectively.

But, out of curiosity we should know some of the arguments experts have made about technology. This will give us a better flavor of the challenge involved in defining technology, to make sure that we have not left out anything and so that we have a decent grasp of the entire field.

Historically, Francis Bacon was one of the important thinkers of the 17th century who brought science and technology closer together. Prior to him, technology was often the domain of the artisans, who built machines based on empirical experience. And science dealt with ideas only. Bacon asserted that scientific ideas could also be practically useful to the mankind. In his book *New Atlantis,* he described a utopian island society exploiting the benefits of science and technology. His society lived in buildings as tall as about half a mile in height, and used pipes and trunks to convey sound over long distances.

Peter Drucker, the most respected guru of business observations and its cogent articulation over a long period of time, pointed out that Alfred Wallace, the co-discoverer of Darwin's theory of evolution, found man unique among all other species because of his ability to make tools (Note 1.1). The "Wallace Insight," according to Drucker, changed the age-old definition of technology from "how things are done or made" to "how man does or makes." This led to Drucker's conclusion that "technology is not about things. . . . It is about (human) work." He also proposed that technology must be considered as a system of "interrelated and intercommunicating units and activities."

Over the years, the boundaries between scientific research and development of a technology have overlapped and merged. Now we cannot do one without the other. Whatever scientists do or discover are continually monitored, used, or adapted by technologists. Also, scientists keep an eye on (a) the emergence of new technologies and (b) the need for the new concepts by the technologists.

Consultant Lowell Steele made an elaborate point about how many of the popularly known definitions of technology fail to appeal to a wider group of people, including the senior executives. He offered a more general definition of technology as "the system by which a society satisfies its needs and desires." (Note 1.2). He admitted that "the definition is pretty sweeping." Professor Michael Badawy of Virginia Polytechnic Institute and State University, in his book entitled *Management as a New Technology,* also acknowledged that "technology is an elusive concept to define" (Note 1.3). He then defined management technology as "the architecture or configuration of management systems, policies, and procedures governing the strategic and operational functioning of the enterprise in order to achieve its goals and objectives. . . a form of social technology."

Professor Albert Rubenstein, who trained many future professors of management of technology at Northwestern University, stated that technology "covers a wide range of activities and functions in the firm which are devoted to producing new and improved products and processes, materials, and know-how for the firm" (Note 1.4). Robert Burgelman of Stanford University, Modesto Maidique of Florida International University, and Steven Wheelwright of Harvard University have proposed a similar definition of technology too. To them, technology "refers to the theoretical and practical knowledge, skills, and artifacts that can be used to develop products and services, as well as their production and delivery systems." They also pointed out that "Technology can be embodied in people, materials, cognitive and physical processes, plant, equipment, and tools." (Note 1.5).

We can list many more definitions offered by other experts. The only consensus we have is that technology is hard to define.

DEFINING BY ENCOUNTERS WITH TECHNOLOGY

Another way to develop our definition of technology is to use the different forms of technology with which we interact on a regular basis. There are at least eight forms of technology we can easily recognize. To remember them easily, we will refer them as the "8Ps" of technology (like the 4Ps of marketing).

Defining "8Ps" of Technology

P1: Technology Is Products

Technology can be defined by the products it is embedded in. These are the physical, useful products and appliances which enhance quality of our life. They include cars, computers, cellular telephones, TVs, VCRs, CD players, and so on.

P2: Technology Is a Production Process or an Operation

Another way to look at technology is as a set of operations involved in a production process. Examples include refining crude oil, fermenting beer, cooking, painting, and so on. At one end of a production process, raw materials of lower value are fed in. At the other end finished goods of higher value come out.

P3: Technology Is Proprietary Intellectual Know-How

Technology can also be defined as a body of knowledge. The recipe for Coca-Cola or for a pharmaceutical formulation is nothing but a set of weights, temperatures, and other conditions which produce something that gives pleasure in the case of the former and relieves pain in the case of the latter. Technology also comes in the form of intellectual property or legally protected proprietary knowledge in a patent, trademark, or copyright. These give a person the exclusive rights to commercially exploit that knowledge for a prespecified period of time.

Thus, if Polaroid has received the rights to exploit instant photography technology, then Eastman Kodak cannot use the same technology during Polaroid's patented period, unless Kodak seeks permission and pays an adequate royalty to Polaroid and vice versa.

P4: Technology Is Processing of Information

In the Information Processing Age, many people equate technology to the computer platforms and other information processing appliances and services. To them technology is nothing but a set of semiconductor chips and intricate databases, or the Internet that allows surfing in Cyberspace. These technologies help one to browse from a web page in Boston to another web page in Beijing and a third one in Bangalore.

It is amazing how many common people and some journalists use the term "technology" in these situations without the suffix "computer" or "information." By doing so, they restrict the meaning of the term "technology." One wonders if the technology used in a computer has a greater right to be called a technology than the earlier modes of technology we used for processing information—for example, a ball point pen, a pencil, or a solar calculator.

P5: Technology Is a Promise (of Expected Quality and Guaranteed Reliability)

Technology may also be defined as a promise—that is, the promise from the producer of goods and services to the customers buying those goods and services. The customers bought the goods with the assumption that the price they paid for the products or services included the high quality of goods produced by reliable processes.

The technology is like the Timex watch promise that it will keep on ticking no matter how bad the beating. The overnight service technology is the Federal Express (FedEx) promise that no matter how many planes are grounded for repairs, the FedEx package will positively, definitely be delivered the next day. Another promise is that the air bags and seat belts in automobiles will save lives rather than trap people, and women are promised that the silicone breast implants are to add quality of life rather than make life more miserable.

P6: Technology Is People (and Their Skills)

Another way we encounter technology is when we see technology as a person with a set of skills and abilities to do certain things. A programmer puts together a few hundred thousand lines of codes of instruction in a certain order so that we can conveniently word-process our ideas. A hairdresser moves her hand a few times, with a pair of scissors and a comb, and our head looks a lot better than before. A fashion designer takes a few rags of fabrics, in different color combinations, and creates an exotic ensemble for an expensive "look." A medical doctor embodies the training to prescribe a medicine when we explain to her which part of our body hurts.

P7: Technology Is a Project

Sometimes we encounter technology as a project. For example, the technology of the Brooklyn Bridge involved a one-time multiyear project to link the island of Manhattan to the land mass of Brooklyn. Projects include: President John F. Kennedy's commitment to send a man to the moon before the end of the 1960s and the project to build the Apollo Lunar module, which made the lunar landing possible; or the Erie Canal project, called Dewey's Ditch, which helped America's westward expansion by connecting the Midwest and the Great Lakes to the Atlantic Ocean; or the technology required in designing and constructing a petrochemical complex that would help refine petroleum oil and produce plastics, rubbers, and synthetic fibers.

P8: Technology Is Pioneering for Profits

Finally, technology is also a pioneer's ability to perceive how certain known things can be put together in an entrepreneurial new way, so that profits are generated to compensate for the long years of the persistent efforts.

Fred Smith, the founder of Federal Express, used a few airplanes, a few vans, and some conveyor belts—very common things available to all of us. But he then arranged them in a hub-and-spoke arrangement to collect packages and deliver them across the nation for a "definitely, positively, next day guaranteed" delivery. After a few initial teething difficulties, the pioneer of "overnight delivery service" reaped rich profits for his efforts and resources invested. Bill Gates did the same with a few lines of programming codes developed by himself or acquired from somebody else.

Technology to some comes in the form of the pioneer that takes the risks to develop and commercialize a new technology to the market, thereby developing the potential to earn profits. George Eastman, while working as a bank clerk for his day job, tinkered with the coating of chemicals on flexible films. Finally he loaded a photographic film in a Kodak camera to help common people enjoy the pleasures of photography—in exchange for $25 they got 100 "Kodak Moments."

So, we have the eight Ps of technology, from the proprietary intellectual knowledge to the pioneering entrepreneur who puts it all together for a profit.

Our Working Definition of Technology

For reasons illustrated earlier, we were looking for a working definition of technology. Based on our own encounters with technology and what the experts have suggested, we will define technology simply as "a transformational system, from proprietary know-how to the commercialization of products. This includes production operations, promised quality, people, processing of information, integrated together as projects and a pioneering vision."

This definition takes care of the Drucker criticism of the classical definition of technology focusing only on the making of technology. It also accommodates the challenges cited by Steele and Badawy. It is hoped that the 8-P definition of technology is easy to remember and handy as a working definition for all involved in technological endeavors.

MANAGEMENT OF TECHNOLOGY

In the past, researchers often defined technology primarily by the research and development (R&D) efforts, and therefore they defined management of technology as the management of R&D. The underlying assumption was that the generation of proprietary knowledge is the most important and basic goal of managing a technology (Note 1.6).

The National Science Foundation therefore defined technology-intensive industries (and organizations) as those industries which devoted more than 3% of their sales revenue on their research and development. This included the computer, pharmaceutical, imaging and office equipment, and scientific instrument industries.

In 1988, the U.S. industries spent $59.4 billion on their corporate R&D. In 5 years, this increased to $83 billion in 1993, with an average increase of 6% per year over the 5-year period. The annual increase over 1992 was a lower 4.2%.

This was great news for the scientists and engineers who worked in the research and development laboratories. The underlying assumption was that a higher allocation of resources would lead to more proprietary new technology and higher revenues. But this operationalization of technology by R&D alone did not make much sense to others, particularly the senior executives that consultant Steele referred to earlier. Further, for multiple industries, macroeconomic studies of correlations between allocation of resources to R&D and the competitive performance showed relatively low correlations (between 0.20 and 0.50) (Note 1.7). Those who asserted that management of technology should be defined by a firm's R&D intensity (ratio of R&D$ to total sales$) had a hard time explaining such weak correlation.

The weak effect could be partly explained by the averaging taking place across different industries. There is now enough evidence that an intensive allocation of resources to R&D alone does not always translate to improvement in competitive performance or profits of an organization. Many Japanese corporations, which devoted less resources to R&D but allocated resources much more intensely to new product development and capital investments for automation and expansion of capacity, earned significantly higher returns in the short-term. An extended definition of management of technology seemed necessary. Therefore it was postulated that the competitive performance of a firm depends not only on the allocation of resources to its R&D capability, but also on its commitment of resources to other complementary capabilities in the firm's total technology value chain (Note 1.8).

According to Dr. John Armstrong, the head of IBM's research and development center, "The traditional academic view is that first you do basic research, then applied research, then product development, and then give it to manufacturing. But . . . the Japanese had been so effective at rapidly translating [their own and others'] research into successful commercial products that IBM had been forced to try to [match their practices to] generate new products more quickly" (Note 1.9).

This observation by a manager of a leading technology-driven firm indicates that the management of technology does not always begin and end with the R&D capabilities of a firm.

Sony's Sponsorship of Transistor Technology

In the case of Sony, a leading Japanese enterprise driven by technology, the management of technology sometimes started with allocating resources to the new product development capability. Sony often licensed the needed proprietary knowledge from outside developers of know-how. (See Case 1A at the end of this chapter, describing the sponsorship of silicon transistor technology from the Bell Telephone Laboratories.)

Resources Required for Managing Nylon Technology

In another case study of E. I. Du Pont's development of nylon technology (described as Case 3A in Chapter 3), Wallace H. Carothers discovered the fiber formation of nylon polymer in a hypodermic syringe after spending about $50,000. It took another four years and $1 million of R&D to synthesize nylon polymer with predictable and stable properties. Then Carothers handed over the nylon technology project to Du Pont's development department, where an additional $44 million worth of resources were invested.

This involved setting up a pilot plant to test the alternate processing conditions for nylon's large-scale production process. Many more people were allocated to the development of production process for nylon technology. Thereafter, expensive plant and equipment had to be fabricated, installed, and commissioned. By then, Du Pont's allocation of resources to the development and production stages in the nylon technology "dwarfed the expenditures on R&D" (Note 1.10). Later, Du Pont allocated more resources to new product development, and education of customers for adopting the nylon products to meet their needs. With the outstanding market success, more resources were allocated by Du Pont to increase the production capacity and make the production process more productive.

Over the years, the focus of management of technology shifted from research in the early years to product development, process design, and new market development in the latter years.

Xerox's Alto Computer Technology

The case of Xerox Corporation's management of Alto desk-top computer technology is quite the opposite. (See the Xerox's Palo Alto Research Center section on page 92.) It is an example of unrealized returns from resources allocated to R&D at Palo Alto Research Center (PARC).

During the 1970s and the early 1980s, Xerox Corporation invested substantial resources to its basic research for developing office computer technology at PARC. After some breakthroughs, the researchers at PARC proposed Xerox's higher management in Rochester, New York, to pioneer the personal computer technology under the office automation mission. They had successfully developed an experimental computer language, a user-friendly computer interface, and basic computer-to-computer communication technologies. But Xerox could not successfully transform its R&D discoveries into commercially marketable products. On the other hand, other companies like Apple Computer, Adobe Systems, and Grid Computer built sizable businesses based on the technological inventions funded by Xerox. These other firms acquired Xerox's technological research inventions (say, by hiring Xerox employees) and coupled them with their own complementary technology-related capabilities. Thus they were able to manage office computer technology successfully, and earned significant profits.

This shows that the effective management of technology must include management of R&D capabilities for creation of proprietary know-how and couple them with an equally significant management of other complementary capabilities (Note 1.11).

MULTICOMPETENCY MANAGEMENT OF TECHNOLOGY

In today's globalized markets, technology provides the competitive edge for organizations' growth and survival. However, management of technology requires allocating resources to complementary value-adding competencies as well. This is almost as crucial as the creation or development of new technology by scientific research and systematic development. Business practitioners and students need to understand how technological issues related to innovation, efficient manufacturing, responsive new product development, skilled human resource, and so on, are linked to the organization's business strategy for gaining sustained advantage over competitors.

To decide how to manage technology, we will use the "8-P" definition of technology.

The V-6 Engine Model of Management of Technology

We will divide the management of technology into three sections. Each section covers a subsystem of the overall technology management system for a technology-driven enterprise. The first two subsystems with six competencies correspond to the six cylinders of a turbo-charged V-6 engine for a yacht sailing in the ocean. The goal is to fight the pirates [read *competitors*] and the stormy weather [read *economy*] to get to an unclaimed "treasure island" of profits. The third subsection of management of technology deals with cross-functional integration and pioneering of the technology. These competencies represent the transmission and the steering of the technology-driven yacht. The three subsections of management of technology will be explained next.

A. Core-Transformation Subsystem. This technology subsystem includes the management of primary competencies in the core-transformational chain of a technology system. Here the transformational process for a technology-driven enterprise converts proprietary intellectual property to production operations and new product development and customer trust. It consists of three primary competencies. These competencies will be discussed in a different sequence. The sequence is based on the historical evolution of industrial technology since the Industrial Revolution. The core-transformational competencies are:

A1. Management of *production* operations and engineering automation.
A2. Management of *proprietary* know-how and intellectual property.

A3. Management of *product* development, customer trust and market diffusion.

B. Supporting/Resource Related Subsystem. This second subsystem deals with the management of three key supporting resources required in the management of technology. The three resource-related competencies are listed below.

B1. Management of *promised quality* of products and services, along with management of production reliability.
B2. Management of *processing* of information and communication in technology.
B3. Management of *people* of technology including teams, culture of change, and innovation.

Management of financial resources for technology is covered indirectly in a later chapter on pioneering. Finance and accounting competencies are subelements of the larger business system, but they involve many decisions not directly related to the management of technology. For example, these competencies deal with interest rates, tax accruals, time value of money, and so on.

C. Integration and Visioning Subsystem. The third subsystem deals with the management of integration of core-transformational competencies and the resource-related supportive competencies. There are two competencies in this subsystem:

C1. Management of *project* technology and integration.
C2. Management of *pioneering* and visioning of leadership for profits.

Finally, we will conclude our discussion with a chapter on the management of *projected* future of technology in society.

Management of Competencies: The Cylinders of V-6 Engine of Technology

For management of each competency, some historical evolution, a few key theoretical models, and principles of management of technology will be provided along with the real-life practices in leading technology-driven organizations. To inspire and learn from the lives of pioneers of technology, their brief profiles will be provided, where appropriate as cases.

This first chapter sets the stage with the definitions of technology and management of technology, along with the various different forms of technology in which we encounter technology in our day-to-day life. A general discussion of global competition in technology-driven industries will be provided later.

The Core-Transformational Competencies of Technology Engine

Chapters 2–4 discuss the primary technology-related competencies in the core-transformational process of a technology. These cover the value-adding chain, from a proprietary know-how and intellectual property in a concept to the commercialization of a product or a service. For a technology-driven enterprise, the primary transformational subsystem of technology covers three complementary competencies. These will be discussed first.

Management of *production operations and engineering automation* will be discussed in Chapter 2. A historical evolution of production operations, from the ancient civilization through the Industrial (Machine) Revolution and onwards, will be reviewed. Why did the mechanization of textile operations start in England in the late 18th century? Why did the machines need to be powered with steam power? Why did the Industrial Revolution start in England and not anywhere else?

The economies of scale and the management of learning curve will be discussed. Other production management practices such as Just-In-Time manufacturing, Kanban inventory management, and Lean manufacturing made an enormous impact on the management of production operations and need to be understood clearly.

The management of proprietary know-how and intellectual property includes scientific research and systematic experimentation. Historically, this was the next phase in the evolution spiral of technology. The Scientific Technological Revolution shifted the center of gravity of the world's economy from Europe to the United States of America. We will review how the freedom and pursuit of happiness in America motivated many pioneers to give birth to new technologies. Soon new industries grew around these. Thomas Edison's "invention factory" and research and development laboratories inspired many other corporations to start their own research and development laboratories.

The role of patents and other modes of ownership of intellectual property, such as trademarks and copyrights, can help readers learn how to protect their know-how. Controversial issues such as the piracy of patents and whether patents are a property or a privilege will be introduced. Management of research and development by acquisition and strategic alliances with universities, collaborators, and competitors are challenging issues in managing this core-transformational competency.

In Chapter 4 we will explore the critical role of *product development and customer trust.* We will cover the role of positioning of products and technologies for *customer satisfaction, trust, and long-term loyalty.* A clear understanding of the new product development strategies, new product development process, and organizational designs for reducing the cycle time for new product development should help a manager of technology gain competitive advantage over the firm's competitors. Building up relationships with customers rather than bargaining transactions with them should add more to the bottom line. Listening to the voice of customers should help build trust and long-term relationships. Customer satisfaction, dissatisfaction, and delighting should be considered in the context of customer loyalty.

Supportive Resource Subsystem of Technology

The core-transformational subsystem of competencies for converting proprietary know-how into commercially successful products in the marketplace requires resources. In the management of a technology system, the supporting competencies are related to the management of promised quality, processing of information for communication, and human resource.

These are the other three cylinders of the V-6 turbo-engine of technology. The first three core cylinders of the V-6 engine must be tuned with the supporting three cylinders.

Chapter 5 will cover the management of *promised quality and reliability of technology*. This includes (a) the quality of goods and services delivered to customers and (b) the reliable production processes used to produce the same. Nine different lives of Total Quality Management (TQM) and their integration over time, the Malcolm Baldrige National Quality Award, and ISO 9000/ISO 14,000 are some of the important topics for this competency.

The management of *processing of information for communication and decision-making* will be the subject of Chapter 6. The goal is to improve the quality of information, and fact-based decision-making. Use of computer-based advanced manufacturing technologies have improved the overall productivity of the management of technology. New communication technologies, such as the Internet and Intranets, are changing the technology-driven enterprise dramatically.

The sixth cylinder of the V-6 engine for a fast racing technology-driven enterprise involves *people building for innovation, creativity, and change*. This will be discussed in Chapter 7. From the historical developments in human resource management, we will also learn practices for management of technology and operations. The most recent trends of massive downsizing, as well as implications of other related issues, will be considered too.

Integration and Pioneering of Technology

The primary and secondary cylinders of the V-6 turbo-engine of technology must be tuned together and guided carefully, just like the efficient transmission and steering of a race car or a high-performance yacht. These higher-level competencies must be managed carefully if the yacht is expected to get to the "treasure island" before other pirates or competitors do so. The project management is like the transmission system, and a pioneer's visioning corresponds to the steering done by the captain of the yacht.

The *project management and cross-functional integration* will be covered in Chapter 8. For competitive success, the six competencies of technology engine must be managed to move in a coordinated manner. Project management does that coordination across different competencies for the one-time initiatives, with clearly defined starting and finishing points. The ongoing processes also require continuous cross-functional integration.

Chapter 9 discusses the role of *pioneering transformational leadership* in developing sustainable competitive advantages for technology-driven enterprises. Analytical tools such as a leader's preferred sources of power to influence others will be briefly discussed. A comparison will be made between the *transactional leaders* focused on reducing the expense lines and the *transformational leaders* excited by improving the top line.

The management of *projected future success of technology in society* will be explored in Chapter 10. A brief historical review of the role of contextual environment on the evolution spiral of technology will help us understand why certain technologies were deployed at certain times. This chapter will explore issues such as the macroscopic roles, rights, and responsibilities of different technologies in the society. Do people regard technology as their friend or as their foe? A force-field analysis will help.

TECHNOLOGY MANAGEMENT AND GLOBAL COMPETITIVENESS

America's global competitiveness in technology-driven industries can be understood by reviewing how the six "cylinders" of the V-6 turbo-engine of technology are run. Some observers believe that American industries seem to be losing their technological leadership in the world. Electronic appliance technology, memory chips technology, and flat-panel display technologies are given as examples. We hope that we won't be too late in fixing the V-6 engine of our technology-driven enterprise and still get to the "treasure island" of profits and prosperity.

Investor Myopia in New Technology

Some "quarter-oriented" managers in the U.S. industries oppose any significant investments in R&D because they will not bring large returns immediately. They are afraid of the lead times involved.

Managers say their hands are tied. If they do not produce quick profits sometime soon, they lose their investors. The truth of the matter is that they also lose their own bonuses and benefits, which are now increasingly tied to stock-market performance.

In a timely new book, *Short-Term America,* author Michael T. Jacobs blamed the entire economic system for the impatient capital in American industries. For the first time, the complete problem was studied as an integrated system. A number of solutions were proposed which have significant implications for the management of technology. The financing of capital by debt and equity has an influence on the sponsoring of new technology. So does the adversarial relationship between the investors and managers. It boils down to trust. When investors can't trust managers, they demand higher returns to minimize their risks, which they want in as short a period of time as possible.

Low Capital Reinvestment

During the 1990s, in the G7 group of the largest economies of the world the American companies invested the least rate of capital reinvestment for its growth and technological advancements even though the American economy was the largest in the world. Production technology is a key source of sustainable competitive advantage for technology-driven enterprises. It must be nurtured with adequate funding and capital allocation. It is hoped that such a trend will be reversed quickly as America's policy makers note a growing gap in its technological advancements compared to some of its global competitors.

Delightful Quality Promise

During the 1980s, America's appliance and auto manufacturers realized how their customers could change their preferences and quality expectations dramatically. They were caught unprepared. Many foreign manufacturers of these goods made deep inroads into the U.S. markets by fulfilling their promises for high-quality goods and services, produced with reliable and robust production operations. Over the past few years, a high-quality promise became a core requirement rather than an additional feature. The threshold limits for such promises are likely to keep increasing further. The captains managing the turbo-engines of technology must keep that in mind.

America's Information Age Advantage

This is the good news. The new information technologies are opening massive doors for future growth in America. The country is in a unique position, with a very significant lead over its global competitors in information processing technologies. The lead is big even over the industrialized competitors in Europe. America has the largest per capita computer installation rate in the world. Internet has become a household name and is particularly a favorite pastime of the young adults in America. American companies could, and should, leverage this technological leadership in information processing and communication technologies for a sustainable competitive advantage.

Going Global Means Learning to Be Local

Increasingly, corporations and captains of technology are quickly realizing that their new opportunities are in the global markets. These global opportunities are strongly influenced by their local flavors. Customers are becoming increasingly "tribal" in demanding that their individual needs must be not only met, but customized. New products must be developed with their needs in mind. Not just by exporting abroad the leftovers from the domestic markets.

Mind Magnet

For many years, America has been a magnet for bright minds. A high concentration of some of the world 's smartest people, blended with good technologies, was the envy of many other emerging nations, particularly the fast-rising economies of Japan, Korea, Taiwan, and so on. These Asian countries primarily gained competitive advantage because of their low relative wages. But, they can foresee the future slowdown of their rapid rise. They are therefore investing enormous resources in their scientific research to generate new proprietary technologies of their own. New emerging economies of the world are striving hard to catch up and fill their technological gap with the industrialized West.

If the American corporations wish to continue to lead the technologies of the world, they must treat their human resource differently.

Pioneers or Poachers

Captains of American industry are becoming increasingly myopic in managing their new technologies. They are reluctant to make investments that risks not giving them large returns sometime soon. This sometimes comes from their lack of understanding of the business technologies which drive their businesses. Yet they can't delegate such strategic decisions to the lower levels of managers. To play safe, they demand risk-free guarantees from investments in new technology, yet expect high levels of returns (which are justified only for the risky investments).

These leaders, unlike their predecessors, are not thinking much about pioneering of new industries. The pioneers like George Eastman, Thomas Edison, Chester Carlson, and Charles Goodyear did not pioneer new technologies for quick short-term profits or discounted cash flows. There is too much expense-oriented-ness in the upper echelons and not enough visioning to expand the top-line horizon. Can American companies become great by shrinking all the time?

A BRIEF SUMMARY OF LESSONS LEARNED

In this chapter we learned about the overall model of technology engine and management of technology. We reviewed our images of technology and developed a working definition by the eight different forms in which we encounter technology. Various examples of management of technology helped the development of the engine model of technology. The core transformation subsystem, the supporting resource related subsystem, and the integrative and visioning subsystem were discussed.

SELF-REVIEW QUESTIONS

1. Are there other "Ps" of technology that you can add and elaborate?

2. Compare and contrast the focus of management of technology with the focus of production management and the marketing management perspective.

3. How can American corporations retain and gain their technological leadership?

NOTES, REFERENCES, AND ADDITIONAL READINGS

1.1. Drucker, Peter F. 1959. Work and tools. *Technology and Culture,* Winter; and in *Technology, Management and Society.* New York: Harper & Row.

1.2. Steele, Lowell W. 1989. *Managing Technology: The Strategic View.* New York: McGraw-Hill.

1.3. Badawy, Michael K. 1993. *Management as a New Technology.* New York: McGraw-Hill.

1.4. Rubenstein, A. H. 1989. *Managing Technology in the Decentralized Firm.* New York: John Wiley & Sons.

1.5 Burgelman, R. A., Maidique, M. A., and Wheelwright, S. C. 1996. *Strategic Management of Technology and Innovation.* Chicago: Irwin.

1.6 Parisi, Anthony J. 1989. How R&D spending pays off, *Business Week,* August Bonus Issue:177–179.

1.7. See McLean, I. W. and Round, D. K. 1978. Research and product innovation in Australian manufacturing industries. *Journal of Industrial Economics,* 27:1–12; Comanor, W. S., and Scherer, F. M. 1969. Patent statistics as a measure of technical change. *Journal of Political Economy,* 77:392–398.

1.8. An earlier version of this discussion was presented as a paper by the author at TIMS/ORSA Joint National Meeting, Boston, April 24–27, 1994, in COLIME/COLTEM cluster. This research was partially funded by the J. Warren McClure Research Award for 1994 received by the author. In this paper, empirical research and teaching implications of this model were discussed by considering the contingency effects of context influences.

1.9. Markoff, John. 1989. A corporate lag in research funds is causing worry. *New York Times,* January 23:A1, D6.

1.10. Scherer, F. M. 1984. *Innovation and Growth: Schumpeterian Perspectives.* Cambridge, MA: The MIT Press.

1.11. A few authors have suggested and researched with the use of multiple technology-related competencies. For example, see Teece, D. J. 1986. Profiting from technological innovations: Implications for integration, collaboration, licensing and public policy. *Research Policy,* 285–306.

CASE 1A: TECHNOLOGY MANAGEMENT PRACTICE

The Tale of Transistor Technology: Gateway to the Electronic Age

The transistor, the small semiconductor device, has often been called mankind's entry point into the electronic era. The transistor was developed in December 1947 at Bell Research Laboratories of AT&T by John Bardeen, Walter Brattain, and William Shockley. Understanding the underlying principles of the semiconductor effect must have required a good knowledge and understanding of solid-state physics. The three inventors made some major contributions in that regard, and they were given the Nobel Prize in Physics in 1956 to honor their expertise.

The discovery of transistor technology at Bell Labs makes the undisputed argument in favor of corporate funding for scientific research and laboratories to discover new technology.

Continual Changes in Solid-State Technology

The development of transistor technology is also often cited as a discontinuous radical departure from the vacuum tube devices used earlier in the electrical appliances. The vacuum tube used a big bulky glass enclosure under vacuum, with a heated cathode. The semiconductor transistor eliminated these inconveniences.

The development of transistor technology, however, can be linked back to a continuous development of other electronic devices dating back to the 19th century.

In the 1870s a German physicist, Ferdinand Braun reported that certain crystalline substances conducted electric current only in one direction. These substances were used as crystal rectifiers to detect electromagnetic radiation, and they helped to usher modern radio reception around the turn of the 19th century. The crystal radio set used a holder with a semiconductor crystal of silicon carbide or lead/molybdenum sulfide. A flexible thin wire (called cat's whiskers) was moved over the crystal to receive a clear signal. It required trial and error, and the signal could not be amplified.

These drawbacks of the crystal radio receiver were eliminated by John Fleming and Lee De Forest 's invention of the vacuum diode and the vacuum triode in 1904 and 1906, respectively. The vacuum tube device, which had evolved as a by-product of the light bulb, allowed amplification of the received signal and made the crystal radio sets obsolete by the 1920s.

The interest in crystal detectors, however, persisted for military applications during the World War II years. Vacuum tube radios were not good at catching short wavelengths, unlike the crystal radio detectors.

Then radar technology was developed, and crystals were noted to detect microwaves as well. Crystalline materials such as germanium or silicon/tungsten probe (cat's whiskers) were used to develop point-contact crystalline rectifiers for microwaves.

The Bell Laboratory scientists are credited with the jump from germanium microwave detectors to the first workable germanium transistors. Similar research was going on in many laboratories across the United States. The new transistor used cat's whiskers, just like in Ferdinand Braun's crystal rectifier radio. The new transistor, however, acted as an amplifier instead of a rectifier, which has less commercial potential. The point-contact transistor also replaced the vacuum triode tube, as well as its use in electric circuits. The whisker ends of transistors were called the emitter and collector, just like in a vacuum tube. Transistors were also eventually sealed into glass casings, like a vacuum tube. Shockley subsequently replaced the point-contact transistor by inventing a junction-type transistor that led to the development of solid-state electronics technology used extensively in household appliances.

Many transistor-based electronic devices were incorporated together to form the integrated circuits (ICs) and very large size integrated circuits (VLSIs).

Let us pursue the evolution of transistor technology beyond its invention. Even though the transistor technology was invented by the American and European scientists, it was supported and commercially exploited first by the Japanese.

Sony's Sponsorship for Semiconductors. After the devastating defeat precipitated by America's atomic technology, McArthur's Occupation Forces disbanded the Japanese engineers working for the Imperial Army. A number of them, including Akio Morita and Masaru Ibuka, who had worked together earlier, decided to get together to use their telecommunications expertise for civilian purposes. They formed Tokyo Telecommunications Engineering Company. They produced a vacuum tube voltmeter, an electric rice cooker, and a magnetic tape recorder for schools (1949–1951). But they were still on the lookout for other consumer products to keep their technical staff busy.

In 1953 Masaru Ibuka, while on a trip to the United States, learned that Western Electric was about to sell the rights to use their patent for transistor. While none of the large electrical or electronics companies in the United States or Japan showed much interest, Ibuka decided to buy the license for transistor technology in 1954.

He then sent Japanese technicians to America to collect all available published or not yet published information on transistors. They visited laboratories and talked to the researchers, engineers, scientists, and technicians who had worked with the transistor technology.

This know-how was assimilated to use transistor technology to manufacture a pocket-sized radio receiver. In 1955, as the newly developed mini radio was ready to be marketed, the producers changed the name of their company to Sony, derived from the sonic sound they wished their company to be associated with. Sony's transistor success was the first of a string of many successful electronic appliances produced based on semiconductor technology—originally invented far from Japan.

The intellectual exploration of semiconductor technology and its successful commercial execution and exploitation could not have been farther apart. This confirms our call for an expansion of the definition of technology far beyond the confines of its intellectual or technical birth. The nurturing parent enterprise of a baby technology deserves as much credit as the natural parent inventor (or firm) responsible for the birth of a "new" technology. The resource constrained Japanese in the 1940s could not have invented the semiconductor technology. But the postwar Japanese reconstruction of the 1950s seemed to be the right time to take up the challenge of nurturing an untested technology desperately looking for a sponsor.

Reluctance to Change. In the 1950s the victorious American electronic device manufacturers, relying on vacuum tube technology, were reluctant to change to semiconductor device technology. Thus they allowed the fledgling Japanese electronic companies an uncontested world market for semiconductor-based home appliances. In the meantime, Americans saw an enormous potential of semiconductor technology in the computers, defense, and space sectors.

Transistor-based radios first appeared in the U.S. markets in a big way in 1956. In 1955, portable radios, with demand for about 2 million units and all made in America, used vacuum tube devices which on average weighed 6 pounds. The new transistor-based radios weighed only 20 ounces. Their average cost was $57. Most of them were produced by U.S. manufacturers. Regency was one of the first producers of transistor-based portable radios.

By 1959, there were 25 U.S. portable radio models using transistors, weighing around 20 ounces, with prices somewhat lower than those from 1956 when transistors were first introduced in portable radios. But by 1959, a new pocket radio market segment was created by 11 Japanese miniature portable radio sets using transistors. These weighed only 10 ounces (half the weight of the U.S. models), and their cost was 10% lower.

Developing New Market Segment. The Japanese radio models entered into a market segment that was not competitive with the American producers. The Japanese producers did that not by innovating the basic transistor technology, but by reducing the size and weight of tuning capacitors, loudspeakers, battery supplies, and antennas. They ushered in the core transistor technology along with these other enabling technologies that created the new pocket radio market, not known to exist before. This helped the Japanese make inroads into the American consumer markets, thereby breaking into the customer barrier of Japan's past reputation for producing cheap and shoddy goods. The proprietary American transistor technology, by itself, was not adequate to make that quantum leap into the market. By the year 1962, the Japanese transistor radios captured over 58% of the U.S. market.

Diffusion in Automobile Radios and TVs. Until 1955, car radios were tube sets. In 1956 these were replaced by improved hybrid car radios using germanium transistor outputs. They would operate from a 12-volt car battery. Then in 1957 they were replaced by a fully transistorized car radio. The use of transistor technology in car radios was facilitated by eliminating the use of a vibrator and by using a car battery as power source. The Motorola Corporation, a car radio and military communications manufacturer and not a vacuum tube device producer, grabbed the business opportunity by effectively integrating backwards to manufacture germanium power output transistors. They did this for use internally, as well as for sales to others.

The transistor diffused into the color television sector at a much slower rate. Vacuum tubes were used from 1955 until 1967. In 1968 the first transistorized television consoles were available. But this became the industry standard only in 1974. Some argue that the reason for this slow diffusion of transistors into the television sector was the lack of significance attached to the portability of television.

Source: Adapted from White, George R. 1978. Management criteria for effective innovation, *Technology Review,* February:21–28; Basalla, George. 1988. *The Evolution of Technology,* Cambridge: Cambridge University Press.

CHAPTER 2

Production Operations and Automation Engineering

PREVIEW: WHAT IS IN THIS CHAPTER FOR YOU?

Production operations and automation engineering have been historically evolving from the ancient civilizations. Production operations involve the physical transformation process of technology, converting the basic raw materials of low value to the finished goods with higher value. In this chapter we will learn from production operations conducted at different times in human civilization. These antecedents define the way we manage our production operations today.

The different significant milestones in this journey include (a) the princely production of pyramids and the Indus Valley public sewage system, (b) the rise of empirical reasoning in the Greek city-states, (c) state-run Roman roads, deficit-financed projects, and centralized Church constructions, (d) the household production system in feudalism, (e) the Renaissance of reasoning, (f) industrial factory production, (g) production operations in market-driven capitalism, (h) automation of production machines, (i) interchangeable parts for mass production, and (j) just-in-time lean production.

In this long evolutionary journey, production operations grew primarily by vocational experimentation and trial and error. This technique prevailed until Eli Whitney introduced the method of mass production which he based on the use of interchangeable parts. Subsequently, theoretical models such as the experience curve, as well as empirical practices such as the Scientific Management, Just-In-Time and lean production, were developed.

EVOLUTION OF PRODUCTION OPERATIONS: FROM STONE WHEEL TO CYBER-WEB

Production operations matter. They have often defined the state of human civilization. Therefore, before we grasp how to manage our production technologies today in a highly integrated and shrinking world, we must first look into the lessons hidden in the pages of history of production and operations. A brief review of our historical heritage will help us obtain a better understanding of some of our nagging search. For example:

1. Why do we produce goods in factories and not homes?
2. How did plants and factories get to be so concentrated as the preferred places of work?
3. How do we react to changes from one production technology to the next one?
4. Does new technology add or shrink jobs for working men and women?
5. Does manufacturing matter in a service economy?

In this chapter we will focus on the management of production technology and operations. Production operations are defined as the transformational processes to convert raw materials into desirable and useful finished products or services. However, we will soon see how these transformational processes are closely related to developments in many other aspects of human civilization, such as (a) the degree of freedom permitted to common masses for the pursuit of happiness, (b) flow of information, and (c) access to financial resources.

Learning from History

Even though much of our modern production technology at the end of the 20th century is anchored deeply in relatively recently developed scientific research in computers and semiconductor chips, the story of technology goes way back to ancient civilization. History holds many lessons for management of technology. It tells us how workers worked, or how rulers ruled and organized their workers. It also tells us the values that prevailed at a time and changed with time.

Man's control over nature depended on his ability to develop (a) tools to produce and gather food and (b) weapons to fight. The ancient human civilizations such as those in the Stone Age, the Iron Age, or the Bronze Age were classified by the technologies they learned to use. The technologies for basic human needs, such as for farming and gathering food, for clothing to protect body from nature, and for building a shelter for adverse weather conditions, have evolved empirically, with trial and error. The lessons learned have been transferred from one generation to the next for the past 6000 years. In today's fast pace of life and work, very often we forget the historical antecedents of our beliefs and practices. A quick review would give a broad perspective of how production and operations changed with new developments, and how they influenced the way we manage our technologies and production operations today.

The Stone Wheel

The wheel is a symbol of man's first significant technological invention. The first wheel was perhaps not invented in its present-day form, and its initial use may not have been for transportation. The early uses of the wheel may have been ornamental or ceremonial instead of utilitarian. The first productive wheels were potter's wheels. The oldest representation of a wheel was found on a clay tablet in Mesopotamia in 3500 BC. It is estimated that by that time they were also in use in the Indus Valley civilization and in China.

Once Paleolithic man got past the early Stone Age, his dependence on hunting and gathering declined. In the new Neolithic Age an Agricultural Revolution took place and man's focus was drastically diverted to herding of cattle and controlled farming of the agricultural food crops. About 3500 BC, the Sumerian civilization flourished in the fertile Mesopotamian delta between the rivers Tigris and Euphrates.

The Egyptians learned about the wheel around 2500 BC, almost a thousand years later. For these ancient civilizations the wheel helped improve transportation by carts and also helped improve farming.

Stone was gradually replaced with copper, which was easy to form and hard to blunt. Metal working and smelting technologies emerged in the copper ore-bearing mountains of Asia Minor and Cyprus (the word "copper" is derived from the Greek word for Cyprus, *Kyprios*). By 3000 BC, Mesopotamian man learned to mix and alloy copper with tin, to make more stable bronze.

At this time a number of factors converged which changed man's life and operations drastically. For example, improvements in wheel-based transportation reinforced the need for nonagricultural skills such as metal-working. This was coupled with a better understanding of the role of irrigation, climatic conditions and use of the plow in agriculture. The combination of these important factors gave birth to a high concentration of population in small communities. The cave man was ready to be civilized in the cities of the future.

EVOLUTIONARY SPIRAL OF PRODUCTION AND OPERATIONS

Mankind produced things long before we learned to industrialize our production operations in factories. Compared to human civilization, the modern form of factory-based industrialization is a relatively recent phenomenon. In the past, most of the goods and services were provided by craftsmen who worked from households. They were not centralized and specialized the way they are now. Yet production and commerce of produced goods have been thriving for about 6000 years. Man produced and traded even when roads and telecommunication services were not available. In this chapter we will learn about the evolution of production and operations since the early civilizations.

Producers in Early Civilizations

Like Mesopotamia, the Egyptian civilization also rose from the valley of another great river, the Nile. For its prosperous economic growth it depended on the rich fertile soil brought down by the river floods. Surplus agricultural produce helped Egyptians pursue luxuries and other refinements of life.

But unlike the Egyptians, the Babylonians did not develop objective thinking. They attributed all natural occurring events, such as rains, floods, and so on, to supernatural forces and divine beings. However, in mathematics they advanced and developed a hybrid system combining numbers with 10 base and 6 base. They did not have a representation of zero, but they did use numbers representing quantities such as 1, 10, 60, 3600, and so on. For multiplication they were forced to use multiplication tables. Even today we use the Sumerian sexagesimal system when we divide a circle into 360 degrees, an hour into 60 minutes, and a minute into 60 seconds. Their calendar was based on lunar movements. One year consisted of 12 lunar months of 29.5 days each. From time to time they added another lunar month to catch with the changing seasons.

THE STATE-RUN PRINCELY PRODUCTION OF PYRAMIDS

The Egyptians lacked abundant supplies of timber or metal for constructing their buildings. Yet, Egyptian pyramids that are more than 4000 years old are testimonials to the organization and management of their technological ability to carry out large projects. In about 2575 BC, the Egyptian pharaoh Khufu started the Great Pyramid of Giza, with a perfect square base. Much later, the 19th century explorers were amazed to discover that the deviation from a right angle was only 0.05%. The sides of the pyramid were oriented exactly in the east–west and north–south direction. The pyramid used granite slabs weighing about 50 tons. It probably took about 25 years to build.

According to the archeological findings of the British explorer Piazzi Smyth, who traveled to Egypt in 1864, the mysteries of the Great Pyramid of Giza were revealed only by the use of pure and applied mathematical concepts (Note 2.1). Smyth's careful examination revealed that "the ratio of the vertical height of the Great Pyramid . . . to twice the breadth of its square base [was], as nearly as can be expressed by good monumental work, [the same as the ratio of] the diameter to the circumference of a circle" (see Note 2.1). This number, the first "key" to resolving the mystery of the Great Pyramid, was 3.14159, which is nothing but the value of pi, or 22/7, or the ratio of circumference and diameter of a circle.

The Egyptian pyramids and canals were amazing engineering projects. These as well as other quarrying and mining operations were managed by a state-run bureaucracy. The state administrators paid out money for the contracted services of the freemen and the slaves. This required careful counting and resulted in the development of certain modern practices of accounting.

In these administrative bureaucracies the pharaohs were the chief executive officers, and they appointed a vizier or a prime minister as the chief operating officer. Then below them came the various ministers. The Egyptians also used forecasting techniques to estimate the periodic flooding of the Nile.

The power of the pharaohs collapsed around 1100 BC. Foreign wars and a gradual concentration of power in the hands of the priests took a toll on the welfare of the common people. The Pharaohs could no longer exercise their control. Foreign invasions and civil wars weakened Egypt further. Finally, Egypt was incorporated into the Persian Empire in 525 BC.

The Urban Planning Projects of the Indus Valley Authorities

Around the time when the Great Pyramid project of Giza was coming to its completion in 2500 BC, not far to its east, in the Indus River Valley, a number of well-planned urban centers such as Harappa and Mohan-jo-daro were thriving along the banks of the Indus River and its tributaries. These highly planned urban cities had sewerage systems. These first cities inhabited by humans had orthogonal road systems (like in midtown Manhattan) with broad avenues and narrower streets crisscrossing each other. The Harappan people also built a huge citadel, 1400 feet (425 meters) long and 600 feet (180 meters) wide. The modularized houses in standardized lot sizes were built with fired bricks.

Further east, along the banks of the Huang He ("Yellow") river, the Longshan people settled the first large Chinese community with distinctly different social classes. The Chinese had by then well established the silk industry. This was started accidentally around 2600 BC by Se Lingshe, the wife of Emperor Huang-ti, when she dropped a silkworm cocoon into hot water.

THE GREEK CITY-STATES AND THE INDUCTIVE METHOD OF REASONING

In many ways, the Greek and the Roman civilizations form the backbone of the modern Western civilizations. For the next 2500 years these civilizations established not only the way managers would think and the way businesses would be run, but also how the roads and other public projects would be undertaken.

The Greek civilization emerged around the Aegean sea when the pastoral people settled down into small communities dispersed in the different islands, mountains, and valleys of the region. The new Greek city-states such as in Athens and Sparta, which they called "polis," were small towns clustered around a place of high elevation. The Parathenon-like buildings in these city-states were fortified for refuge to the people in the face of an outsider's attack. By 500 BC the city-state of Sparta emerged as politically the most powerful city-state in the region. Sparta's strong infantry, with a strong navy from Athens, defeated many invasion attempts by Persians.

The Rise of Empirical Reasoning

The Greek civilization made more contributions to the development of intellect than to the advancement of technology. In the Greek value system, manual work and trade were for slaves, while the respectable men contemplated on higher ideals. During the late fifth and fourth centuries, Socrates developed the inductive method of reasoning to explore social inter-relationships between fellow human beings instead of exploring the natural world. His famous pupil Plato followed in his teacher's footsteps. Plato proposed that God was the ultimate source of all life and matter. Plato's pupil Aristotle then developed a comprehensive philosophy to integrate the knowledge mankind possessed until then. In his *Metaphysics* he proposed that man's reality is best understood not by mysticism but by reason and physical senses. Collectively, this dynasty of Greek teachers and their students gave shape to the scientific foundations of reasoning.

A Vision for Leadership

Socrates, Plato, and Aristotle provided a scientific attitude and reasoning in their thinking. People like Aristotle and Socrates were concerned with the leadership and administration of the public affairs. Direct democratic participation through elections of leaders, along with job rotation of administrating ministers, was a common practice. In his treatise *Politics* (which meant everything concerning *polis,* or state), Aristotle discussed the relation between the management of the household and the leadership of the state. In another famous book, *Republic,* Plato portrayed an ideal state run by justice. In his book *Laws,* law ruled a community. With the intellectual development of reasoning, the Greeks gradually rejected mysticism in the face of empirical observation and deduction.

The Greek eminence declined in the fourth century BC because of ecological devastation of natural resources and forests. Alexander the Great marched to the edges of India, but could not enter. The fall was accompanied by a moral decay of the people's values. With isolationism and economic weakness, the Hellenic city-states started dissipating in the fourth century BC. The Greeks had funded their city-states only by the limited resources they had available at any time. They did not have budget deficits. Borrowing money from the public was not conceived yet.

RISE OF THE ROMAN EMPIRE AND DEFICIT-FINANCED ROADS

The Roman Empire rose and overwhelmed the Greek city-states with alliances and more centralized organization of labor. The Romans constructed paved highways, radiating from Rome to the distant areas acquired by Romans. They were used for faster transportation of not only military personnel, but also goods and supplies needed by the marching armies. In about 312 BC the famous Appian Way project was completed.

The Romans produced arms and weapons in state-run factories. Textile goods and pottery were produced by craftsmen for international trade. Such goods for export were often taxed with imposed state-tariffs. The export trade was primarily run by the Orientals and the Greeks. A system of standardized measures of weights, size, and currency was adopted. The Roman governments funded their public projects by floating the early forms of joint-stock companies, with stock shares issued to Roman citizens.

Workers and craftsmen specialized themselves and organized themselves under *Colegia,* or vocational guilds. Their primary goal was to gain respect in the society rather than to do collective bargaining.

The Roman rulers laid out strict laws and enforced order. Executive and legislative powers were separated to curb the runaway control of dictators. With the spread of the Roman Empire, monarchy was replaced with a well-developed judicial law system. This was Rome's major original contribution. An imperial form of government emerged to rule the widespread empire. Acquired territories were administratively grouped under province districts and were governed by a governor. These governors in far-off lands exploited their subjects, who were also abused by *publicani* or tax-gathering public officials. However, once people got elected to high places, they abused power and did not want to give up their powerful revenue-generating positions.

By the end of the second century BC, a period of civil war broke out and lasted for a century. At the end of the civil war, military party chiefs took over. Julius Caesar, the leader of a popular faction, took the power and gradually became a dictator. In 44 BC he was assassinated on the senate floor by his fellow aristocrat members of his governing body. A new power struggle began in Rome. Eventually, Caesar's adopted son Augustus, also known as Octavian, took over in 27 BC and brought in a new form of government. Pax Romana, a period of peace in the Roman empire, followed for the next 200 years.

The Romans valued character, loyalty, and family ties. They remained primarily agricultural people and did not like the theoretical speculations of the Greek. Hence there was only limited scientific literature written in

Latin. This practical orientation of Romans was best reflected in Julius Caesar reforming the 12-month lunar calendar. It required adjusting every few years with an additional 20–22 days. He replaced it with a solar calendar that has 365 days in a year, with an extra day added to February every fourth (leap) year.

The Roman Empire survived a few ups and downs, and it triumphed in the Western world for about two and a half centuries. Caesar's son Augustus reconstituted the government to avoid the problems that plagued his father's rule. This enabled Rome to flourish for another two centuries. In the third century AD the Roman legal framework collapsed.

The Centralized Construction of Christian Churches

Some historians hold that the rise of the Christian Church was the root cause of the fall of the Roman Empire. Others believe that Christianity may have been the result of the fall of the Roman Empire. Christianity brought hope for the increasing masses of the poor and the slaves (Note 2.2).

Initially there were many different forms of independent church organizations catering to the rising number of worshippers. Eventually, the spiritual effort came under the centralized authority of the Catholic Church at St. Peter's in Vatican City. With this centralization came the well-defined policies and procedures for all church operations. Policies defined how to raise funds, construct churches, train priests, and distribute information.

The centralized church also became the primary builder of elaborate construction projects in the Western world.

THE HOUSEHOLD CRAFTSMAN PRODUCTION

Dark Middle Ages

In the late Roman Age, it became financially uneconomical to maintain slaves. The Roman landowners discovered that hiring freemen to work as part-time tenant farmers was economically more advantageous than bearing the living expenses of their full-time slaves. This was particularly hard under declining economic conditions, such as those after the fall of Rome in the seventh century. The resulting political anarchy and chaos gave birth to feudalism and what is often referred to as the Dark Ages. As a way of life, this prevailed for almost nine centuries, from 600 AD to about 1500 AD. It was a period in the Western civilization when little technological progress was made.

The Feudal Economic System

Feudalism started in Europe about 800 AD and was based on a rigid and oppressive economic hierarchy. At the base of the feudalistic economic system, the burden of maintaining the feudal system was carried by the bonded or free serfs who were economically more oppressed than their slave counterparts. The serf worked day and night on a small and often uneconomical plot of land that was rented from a manor lord. The manor lord shared a portion of his produce (to the knights) in exchange for military protection. A vassal was a noble who was given a fief—that is, property—in exchange for his loyalty.

Feudalism created the order of chivalry in the upper crust of noblemen. The word *chivalry* was derived from the French word for horse, *cheval,* and is related to *cavaliar.* This taught discipline and provided apprenticeship in leadership skills to the sons of the noblemen. This was sort of a training ground for the future rulers. The nobles got knighted by the king or his representative. The journey to knighthood started for a son of a nobleman when he started as a page at the age of 6–8 years. He became esquire at 15 or 16 years of age, and he could be knighted at 20 years of age. Thereafter he was called "sir."

The life for the masses of working serfs, even though they were politically free, was oppressive. With rigid class lines, poverty and ignorance perpetuated among the masses. Excessive controls stifled human creativity and growth. The centralized Catholic Church provided the only hope—in the form of salvation and a better afterlife. This made their miserable lives bearable. Historians consider that the Dark Ages of Europe lasted from about 500 AD to about 1500 AD. During this period, most of Europe was very rural. There was no sign of urbanization in this part of the world yet.

Crusades and International Trade. A firm feudalistic control on human endeavor and growth resulted in the religious Crusades for ambitious expansion of territory. The Crusades, or the Holy wars for the Cross, spanned from 1096 to 1291. The Crusades had many significant side effects. A new class of warriors and disciples emerged. Chivalry bloomed. After 200 years of religious wars in Europe and Northern Africa, the impact of an alternate religion, Islam, and a very different culture of the Middle East was fully realized and accepted by the Europeans. Crusaders also helped open new trade routes to distant, and until now disconnected, markets. Merchants grew by lending funds for financing the Crusaders. Luxurious goods from the East, such as fine silk clothes, spices, carpets, and silverware, not seen in Europe until then, were introduced in the markets. This international exposure broadened the belief system in feudalistic Europe.

Household Textile "Put" Production System. The new international exposure expanded Europe's trade with the newly discovered markets of the East. The new trade was mostly dominated by a back-and-forth flow of textile goods. An early form of "household"-based industrial production emerged. Using a "putting system," merchants procured the required amounts of raw materials. These were issued to individual workers or families on contract wage basis. They processed and produced the finished goods in their own homes, using the machines and equipment they purchased and owned. The finished goods were consolidated by the merchants and were traded in domestic or international markets.

In many ways this system was not very different from the way Wal-Mart and other retail merchants procured their merchandise from other vendors and small businesses in the 1990s. The growth of the Small Office Home Office (SOHO) is reminiscent of the "household" production system. In the diamond industry of India in the 1980s, most of the diamond polishing and cutting was done by the "household" production system.

The Career of a Craftsman. The craftsmen learned their craft through 7 years of apprenticeship under a master. Then they became wage-earning journeymen. They submitted a sample of their work, or a "masterpiece." If it was approved, they became masters.

The master craftsmen from one industry or a profession were members of a craft guild. There were craft guilds and merchant guilds for processors and weavers, bakers, butchers, brewers, cobblers, carpenters, druggists, doctors, lawyers, and many more industries and professions. These guilds got together separately for professional as well as social interactions. They played an important role in the running of the cities, and they defended members from the oppression of feudal noblemen. From time to time, the noblemen from the rural country tried to forcefully extend their jurisdiction over towns which were outside their control.

The European cities were highly concentrated. Lack of sanitation in the cities caused epidemics. The Black Death epidemic of bubonic plague in 1348–1349 decimated about one-half of Europe's population—the buyers as well as the producing craftsmen.

Bookkeeping and Inventory Control Emerges. This "household" production system involved many hand-overs and required detailed accounting of goods flowing in and out. The first double-entry bookkeeping of debits and credits emerged out of this necessity. In 1494, a venetian named Luca Pacioli wrote a manual of accounting practices. This provided the first manually run management information system (MIS) for the trader-merchants of that period.

The trader-merchant also faced a continual challenge to raise more capital needed for further growth. Recall that earlier, during the rise of the Christian Church and feudalism, lending of money and receiving interest was denied religious sanctity. Most states had to pass orders against these sanctions. Bankers of Venice, Genoa, and Florence, such as the Medicis, innovated new ways to lend and collect money.

Universities Emerge for Education. The medieval Cathedral and monastic schools gave birth to universities. The term *university* was derived from the Latin *universitas,* meaning "all (persons) together." Later the meaning became restricted to a limited guild of scholars who assembled to study. Most medieval universities had only four groups of faculties. These were for liberal arts, law, theology, and medicine. The study of law was further divided into (a) common or Church law and (b) civil or Roman law.

As guilds, the universities followed the practices of a craft guild and had apprentices called *bachelors*. They completed their apprenticeship, and at "commencement" they were granted permission to move on to advanced studies. The master's program was to get a license to teach and was used interchangeably with "doctor," meaning a teacher. In later times, a doctorate degree required a public examination and defense of an original proposition or thesis.

Challenges of Household Craftsman Production. The household production system worked very well for the merchant as well as for the household-based craftsman. But it had certain structural disadvantages for its survival and growth. The individual domestic producers had low economies of scale and very limited incentives to invest in new production equipment or development of technology of their operations. The craftsmen did not learn from each others' mistakes. The merchants held back most of the profits. The craftsmen worked at the mercy of the merchants, who made more money without adding much value. The system needed a change. A change brought about the Renaissance or rebirth of Europe.

RENAISSANCE OF REASONING

The Rebirth of Europe

In 1215 the Magna Carta, a long list of laws, was written. It listed the rights of an individual to a jury trial by his peers, and it limited the arbitrary power of the kings. The taxes were to be collected by the English parliament, divided into the House of Commons and the House of Lords. Such events slowly brought back political stability to Europe, which gave birth to the Renaissance and the Reformation of Europe. The dark age of feudalism was replaced in the beginning of the 16th century, when the Renaissance ushered in a fresh bright light of freedom of expression and liberty to pursue happiness.

The progress of ideas and the desire for human exploration was reborn in the Renaissance. The early phase of the Renaissance Age started in Italy around 1300 AD and lasted for about 300 years. *Renaissance* means rebirth of the Roman ideas, after a sleep of about 1000 years. It marked a rebirth of the pursuit of fine arts, music, theater, and philosophy. But the Renaissance mainly symbolized a willingness to once again view reality in a scientific way. Some of the inventions during this time period included the printing press, cloth paper, optical lens, telescope, and microscope. During this period gun powder, saw mills, the spinning wheel, and oil painting were also introduced.

Renaissance Inventors. **Leonardo da Vinci (1452–1519)** has been often considered to be one of the most famous artists, inventors, and representatives of the Renaissance Age. He conceptualized the ideas of the submarine, the airplane, the helicopter, the tank, and many other products and technologies still in use today more than 500 years later. Beside his 'engineering' and technological ingenuity, he is also famous for painting the Last Supper and the Mona Lisa—two of the world's most valuable paintings.

Among the artists, **Michelangelo (1475–1564)** was the most famous and was undoubtedly the best sculptor during the Renaissance. His masterpieces include the ceiling of the Sistine Chapel in Rome as well as the statues of David and Moses. Other famous artists of the Renaissance Age included Raphael and Donatello.

Machiavelli's View on Leadership. The resurgence of new ideas during the Renaissance also permeated to a new look at the leadership and administration of the re-emerging large city-states. For example, in 1513 **Nicolo Machiavelli,** an ex-administrator from Florence, dedicated his book, *The Prince,* to a Medici. The Medicis established libraries in different regions and were also patrons of fine arts. Machiavelli's book was a handbook for the state administrators. It explained to them how to rule and control their subjects. Machiavelli assumed that all men were bad and were always ready to behave badly. He therefore urged the state administrators to lead adaptively by using deception. He recommended that the leaders should prefer to be feared and forceful like a lion, rather than be loved and perceived as weak administrators.

The Printing Press and the Spreading of Information. One of the most significant technological inventions of the Renaissance Age was the printing press using movable metallic print type. This was perfected and introduced in 1444 by **Johannes Gutenberg (1400–1468)** in Germany. The printing press helped print books and religious documents faster, cheaper, and in larger numbers for the benefit of larger masses of people. It also made a radical change in the way inventors and innovators founded their reasoning. Prior to Gutenberg, the medieval man was forced to reason using the thoughts of Socrates, Plato, Aristotle, and the Bible. After the printing press, new thinkers suggested relying more heavily on observations and deductive empirical evidence.

This fundamental technological innovation became a fountainhead of a large number of amazing ideas, inventions, and discoveries in the not-so-distant future.

First came **Nicolaus Copernicus (1473–1543),** who challenged both (a) the prevailing authority of the Church and (b) its sponsored belief that the universe was centered and moving around the earth, which was considered the home of the divine being. Copernicus proposed, based on his direct observation rather than deductive logic from Church-approved beliefs, that our universe including home planet earth was heliocentric—that is, centered around the sun, not around the earth. We were on the right track again.

Soon thereafter, support came for Copernicus from **Galileo Galilei (1564–1642).** He also studied gravitational forces and spent a lot of time observing (a) objects dropped from towers and (b) church bells swinging like clock pendulums.

France produced **Rene Descartes (1596–1650),** who proposed that the universe operated mathematically, and therefore mathematics would describe its movements. England gave us **Sir Isaac Newton (1642–1727).** He was impressed by falling apples and the laws of moving objects.

This new practice of empirical scientific inquiry helped lay a firm foundation for the subsequent rapid development of new technologies that led to the production of swords, silverware, and sweaters during England's Industrial Revolution in the 18th century.

Mercantile Trade and Barriers

In the 16th and the 17th century, the newly emerging stronger nation-states gave birth to state-sponsored exploration of distant lands by discovering untried routes. For example, on December 31, 1599, the East India Company was chartered by the British Crown to trade with India. This charter and the British influence on India continued very profitably, in one form or another, for almost 350 years until India's independence on August 15, 1947. Similarly, other European states also forced other lands to open their shores while protecting their own markets with tariffs and other barriers. This national chauvinism guided the private economic activity of merchants.

In 1651 **Thomas Hobbes** published *Leviathan.* He supported a strong central leadership and suggested that man's innate tendency toward chaos must be curbed.

Western mercantilism crumbled under its own discriminatory bias. Just as the mercantile bureaucrats were able to enact policies to defend the domestic merchants from the competition of their foreign producers, they also ended up sponsoring inefficient and uneconomical but politically influential businesses and industries. Isolationist and protectionist mercantile merchants also passionately promoted wars, thus often destroying the very markets they were seeking to conquer.

The Reformation: Prelude to Industrial Revolution

The Age of the Reformation, from 1500 to 1648, was marked by religious upheaval in Germany in 1517. This led to the establishment of Lutheran, Calvinist, and Anglican Protestant denominations. In 1690, **John Locke** wrote *Civil Government,* and once again he questioned the divine right of kings to rule others. He proposed that most people were governed by reason and were not easily intimidated by a central authority. (He was born too soon to see Germany after World War I.) He urged building a civil society by respecting each other's right to own private property. This had a profound effect on Cromwell's England. Locke's movement brought about a radical change in the British Constitution in 1688.

Such political pragmatism, coupled with the intellectual enlightenment unleashed earlier during the Renaissance and Reformation, gave birth to a unique period in British history. During this period, the accelerated technological progress and growth motivated man's audacity to improve production of goods with an accelerated pace. Around 1750–1751, England experienced a rapid replacement of manpower with machine power. In centralized manufactories, large quantities of coal and other goods were produced for the growing markets of free people seeking to improve their quality of life.

CENTRALIZED MACHINE PRODUCTION

The Industrial Revolution

The Industrial Revolution marked a rapid change from production in households and domestic workshops to centralized factory production—in other words, a transition from "manu-facture" to "mechano-facture."

During the second half of the 18th century, one of the earliest introductions of machinery to manufacture occurred in the spinning of cotton yarn and the making of cotton cloth. During this period, called early Industrial Revolution, the manufacture of cotton cloth was transferred from cottages to factory. The power for centralized production came from water power. The first textile mills were operating for about a generation before steam power entered the cotton textile manufacture. James Watt's engine provided blast in 1776 and drove forge-hammers in 1783, just in time for such conversion.

The cotton industry was the first sector which demonstrated unprecedented growth with the Industrial Revolution. Mechanization and continuous motion reduced the cost of cotton goods drastically. In a quarter century between 1760 and 1785, production of the cotton industry in England, as measured by consumption of raw cotton, expanded by a factor of 10. The cotton industry grew more than 10 times again between 1785 and 1827.

Some of these technological advances in cotton textile technology were either adapted to or were derived from innovations related to woolen textiles. Together, innovations in textile technology resulted in some major geopolitical changes throughout the world, including the rise of England as the new industrial leader of the world.

Three Phases of 18th-Century Cotton Production. The story of 18th-century cotton technology can be divided into three distinct phases. In the first phase, between 1700 and 1740s, the compound rate of growth of cotton textile production was as low as about 1.4% per annum. This was prior to the invention of John Kay's flying shuttle for weaving cotton broadloom. Then in the second phase the cotton production rate changed to an increase of 2.8% per annum until the 1770s. In the third phase until 1800, the cotton production jumped to 8.5% per annum. Textile inventions by Hargreaves, Arkwright, and Crompton were introduced. (See Case 2A on innovations in cotton technology at the end of this chapter.)

It is difficult to infer which came first. Did the textile innovations in textile technology cause the phenomenal growth in cotton production in England? Or, did the phenomenal growth in cotton production gave birth to the pioneering inventions of textile technology? Some analysts argue that the hundredfold growth in cotton production between 1760 and 1827 could not have been made possible by a matching increase in labor force. The growth can only be explained by the improvement in productivity caused by technical progress in spinning machinery and power looms for weaving.

The higher production efficiency, accompanied with higher quality and lower prices of the cotton yarn and fabric thus produced, helped cotton substitute the place of linen and silk for many purposes around the world. The British-made cotton substituted for competing textile fibers in all markets around the world. There was no difficulty in selling the hundredfold growth in cotton production from 1.1 million pounds of imported raw cotton to 44 million pounds of retained raw cotton imported to Britain.

Continental Spread of Cotton Textile Technology. The new cotton textile machines developed in England were soon introduced by British businessmen and mechanics in France and Germany. For many years John Kay and his sons lived in France making flying shuttles, card-making machines, and other textile machines. A Jacobite exile, John Holker, settled in Rouen in 1751 and founded a textile factory at Saint Sever. As he became Inspector-General of Factories, he introduced the latest cotton machinery and invited skilled operators from Lancashire to train the French workers.

After the Napoleonic Wars, English experts such as Job Dixon and Richard Roberts introduced modern textile machinery to Alsace. New carding, combing, and spinning machines for wool and shearing of cloth were introduced by William Douglas and John Collier.

In Germany, in the 1790s K. F. Bernhard started textile operations at Hartau in Saxony. At Guben and Grunberg woolen mills were operated by William Cockerill in 1810. Some indigenous textile inventions included the Jacquard silk loom, the Oberkampf–Widmer printing cylinder, and the improved dyeing technique by Macquer and Berthollet (Note 2.3).

Impact of British Textile Innovations. The faster and easier-to-use textile machines, developed during the Industrial Revolution in Great Britain between the 1750s and 1780s, had an enormous impact on the industrialization of Great Britain. The British cotton industry expanded rapidly. The cotton yarn that cost 38 shillings per pound in 1786 fell to 2 shillings 11 pence per pound in 1832.

In the 1840s a cotton-mill used a 100-horsepower engine and employed 750 workers to run 50,000 spindles. It produced a cotton yarn equivalent to that produced by 200,000 operators using spinning wheels. Similarly, the amount of four-color cloth that 200 men could print in one hour by hand was being printed by one man on a calico-printing machine.

Motivation to Innovation. Why did Europe gave birth to so many innovations in the 18th century? What were the incentives to invent? Was it a mere coincidence? Or was it the work of pragmatic rulers of that time?

1. *Crisis Is the Mother of an Innovation.* Many textile innovations were born out of crisis situations. For example, in 1824 Lancashire reeled under a severe strike by cotton workers. This threatened the future of an emerging export industry of Britain. To break the workers' strike, three leading mill owners commissioned Richard Roberts to construct a spinning mule that operated automatically without manual supervision. Roberts developed a self-acting spinning mule, to mill owners' satisfaction. According to Andrew Ure and Samuel Smiles, many industrial machines were invented in response to labor disputes and to replace striking workers.

2. *Try, Try, and Try Again.* Many innovations of the Industrial Revolution in the 18th-century textile industry of England were the result of trial and error rather than the application of precise scientific principles. Crompton was a spinner and Hargreaves was a weaver—not engineers.

3. *Interindustry Transfers of Inventions.* Sometimes a textile invention resulted from innovative application of an invention developed and successfully used in some other industry. William Fairborn introduced many improvements in the textile machines and hydraulic machines. He also established a shipbuilding yard at Millwall (London) to build iron ships. Matthew Murray in Leeds innovated flex-spinning machines as well as a heckling machine. He also built locomotives (Note 2.4).

4. *Race to Invent.* Very often, different inventors, struggling to solve similar problems, came up with quite similar solutions. They could be in very different parts of the world, with no potential for piracy.

PRODUCTION IN MARKET-DRIVEN CAPITALISM

The Industrial Wealth of Nations

While the Industrial Revolution was accelerating in England, in 1776, the year when the new United States of America gained freedom from the clutches of the British colonialism, a Scottish political economist named

Adam Smith (1723–1790) published his revolutionary book, *An Inquiry into the Nature and Causes of the Wealth of Nations.*

Adam Smith proposed what became known as the classical theory of economics. He proposed that state-run mercantilism and its selective tariff barriers on produced goods were self-destructive. This "visible" and direct hand of the State on industrial enterprise, according to Smith, was unable to guide and ensure the most optimum allocation of resources in a society. Instead, he proposed that an "invisible hand" of market forces of "self-interests" provided better regulation of rewards and resources. The "wealth" and prosperity of a nation, according to Adam Smith, lay mainly in the free and unregulated pursuit of maximum profits by the entrepreneurial individuals and industrial enterprises.

Reforming Protestant Ethic and Capitalistic Production

The spread of trade and commerce, along with the exchange of new ideas with distant lands, resulted in an implosion against the centralized control of the Catholic church. In Germany, Martin Luther and John Calvin gave birth to a Protestant movement for religious freedom and choice. They did not, however, sponsor an uncontrolled pursuit of happiness and commercial business.

In 1905 Max Weber published *The Protestant Ethic and the Spirit of Capitalism,* later revised in 1920. He tried to link personal freedom and pursuit of prosperity. He redefined the concept of spiritual "calling" in the Reformation movement to the pursuit of personal perfection in one's occupation of choice. This was to glorify God and was called "Puritanism." This Reformed approach had markedly different implications on human work practices when compared with those from the alternate value system which asked man to wait until the Day of Judgment for a better afterlife.

As a corollary to the glorifying call for intense personal and specialized vocational activity, the unequal distribution of wealth among people was accepted as natural. Each individual was expected to, and accepted for, developing his or her own specialization. This resulted in a high priority for the value of individualism in Protestant beliefs.

This concept of individualism fit very well with Adam Smith's specialization of labor in a market-driven business enterprise. Adam Smith had illustrated the benefits of specialization with the simplistic example of a manufactory producing pins. He had proposed that the "wealth of nations" depended on specialization of labor, and this resulted in economies of scale.

For example, an individual pin-maker could hardly produce more than 20 pins per day per person. On the other hand, a pin manufactory, with specialization of workers operating different pin-making operations, could produce up to 48,000 pins per day. This significantly large increase in production output was due to three important factors.

1. Increase in the dexterity of each workman.
2. Reduction in downtime due to switching from one type of work to a different type of work.
3. Need and resultant invention of a large number of machines, which help as well as reduce labor.

The Founding Pillars of Manufactory Production System

The above-mentioned three factors helped one specialized worker do the task of many workers. Here is the foundation of the concept of "experience curve" or "learning curve." This model will be discussed in greater detail later.

Adam Smith's specialization of labor was particularly well-suited to the booming growth in economy and demand. The domestic production system, under mercantile economy, did pretty well when the total demand was low and the increase in demand was limited. But when the demand started increasing rapidly, economies of scale with specialization of labor was better suited and likely to be more profitable. This is the earliest use of (a) functional specialization of production processes and (b) the manufacturing–marketing integration. These concepts will be developed in greater detail too.

Pioneering in Capitalism

Some observers (Note 2.5) have linked the religious spirits of individuals and their psychographic needs for high achievement. McClelland hypothesized that Protestant individualism produced an entrepreneurial personality with an attitude favoring risk-taking, innovation, decision-making, and hard work. This once again fit well with the emerging concept of capitalism. We can tie this to pioneering and visioning in modern context.

Finally, we must also consider the implications of an individual's individualism, task specialization, and pursuit of happiness, with the ease or difficulty of managing and organizing a large number of such men in a centralized factory. Recall that Nicolo Machiavelli urged the leader to assume that all men working for "the prince" are bad troublemakers. He therefore recommended the ruler to rule them with deception and fear.

Thomas Hobbes also proposed a strong central leadership to suppress the natural tendency of men toward chaos. But John Locke went the other way around, urging leaders to believe the reasonableness of people under their rule. He instead suggested to offer people the incentive of owning property and a pursuit of happiness to bring about a civil order in the society.

All these proposals have a strong bearing on the way the production operations are organized in a modern corporation.

AUTOMATION OF PRODUCTION MACHINES: THE SECOND INDUSTRIAL REVOLUTION

Reasons for Automation

Production machines had to learn automatic control before they could be widely used. Automatic control or self-regulation, a key feature of many production processes, was adapted from the processes in nature. The early introduction of automatic control in production was to reduce the dependence on labor and save the rising labor costs. Later, automatic control became a necessity with increasing demand for goods of high uniformity and quality—for example, for producing interchangeable parts of standard shapes. The automation of machines became unavoidable as the production speeds increased or the work had to be done under hot, humid or toxic conditions. Workers were too glad to adapt the use of automation under such adverse conditions.

Automation often increased productivity. Over the years, workers' average working hours have decreased. Yet, the industrial production increased because of increasing use of automatically controlled production machines.

Workers often feared that automation would lead to large-scale unemployment. This could result in economic distress and even social upheaval, as it did from time to time, in different parts of the world.

In the past, some analysts have feared that automated production might rob mankind the opportunity to apply its creative genius. Or, automation would reduce craftsmanship.

But, automation has often given birth to new professions—for example, the automatic control engineer or time-and-motion analyst. Automation often requires reskilling.

Self-Regulating Open Systems

The core principle of self-regulation is simple. Every stable operating system shows a typical pattern of behavior. It requires a supply of energy and other input resources. Open systems operate with equilibrium under changing environmental conditions. The changes in external conditions must be adapted by changing the input resources or the operating process of the system. An automatically controlled system maintains its typical processes even under excessive variations in environmental conditions.

Steam Engine Regulator. In 1788 James Watt invented a flyball governor for the automatic control of the output of a steam engine. This was a major critical contribution by Watt to the development of the steam engine. Subsequently when the reciprocal steam engine was substituted, the spinning flyballs also disappeared.

Some machines used more automation than others, depending on the function and processes a machine was expected to perform. Feedback arrangement was the key driver of a self-regulated automatic operating system.

Windmill Self-Regulator. A very early use of feedback mechanism was made in a simple device used on windmills. It helped keep the sails of a windmill always facing the wind. The device consisted of a miniature windmill, attached at right angles to the main windmill. The sails of the mini-mill were at right angles to the sails of the main windmill. Whenever the main sails faced in the wrong direction, the wind was caught by the mini-mill sails. This rotated the main windmill back to the correct position.

With industrialization and industrial progress, more production machines and processes used automatic control devices.

Late Automation of Machine Tools. Surprisingly, the metal-cutting machine tool industry was one of the slowest industries to adopt automatic control. As late as in the 1950s, skilled machinists believed that their speed, judgment, and flexible dexterity to use machine tools could not be duplicated by the automatic machines. They considered their tasks too sophisticated and very different from the mass production of automobile parts. Many more sophisticated developments in automatic control had to take place before they were convinced otherwise. To fully understand this resistance to adopt change, we must first look into the history of the machine industry. It is a fascinating story.

History of Machine Tool Technology

Until the last quarter of the 18th century, the machinists (or millwrights, as they were called then) mainly used a hammer, a chisel, and a file. They measured dimensions using a wooden rule and crude calipers. They prepared materials either by hand-forging or by rudimentary foundry casting. Crude hand-powered lathes were primarily used for wood-turning or making clock parts.

The first major machine tool innovation was a cylinder-boring device invented in 1774 by John Wilkinson. This is what enabled James Watt to build his full-scale steam engine. Wilkinson's boring machine is what made the Watt steam engine a commercial success. Prior to this, James Watt struggled for 10 years to turn a true cylinder to do his intended job. He once wrote about his cylinder of 18-inch diameter that "at the worst place the long diameter exceeded the short by three-eighths of an inch." By 1776, Watt's assistant Matthew acknowledged that "Mr. Wilkinson has bored us several cylinders almost without error; that of 50 inches diameter, which we have put up at Tipton, does not err the thickness of an old shilling in any part."

The Wilkinson boring machine was the forerunner of the precision metal-working machine tools of latter years. Another machine-tool technology pioneer Joseph Bramah and associate Henry Maudslay invented a screw-cutting lathe with a slide rest and change gears like many modern lathes.

They invented successful locks, the hydraulic press, many woodworking machines, the four-way valve, a beer pump, and the water closet.

INTERCHANGEABLE PARTS FOR MASS PRODUCTION

Whitney's Interchangeable Musket Parts

The next major milestone in production operations and machine technology was pioneered by the American inventor **Eli Whitney (1765–1825).** He is more often remembered as the inventor of the cotton (engine) gin. His greater contribution was the use of interchangeable machine parts.

In 1789, after earning some money from the invention and sale of the cotton gin, Whitney started the manufacture of muskets for the U.S. government in a New Haven plant. For this defense contract Whitney planned to use interchangeable parts. At that time, most industrial experts questioned if the same could be done for muskets. In 1791 Whitney had to go to Washington to reassure the U.S. Secretary of War and other Army officers about his progress. He took with him 10 muskets produced with interchangeable parts. He demonstrated to his sponsors that the gun parts could be interchanged, without adversely affecting the gun's firing performance.

Whitney also argued that the interchangeable parts were manufactured in his New Haven factory using precision machinery that could be operated by relatively unskilled workers. He asserted that this would save on the cost of hiring skilled machinists, who were in rare supply during that period. His critics refused to be convinced. To prove the usefulness of the interchangeable parts, different muskets were disassembled and their parts were mixed. Then muskets were reassembled from the heap of parts. Eventually, interchangeable part manufacturing became the common practice in production of all mass-produced goods.

The production with interchangeable parts used two primary tools. These were the lathe and the milling machine. The lathe was the same as the one invented and used by Maudslay earlier. Around 1854, the tool-changing turret was added on the lathe, and the automatic screw-cutting machine was developed.

In 1818 Thomas Blanchard invented the "copying" lathe for turning gunstocks. These types of tools were called "cam-following" machines. These machines automatically oriented to machine irregular shapes according to the contours of a prespecified cam. Sometimes with use the cams wore out, and they had to be carefully monitored and machined to account for that.

Whitney developed the first milling machine for producing his oddly shaped interchangeable parts in 1818. It could be used not only for muskets, but for interchangeable parts for many other devices too. The milling machine consisted of a power-driven horizontal table which passed under a rotating cutter at right angles. According to the legend, Whitney did not patent the milling machine and freely shared its design with many other manufacturers of muskets for the U.S. government. During Whitney's time, the milling machine played a more significant role than his cotton gin in America's industrial development.

The All-American Production Method. There is no doubt that Eli Whitney's cotton gin radically changed the life and agricultural production on plantations, particularly their use of slaves. It helped the South economically. These changes in the South, some say, in the 1860s caused the American Civil War between the agricultural South and the industrial North.

But Whitney also made a big impact on the industrially emerging northern parts of America. By developing specialized machine tools, using standardized parts, gaining government contracts, and many other first-time industrial practices, Whitney almost single-handedly created the "American Production Method," as we know today. Prior to Whitney, Americans made muskets one at a time. Some crafted all the different parts of a musket individually and by hand. Each part was different in different muskets. Getting Whitney's progressive ideas accepted by the rest of the industry was not easy.

Later in 1862, the Providence inventor Joseph R. Brown modified Whitney's milling machine to the universal milling machine used in America's industrialization. The 19th century also added other machine tools such as drilling, punching, sawing, and metal shaping machines to industrial production.

Machine Control. Machine tools and machines used to make interchangeable parts used guides, tracks, and other control devices. These helped improve the precision and increased manufacturing speed while reducing the need for human skills or intervention.

MASS PRODUCTION, ECONOMIES OF SCALE, AND THE EXPERIENCE CURVE EFFECT

The Experience Curve Effect

This is a useful model for managing mass production of standard products. The experience curve is also known as the learning curve, the progress curve, or the manufacturing progress function. The experience curve expresses the expected rate of improvement in productivity as more units of a standard product are produced, and accumulated experience is gained as a result of it.

Let us try to understand what this experience curve stands for. Consider that we want to learn a new skill, such as how to play piano, swim, or type. As we practice more, we gain experience as a result of doing the activity, and we get better and better. A person learning to type will first type the first page in one hour or even longer, say 100 minutes. But, as the individual learns where to position fingers, and how to type different letters or letter combinations to make different words, the speed of typing improves significantly with practice. Thus typing one page would require much less time. The second page may take 80 minutes, and the fourth page may take only an hour.

Origins in an Air Base. The experience curve was first postulated at a U.S. Air Force base. In the 1920s, at Wright-Patterson U.S. Air Force Base, production managers in charge of building airplanes noticed a trend. They found that the number of labor hours required to build the second plane were only 80% of the labor hours required to make the first plane. In other words, the air force production staff noted that as they doubled their cumulative production, the number of labor hours required to build a unit plane reduced by a certain fixed ratio. In the above example, the pattern followed the ratio of 0.80. This was called the *learning factor.*

The value of the learning factor depended on the technology used and the operation involved. If an operation followed the "80% experience curve," then as the cumulative production doubled from N planes to $2N$ planes, the labor time required for production of the $2N$th plane was reduced to 80% of the time taken for the Nth plane. If our actual measurements indicated that the 10th plane we produced took 1000 hours to build, then the 20th plane should have taken only 800 hours. If we followed the same experience curve, then the 40th plane was expected to take only 80% of 800 hours— that is, 640 hours. And the 80th plane would have taken only 80% of the time taken to build the 40th plane—that is, 512 hours.

If the production operation used a different technology, then it would follow a different experience curve trajectory. If the new technology empirically followed a different 90% learning factor, then the 20th plane would take 90% of the labor hours required to build the 10th plane—that is, 900 production hours per plane. The 40th plane should have taken 810 hours with the technology, and the 80th plane was down to only 729 man-hours per plane. Notice that the economies of scale for the 90% learning factor are less steeply declining than what we observed with the experience curve following the 80% learning factor. These values are illustrated in Exhibit 2.1.

In general, this relationship holds true for different measures of input resources. The trend also applies to the cost, man-power required, and so on. Thus, if the 10th plane costs $10 million to build per plane, then the 20th plane on an 80% experience curve should be expected to cost $8 million. The 40th plane was expected to cost $6.4 million.

Causes of the Experience Curve Effect. The experience curve effect does not occur automatically or "naturally." As the accumulated experience (or production volume of a product) increases, managers should be able to achieve a systematic cost reduction due to their managerial synergy. Some of the sources of such synergy are as follows:

EXHIBIT 2.1. Learning–Experience Curve

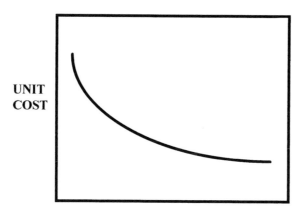

UNIT
COST

CUMULATIVE EXPERIENCE

1. *Improved Labor Productivity.* With experience, the same worker can produce more in the same amount of time. Practice makes a person produce more. Workers, if motivated right, can learn to do their task more efficiently. Sometimes the operator memorizes the steps involved. The hand–eye coordination gets better. Mind almost knows what to do, without thinking about it.

2. *Higher Degree of Specialization Permissible.* As the production volumes of a standard product increase, complex production tasks in its manufacture can be broken down into simple and more specialized tasks. Then these simpler tasks can be assigned to those workers who can do them best, either because they have more experience or because they have been specially trained to do those tasks. In more recent years, specialized tasks were out-sourced to specialized vendors with low overheads and higher expertise.

3. *Innovation in Production Methods.* With increase in accumulated experience and higher specialization, the assigned workers come across new, innovative, and more productive ways to accomplish the same tasks. Much of the improvement in productivity in Toyota's automobile production system resulted from innovative suggestions made by the shop-floor workers. Most of the time these innovations involved the use of simple convenient jigs and fixtures and did not cost the company much money to implement them.

4. *Value Engineering and Fine-Tuning.* As the production floor staff gains experience with the production of a product, added savings in production time and cost can be achieved by using the principles of value engineering. For example, Japanese producers were able to cut significant costs by reducing the number of parts in an electronic appliance, such as a photocopier. Fewer parts also meant lower maintenance costs. Thus, not only cost was reduced, but quality and customer satisfaction were also improved.

5. *Balancing the Production Line.* Higher production volume allows supervisors to achieve large increases in production capacity by selectively adding a few pieces of balancing equipment. The pieces of equipment add capacity to certain "slow" spots in the overall production process. These slow spots slow down the speed of the entire production process. Adding the balancing equipment costs much less investment than that required to duplicate the entire production line to increase capacity.

Caveats for Experience Curves

The experience curves seem simple to use and understand. But they are actually simple approximations of complex interactions taking place between a wide variety of interrelated parameters. Therefore for decision-making, production supervisors must exercise some caution in using the experience curve effects.

1. *Learning Is the Key.* The experience curve economies take effect because learning takes place with accumulated experience. If the work force is such that either they cannot learn because of lack of basic literacy or training, or they do not want to learn because of their mistrust of management, then the experience curve effects are not likely to be noticeable. On the other hand, well-educated and highly motivated workers are able to exploit all the advantages of the experience curve effect, and more.

2. *Distinguish Accumulated Experience from Time.* Even though experience seems to increase with the passage of time, the two are not the same. If no production is done during part of the time, there is no experience gained during that period. It is like taking college courses on a part-time basis. If no courses are taken during a particular semester or academic year, then the graduation is not granted at the end of the prescribed period. Besides, what we learn during each class or work experience is what helps us get better over time. No pain, no gain.

3. *Changing Units of Experiences.* Products produced by a manufacturer do not remain the same over the years. Computers or the telephones produced in the mid-1990s were more sophisticated and had more features than the computers or telephones produced 2 or 4 years earlier. Having more than one or two gigabytes of memory in desk-top computers was common in 1996, whereas a computer model produced a few years earlier had only half or less memory. Intel's MMX Pentium microprocessors ran at 150 MHz speed or more, whereas two generations ago, Intel's top-of-the-line microprocessor processed at 20 MHz speed.

Thus, one of the challenges of the experience curve effect is how to compare the "apple" production units of the 1990s with the "orange" production units of the 1980s.

4. *Demand Elasticity.* The experience-curve-based production works only if the market demand is elastic. That is, as the price of a unit produced is re-

duced, the demand will increase by a certain ratio. If the demand is inelastic, the price reduction will not generate the higher demand needed to justify higher production volumes.

5. *Highly Fragmented Markets.* Similarly, in highly fragmented markets, the individual competitors have only small market shares each. Even if one producer increases its production capacity substantially to gain economies of scale, it is not likely to improve its slice of the market pie significantly.

Uses of Experience Curve Effect

With the above-mentioned caveats in mind, experience curves can be used in a variety of production situations.

1. *For Forecasting Future Budgets and Cash Flows.* Using the experience curve, production supervisors know that their total cost is not linearly proportional to the number of units they produce. It is less than that. There are ready-to-use tables available to estimate the cost of producing, say, the 3000th airplane along an 80% experience curve if the cost of the 150th plane was known.

2. *For Outsourcing and Purchasing.* The experience curve can be used to estimate and negotiate what it should cost for a supplier to produce 1500 widgets, as well as how much more should it cost to produce 3000 or 6000 widgets along its experience curve.

3. *For Competitive Bidding of Contracts.* In contracts which are very competitively contested and are decided by the minimum bid, the experience curve can help estimate the bidding cost more accurately. Thus, one can estimate the accumulated cost and then bid a price close to that without losing money on the contract.

4. *To Benchmark Competition.* Experience curves can help estimate the likely unit cost of production for competitors with too high or too low capacity. This information can be used to price the product in such a way that either higher profit margins are earned or the competitors are priced out of the market.

5. *Experience Curve and Dumping in Global Markets.* The experience curve effect can be used to price products for export markets. These are often priced differently from the way products are priced for the domestic markets.

Let us consider a hypothetical example. The Acme Company manufactures cyber-widgets in La La Land. The domestic demand is 2 million cyber-widgets a year. If Acme produces 2 million units a year, its unit cost is $100 per piece. Acme retails the cyber-widget at $120 and splits the margin with the wholesaler. Acme gets a decent 10% return. Acme is a late entrant and does not produce for international markets yet.

In Liberty Land there are the Big Three makers of cyber-widgets. Cyber-widgets sell for $110 each, and the three producers, Alpha, Beta, and

Gamma, produce 4 million, 2 million, and 1 million cyber-widgets, respectively.

John Smith completes his MBA from the University of Acorn and joins Acme in the Business Development Department. His boss asks him to develop a global strategy for the next 7 years. John recalls his lesson on the experience curve effect and suggests to his management to expand capacity to 4 million cyber-widgets a year. With 0.8 learning factor, this would reduce their unit cost to $80 per cyber-widget. They will continue to sell 2 million cyber-widgets in their regulated domestic market at $120 each, thereby earning $30 profit per cyber-widget.

They can export the rest of the cyber-widgets for a retail price of $100 per widget, undercutting the Big Three cyber-widget makers in Liberty Land. If they add 10% commission to the retailer and add another $5 shipping and handling cost from La La Land to Liberty Land, their landed revenue is about $100 − $10 − $5 = $85. They still are making $5 profit per cyber-widget in the international markets. They already have $30 profit margin per cyber-widget in the domestic market. These profits should be enough to break even in less than 5 years and recover the cost of the new plant to produce an additional 2 million cyber-widgets a year.

The Big Three cyber-widget makers in Liberty Land are fuming mad. They accuse Acme of dumping charges. That is, Acme is selling goods below cost. How could Acme sell cyber-widgets for $120 in Acme's domestic market in La La Land and afford to sell them for $100 in Liberty Land, after adding shipping and handling costs?

Should John Smith get worried? Not really, he has his notes on the experience curve effect. He remembers from his MBA classes that he can charge marginal cost. He has coupled it with the experience curve effect from the Management of Technology and Operations class. Is he dumping?

LEAN PRODUCTION PARADIGM: TOYOTA'S JUST-IN-TIME KANBAN SYSTEM

The Oil Shocks and Auto Production

During the 1970s, U.S. auto production changed radically. For the first time, American production and markets were governed by the producers of Japan, the nation that was defeated in World War II less than three decades ago.

The auto industry in the United States was hit hard three times: two times by developments in the Middle East, and the third time by Japan's producers in the American markets. The first shock came in 1972–1973, when the Middle East oil crisis increased the price of one barrel of oil many times its previous price. This impacted many industries. Most of them were heavily dependent on cheap petroleum products. They needed a bigger budget for larger fuel payments. Many downstream products, such as steel, metal fabrication, and so on, became more expensive to use. There was a spiraling effect on downstream industries like the automobiles and motorcycles. Assembling also became more expensive.

American producers shuddered. But most thought this was a temporary phase and that it would go away. After a few months of belt-tightening and gasoline lines, life returned to normal. America had its own reserves of petroleum in Alaska and Texas.

Thinner, Lighter, Stronger Production in Japan

On the other hand, the Japanese producers and consumers did not have the luxury of domestic oil resources. They had to import 100% of their petroleum and fuel requirements from any available source they could find. The First Oil Shock brought negative growth in the Japanese economy. The Japanese confronted their vulnerability. Afterwards they had double-digit annual increases for almost two decades.

The Japanese government and producers started war-like efforts to reengineer the production processes. They knew that their growth and survival depended on exporting their products to the world markets. These markets and customers determined the prices of produced goods. The Japanese makers had to absorb the increase in the cost of their fuel and oil bill by improving their production efficiencies.

There was a nationwide movement in Japan to conserve energy everywhere they could. In production they saved energy by making the parts and the products thinner, lighter, and stronger. Patiently they reduced the number of parts and made their appliances as well as cars less energy-intensive. After much hard work and patience with a *kamikaze* spirit, the producers started seeing some significant savings.

The Second Oil Shock

Then came the unexpected Second Oil Shock in 1979–1980. The world's producers were not prepared yet. But the Japanese were. Their products were much less energy-intensive than any other competing brands. They were lighter, stronger, and more compact too.

The skyrocketing increases in the price of crude petroleum oil, along with the mile-long lines for gasoline, shifted the customer preferences rapidly in America. They could no longer afford to own and maintain large gas-guzzling cars with heavy metal and chrome. American consumers, who had earlier rejected Detroit's compact cars over large cars, were now desperately shopping for smaller cars.

The Japanese auto makers were ready to offer American consumers what they wanted. They had developed small fuel-efficient cars for their own consumers more heavily crushed by the rising gas prices. They lived in crowded cities with narrow streets. The Japanese automobiles and appliances started penetrating the hard-to-penetrate American markets. Companies like Toyota and Honda had tried hard for the past two decades to improve their products and make them more acceptable. But their small products could not take the beating required by the American highways and turnpikes, spanning from one end to the other end of this continent-sized country.

THE LEAN MEAN PRODUCTION AGE

How did the Japanese auto producers gain sustained market share in the U.S. car markets? Besides getting lucky with the consumers, they had to work hard on another bias. The American consumers had always seen cheap and shoddy products from Japan. How could they trust their lives in the Japanese cars, and how could they trust their foodstuff in the Japanese refrigerators? Over the past two decades, Japanese auto makers like Honda and Toyota had been tinkering with their compact cars to adapt to the American road conditions. The lucky break in the American markets meant that they had to do more.

To improve their image, the Japanese producers like Toyota had gone back to their shop floors and reengineered their entire production processes. They invented a new production system called the Kanban method or the Just-In-Time pull production method.

The Just-In-Time Pull Production Method: The Kanban Method

The automobile may be considered in the mature stage of its production life cycle. Many of the major technological innovations in auto technology were made decades ago. The Japanese auto producers entered the auto business much later than their American or European counterparts. The Toyota Motor Company was an "independent" late entrant. Their auto production did not start until the 1930s.

The Japanese auto makers also knew that their own domestic markets were too small to grow. Japan's islands were covered with mountains, and people lived only on one-twelfth of the total land. More than 50% huddled in the eastern part of Japan, stretching from the Kanto region around Tokyo to the Kansai region around Osaka. They had to depend on global markets. This meant making not one standard product, but a wide variety of car types, to meet the varied demand in different international markets.

The Japanese producers also knew that their cost of raw materials was much higher than similar costs for their U.S. counterparts. The Japanese manufacturers recognized these given "external" factors. They were forced to cut costs. The corporate focus became, "keep production costs down."

Japan's War Against Waste. The Japanese auto makers could not use new technologies to improve their productivity. Not many new technologies were emerging in the auto assembly process. Further, the worldwide customers held a very low image of Japanese products. Toyota decided to pursue "low-cost leadership" strategy. They had to cut out waste. Every production operation was rationalized. Large corporations used paper on both sides even for the business documents. Small ends of steel at the end of the spools were collected and recycled. This resulted in a well-documented saving of 2.5 tons of steel per month at Toyota Motor Company. Use of office air-conditioning was minimized, even during the hot and humid days of Japan's summer months.

The Push Production

The engineers at Toyota noticed that one major source of waste in their auto production process was in the large amounts of resources held up in their inventories. As Henry Ford had learned many years earlier (see Case 2B on Henry Ford at the end of this chapter), the challenge of auto production was in managing the supply lines. For the moving assembly line to keep moving, thousands of parts, supplies, and subassemblies had to be delivered to the relevant workstations ahead of time. The auto makers made that happen by stocking large volumes of buffer stock at each stage in the production process.

The Economic Order Quantity. In the push production system, the production planning group determined the optimum quantities and schedules for different parts needed for different work centers. This group released the raw materials at the start of a work line. As the operation of a work center was completed, work was pushed to the next work center. The work waited until the next work center was ready to operate on it.

The optimum lot size of inventory was estimated by the Economic Order Quantity (EOQ) formula. A simple formula was developed by F. W. Harris in 1915. A stochastic model, with statistical variations and distributions added, was developed in 1934 by R. W. Wilson. The optimum lot size traded off the cost of ordering a new consignment of parts for the cost of holding a unit part in stock. The optimum inventory lot size for independent demand was determined to minimize the total production cost and the inventory cost. This kept the inventory levels down to their minimum acceptable levels.

Materials Requirement Planning (MRP). After the introduction of computers in production, the planning for material requirement was done using computer-based models. The Material Requirement Planning (MRP) model used production master schedules, product structure or a bill of materials, and inventory status files to produce schedule plans for placing orders for parts and components.

Later versions of MRP models converted to "closed-loop MRP." These helped simulate the dynamic effects of changing customer requirements on the manufacturing operations.

Finally, fully integrated MRP systems for business planning emerged. These were called "manufacturing resource planning, or MRP-II."

In all these systems, the primary production flow was by a push forward, with buffer stocks of inventory to tide over fluctuations and other factors. The control of this flow was not in the hands of the production operators, but with the production planning experts.

Market Conditions for a New Production Paradigm

A number of market conditions warranted need for a new paradigm for radical cost-cutting. This was required to survive and grow in the auto production business.

1. Auto products were fast becoming standardized products with well-established technology and core production processes.
2. Automobiles were produced in large numbers.
3. Automobiles were produced from a large number of parts. These parts were either (a) produced internally but elsewhere or (b) procured from outside.
4. Automobile entered a mature stage in its life cycle.
5. Auto markets around the world were developing a high price elasticity. Thus, lowering the production costs and prices was critical to increase the demand or keep it from shrinking.

The Pull Production System

Toyota decided to change the prevailing paradigm for management of inventory in auto production. Instead of maintaining certain minimum levels of buffer stocks all the time, Toyota decided to stock inventory only when needed. The Japanese called this the Kanban system, meaning the "notice board" approach. The Western production managers called it the Just-In-Time (JIT) production system.

Kanban or Just-In-Time System. The Kanban system or the Just-In-Time (JIT) system are both zero inventory systems. It was developed by the Toyota Motor Company to control their cost of production. Basically it is a production practice whereby the production of a part is authorized at a work station based on the requirement of that part at the next work station. If the part is not needed at the next station, it is not produced. Parts are produced directly to use. Just in time, when they are needed. They are not produced to stock, as in the case of the "Push" production systems.

The "Pull" system used simple cards and bins for storing the parts and requisitioning their production. Just in time, when they were needed.

Each work center held certain prespecified number of bins for items. A work center X produced new parts only if a full bin of parts was used up by the next work center Y. The work center X then produced only enough parts to replenish what was used by the work center Y. The "Pull" system worked with a set of chain-reactions. These were set in motion by the actual requirement and consumption rather than some prespecified plans or optimal numbers.

The Kanban system was a simple information method using "cards" and a "display board," called Kanban. These were used to convey relevant information to the work centers—for example, the amounts they needed to produce to replenish the "pulled" bins. In a modified version, two types of cards were used. Card P was used for production, and card M was used for material movement. The two sets of cards were continually reused.

The "Pull" production method was an "advanced manual" version of MRP. This suited Japan, where the computers were not as commonly used as in the United States. In the JIT system the rapid throughput rate reduced the lead times to almost zero. Compared to MRP, JIT used fewer levels of hierarchy in the bill of materials.

Prerequirements for Pull Production. The Pull production system was based on the following parameters:

1. An agreement on average production rate.
2. A consensus on the number of full bins of items between different work centers.
3. Early sharing of information with suppliers about the company's production/purchase plan.

The Pull production system worked only if certain prerequirements were met:

1. Greater production flexibility, with small production and delivery lots.
2. More and frequent setup changes, with very low setup times and costs. Toyota reduced their changeovers to single-digit minutes, while die changes were made in under a minute.
3. More skilled and flexible work force.
4. Reliable production and deliveries, with well-maintained machines.
5. Quality given a genuine high importance by top management. Product quality and customer needs were non-negotiable.
6. Everyone strived for continuous improvement.
7. There was a high level of trust between workers and managers, between procurers and suppliers, and between producers and consumers.

The Benefits of Pull Production. The Pull production system, when implemented properly, generated many benefits internally, externally, and competitively.

Internally, inside the production organization, the Pull production system increased inventory turnover dramatically, reduced the inventory and warehousing costs, and reduced the cost of production.

Externally, outside the production organization, the Pull production system improved the on-time delivery of quality goods, which generally improved the customer satisfaction and their loyalty.

Competitively, the Pull production system could help a producer gain market share and improve profit margins.

Applications of the Pull Production System

Toyota derived great benefits from using the Pull production system. Many other production companies, including some in the United States, also adopted the JIT paradigm.

A Turnaround at Harley–Davidson (1982–1986). During the 1970s and early 1980s, Harley–Davidson, a pioneer motorcycle producer, faced severe competition and market erosion due to the Japanese motorcycle makers, Honda, Suzuki, Kawasaki, and Yamaha. This resulted in staggering losses in 1980 and 1981. Harley's survival was at stake.

In 1982 Harley adopted the Pull production paradigm. With JIT, phenomenal improvements were achieved:

1. Inventory turnover per year increased from 7 to 20.
2. The productivity per employee increased by 50%.
3. The rework costs were reduced by 80%.
4. The warranty costs were reduced by 45%.

The JIT Worm in the Apple Computers. Not all companies have benefited from the Pull production paradigm all the time. For example, researchers have surveyed manufacturing companies and reported that the JIT system would fail without suppliers' trust. The failure could also take place because there is only one sole supplier, located thousands of miles away. That is what happened at Apple Computers in the 1980s.

At Apple's $20 million Fremont, California factory, 1000 Macintoshes a day were assembled using the Pull production system. The plant used JIT delivery of parts the day they were needed for assembly. According to Apple's Planning Manager, Peter Barron, the Planning and Scheduling department was always "on the dot." There was no 30-day inventory of parts, as in other plants.

One time the Mac plant got one defective container full of hundreds of cathode ray monitor tubes (CRTs). These were shipped to them from the Far

East. While in use the CRTs turned brown or yellow after a while, instead of remaining crispy gray. But, there were no buffer stocks available to replace the defective parts.

The day the problem was discovered, Apple dispatched to the Orient a team including its Display expert and a Quality Control Engineer. The monitor manufacturer in the Far East had a problem with the phosphor composition used for the Apple's CRTs.

In the meantime, the Fremont plant exhausted their supplies of properly functioning monitors. The Mac line had to be idled for 8–9 days according to President John Scully. The Mac Product Manager said it was down a little more than a week, and the Planning Manager claimed that it was idled only for 2 or 3 days.

Due to the uncertainty caused by the CRTs, Apple's dealers started telling their customers who wanted to buy Macintosh that they would have to wait about 6 weeks for delivery.

Finally, some extra workers were asked to work on a Saturday to catch up with the delayed production. But the Production Department missed their monthly shipment plan (Note 2.4).

FLEXIBLE MANUFACTURING SYSTEM (FMS)

Modern production technology uses flexible machines to produce a wide variety of parts.

In general, the cost of flexible automation depends on the amount and nature of information that an automatic control machine must handle and process. To machine different metal parts flexibly, the machine must be given information about the characteristics of all the metals. This requires substantial initial fixed capital investment. The mass production of a standard part, such as a crankshaft, distributes the fixed capital cost over a large number of crankshafts produced by the machine without any additional cost. If the machine is expected to produce not a single part, but a large variety of parts, and only a few of each at a time, the machine tool must be retooled to a flexible manufacturing center.

The flexible machine systems will be discussed in Chapter 6, in the context of management of information processing for cross-functional communication. The flexible manufacturing systems are driven by the computer and information processing technologies

A BRIEF SUMMARY

We reviewed the historical evolution and antecedents of management of production and engineering operations in the West. Our journey started from the princely production of Egyptian pyramids to the current lean mean production age. We learned about the experience curve effect, lean mean production and other practices.

SELF-REVIEW QUESTIONS

1. Consider the industry you are familiar with, and trace the antecedents of the production practices used in the industry to their historical roots.

2. Can we apply experience curve to cooking? Are there diseconomies of scale?

3. What are the limitations of applying the Just-In-Time pull production method to industries in America?

NOTES, REFERENCES, AND ADDITIONAL READINGS

2.1. Smyth, Piazzi. 1978. *The Great Pyramid.* New York: Gramercy Books.

2.2. Durant, Will. 1935. *The Story of Civilization, Part I: Our Oriental Heritage.* New York: Simon and Schuster.

2.3. Clifford, James. 1956. *Principles of Economics.* New York: Barnes & Noble College Series.

2.4. Scientific American. 1955. *Automatic Control.* New York: Simon & Schuster.

2.5. McClelland, David C. 1961. *The Achieving Society.* New York: Van Nostrand Reinhold.

2.6. Richards, E. 1984. A worm spoils Mcintosh's barrel, *St. Louis–Globe-Democrat,* March 22:6C.

2.7. For more on Henry Ford, see: Staudenmaier, John M. 1994. *Invention & Technology,* Fall: 34–44. and Livesay, Harold C. (1979), *Men Who Shaped the American Economy: American Made.* New York: Harper Collins. See Chapter 6: The Insolent Charioteer, Henry Ford, pp. 139–181; and Chapter 9: Superstar of the Worldwide Car Henry Ford II, pp. 241–266.

CASE 2A: TECHNOLOGY MANAGEMENT PRACTICE

Role of Cotton Production in Industrial Revolutions

Cotton: The Kingmaker. Cotton, the stuff of many legends, has been the kingmaker for more than 50 centuries of human civilization. It has clothed many different civilizations in different parts of the world. Cotton has helped many enterprising traders prosper and created new trading centers. At the same time, cotton has driven men against men, capturing and trading other men as slaves, to be employed in cotton plantations in distant lands, far away from their homes. Cotton has produced the economic miracles of the Industrial Revolution and has led to the colonization of huge empires. Cotton also gave birth to new industries (such as cellulose acetate and rayon) that caused the demise of its own cotton textile industry.

Cotton, a fiber produced by nature, has a length of 1 inch to 1.75 inches. It is strong, yet soft and easy to bend. It has a natural curl or slight twist that holds the filaments together. Most of the fibers that are to be woven into fabrics need to be twisted into yarns.

To grow well, the cotton plant requires a warm climate, lots of sunshine, rich soil, and a large water supply. For proper cultivation, cotton requires a long (about 6–7 months), warm or hot growing season, and 3–5 inches of rainfall during its growth. The cotton plant takes 80–110 days from planting to flowering, and another 55–80 days from flowering to opening of the cotton boll.

Competing Textile Fibers. Until the birth of the man-made and synthetic fibers in the late 19th century, for 70 centuries the history of human civilizations has been marked by the use of four natural fibers. These gifts of nature—flax, wool, cotton and silk—appeared in that order and competed with each other for the clothing needs of mankind.

1. *First Came Flax.* Flax fiber and linen-like fabrics were produced and used by the people of primitive Ancient cultures along the Nile River around 5000 BC. These primitive people lived in the region now known as Egypt, and they separated the strands of the bast fibers, including flax, from their stalks. They then plaited or wove them into simple textiles to cover their bodies. In the Later Stone Age, the Swiss Lake Dwellers around Lake Robenhausen in Switzerland used bundles of flax fiber, spun threads, and fragments of fabrics. These were found preserved in mud and waters of this lake.

2. *Wool Came Next.* In the Later Stone Age the Neolithic Man of the Swiss Lake Dwellers also domesticated sheep, and they used wool along with flax. However, in ancient Mesopotamia, on the banks of Euphrates River, around 4000 BC, sheep were raised and their wool was used in primitive fabrics. The early Babylonians and Assyrians perhaps also used woolen robes.

3. *Cotton, the Next Fiber of Civilization.* Around 3000 BC, according to the archeological excavations in the Indus River Valley, in the Sind region of undivided India, use of cotton and its weaving was clearly evidenced. This fiber will be discussed in detail later.

4. *The Fourth and Finest Fiber: Silk.* Silk originated in China probably by an accident. According to a popular legend, around 2640 BC, the Chinese Empress Hsi-Ling-Chi accidentally dropped a mulberry cocoon into hot water. She then noticed the separation of silk filaments. Realizing the textile potential of her findings, the Empress started the practice of cultivation of silkworms and the white mulberry leaves that silkworms ate. By 1400 BC, the Chinese perfected the craft of sericulture and produced woven silk fabrics. These fine fabrics were often used as a means of paying tribute to higher authorities in the Chinese society.

History of King Cotton

Origins of Cotton. Cotton is no doubt the earliest fiber used by mankind, and it is older than even the recorded history. Cotton has been known to mankind for so long that its origin is often lost in legends. The most popularly accepted origin of cotton is in India in 3000 BC. According to the archeological discoveries of the Ancient civilization of the Indus River Valley, Indians were the first ones to spin cotton and weave cotton into fabric. Cotton is mentioned in the historic book of Indians called *Rig Veda.* The cities of the Indus Valley of ancient India showed evidence that cotton was grown and used by the people as early as 3000 BC.

Weaving of cotton into cloth is seen in recorded history, going back to 1500 BC. For example, the ancient Hindu Laws of Manu specified that the sacrificial thread worn by the Brahman had to be made of cotton (karpasi in Sanskrit). Furthermore, rice water, probably used as the first application of starch, was used in the weaving of cotton. There is abundant evidence that by 1500 BC, Indians were clearly raising abundant cotton and spinning it into yarn to be woven into cotton fabric.

Nearchus, an Admiral for Alexander the Great, settled a colony of Macedonians on the banks of the Indus River. He reported the chintz or flowered cotton fabrics, which rivaled and resisted the sunlight and washing.

Hindu and Arab traders took small quantities of cotton fibers and woven fabrics to other parts of the world. How to grow and weave cotton cloth spread through other parts of Asia Minor in Persia, Babylonia, and Egypt. This probably took place with the help of the Arab traders who traveled to India for its spices, silks, and silver. One consignment of cotton changed hands many times. Often, the people who finally bought the cotton fabric did not know where the fabric had originated from.

The Greek historian Herodotus (485–425 BC), who is often considered to be the father of the field of history, wrote about cotton after a trip to India. He wrote, "There are trees in which fleece grew surpassing that of sheep and from which the natives made cloth." He noted that the Indian auxiliaries of the Persian King Xerxes were clothed in cotton and that the fabrics made of this exotic fiber owed their exquisiteness to the 1000-year-long perfection of the craftsmanship practiced by Indians.

Alexander, during his attempted invasion of India around 327 BC, was amazed by the beautiful printed cotton fabrics produced in India.

Some traders took cotton fabric to China about 502 BC. But the Chinese already used silk a lot more. Therefore they were not much interested in cotton. Initially they grew cotton only as a decorative garden plant.

With time the diverse uses and manufacture of cotton spread to places as far away as Latin America. During the pre-Inca civilization, mummies in Peru were found to be wrapped in fine cotton. Often the imported cotton fabric was very expensive, and only the very wealthy could afford it. Even the emperors or princes were proud to have a fine garment made of cotton.

Introduction of Cotton to Europe. By the early Christian era, cotton was known to be grown in Asia Minor. Europe had yet to learn how to do so. Often the Europeans did not know that the exquisite cotton fabric they wore had originated from India. The climate and characteristics of the land soil in India produced cotton with a short-staple fiber.

Cotton came to Europe in great quantities because of two major historical developments. Alexander's armies introduced the cotton culture into Northern Africa. The Nile Valley with fertile soil offered great conditions for the cotton "wool plant" much closer to Europe. Not many years later, Egypt produced a variety of cotton that had a longer cotton staple fiber. This, coupled with the proximity of Egypt to Europe, helped the long-staple cotton become the preferred choice of the people of Europe. First the Hebrew and then the Phoenician traders spread the cotton trade westward.

In the 7th and 8th centuries the Crusades brought the craft of cotton spinning and weaving from Egypt to Spain. As late as in the 10th century, cotton was introduced by Moors to Spain and then spread further to Northern Africa, along the coast of the dark continent—as far south as the island of Madeira. The European word *cotton,* according to some sources, is based on the Arab word "quttan or kuttu" meaning "a plant found in conquered world." The rest of Europe imported cotton from Asia. However, the trade routes between Europe and Asia were closed when Constantinople was conquered by the Mohammedan rulers.

Cotton has been the cause of many major milestones of modern mankind. Cotton was one of the main causes for the birth and growth of Venice and Genoa as the great trading cities of the Middle Ages for trade with the East.

For centuries, cotton was processed and woven based on a method derived from the ancient Indian method. The raw cotton fleece was handpicked. The seeded South Atlantic cotton in America involved harder work. Hand cleaning of cotton was too slow and unproductive. The cotton fibers were to be spun into yarns and threads. The Chinese had attached a foot treadle to the ancient Indian hand crank. Leonardo da Vinci's rotating flyer, called the Saxony wheel in the colonies, helped do this in a continuous motion. The picked raw cotton was then "bowed." A workman struck the string of a bow with a mallet. The vibration that was produced opened the cotton knots, shook out particle dust, and fluffed it like a down fleece. This was spun, and the home-made thread was woven on hand looms in the households.

Seven Pioneers of Industrial Cotton Technology. Six Englishman and one American particularly stand out among the millions of inventive geniuses who contributed to the mechanization of processing of cotton and other fibers. This led to the Industrial Revolution.

1. *John Kay's Flying Shuttle.* Cotton cloth used to be woven rather slowly, by manually passing a shuttle in and out of the warp threads, until one Englishman changed that. John Kay's father, who produced woolen goods in England, trained him in textiles. In 1733 (or 1730) John Kay invented the flying shuttle, an important textile mechanical device. For the first time, the flying shuttle enabled English weavers to make cotton fabric as wide as the fabrics imported or smuggled from India. The flying shuttle passed speedily back and forth in one groove. This made weaving much easier and faster than before. The hand-loom weaver thus doubled his output with the use of a flying shuttle.

In England at that time, cotton fabric was practically outlawed because it posed a threat to the workers involved with woolen goods. Mechanization further added to their threat. Thus mobs destroyed the first "bobbin" machine.

John Kay also designed a machine to make the cards used to disentangle the cotton fibers before spinning. In 1759, one of Kay's sons invented a drop box to help weave a cloth in three colors nearly as fast as a length of plain cloth. Speeding of the cloth weaving process caused an imbalance in the cotton textile making process. One weaver needed the output of four to five spinners. This set the stage for improving the spinning operation.

In 1738 Lewis Paul applied for a patent to use rollers to draw out the rovings as part of a power-driven spinning machine. In 1741 this was powered by two asses, and in 1743 it was run by water power. These did not have a significant impact on the textile industry until the 1750s and 1760s when John Kay's flying shuttle was popularly adopted. By then the market demand for cotton fabric was increasing so rapidly that there was a shortage of spun yarn. Weavers had become more than five times faster than the spinners. This created a pressure to improve the productivity of cottage spinning using spinning wheels.

2. *James Hargreave's Spinning Jenny.* In 1764, James Hargreaves, a cottage weaver, made the first practical spinning frame in England with multiple spindles. He called his machine the *spinning jenny.* This could spin eight spindles at a time. Thus eight strands of cotton thread could be spun at the same time. The old-fashioned spinning wheel turned only one strand of thread at a time. Hargreaves' original machine design copied the actions of a traditional spinning wheel without a flyer. He combined multiple spindles in such a way that one spinner could operate all eight spindles. Later Hargreaves continuously improved his invention, perfected his design by 1768, and designed a Jenny with 100 spindles. This was designed for manual labor and could not be adapted to use power drive.

But the yarn spun by Hargreaves' spinning jenny was only suitable for use as a weft in fabric weaving. The warp yarn for fabric weaving still had to be spun by a hand spinning wheel.

3. *Richard Arkwright's Roller-Spinning Method.* In 1771 Richard Arkwright, a barber, introduced the roller-spinning method. The spinning frame was patented around 1769 as "water frame" or "thostle." The machine was specifically designed for use with animal power or water power. It used rollers to draw out the rovings, similar to Lewis Paul's machine. Otherwise it was almost the same as the late medieval spinning wheel, using a flyer with a power drive added to it. Initially, Arkwright's roller-spinning frame used pins to guide even distribution of thread, just like the traditional spinning wheel. The flyer gave the cotton thread a twist while drawing it through rollers. The twisted thread could be used for warp as well as weft. Two years later, in 1773, the roller-spinning method was used by the first cotton mill in England to draw finished spun yarn from raw fiber stock. Arkwright also developed the carding processing method using cylinders.

Hargreaves' spinning jenny and Arkwright's roller frame were constructed of traditional materials that also made the medieval spinning wheel. More parts were interlinked. Such devices had been already developed by the end of the 18th century, and they were popularly used by textile industry.

Initially, Arkwright's spinning frame was driven by horsepower turning a treadmill. Later England started using water power to drive Arkwright's rolling "spinning frame" machines. Subsequently, steam powered the roller-spinning frame. With these improvements, a single power source could simultaneously run many textile machines. Thus, Arkwright's inventions marked the end of the traditional domestic system of manufacture, and it ushered in the start of centralized factory system of machine based manufactory.

4. *Samuel Slater's Transatlantic Transfer of Cotton Technology.* Samuel Slater was not a creative inventor, and he worked as a machinist in an Arkwright mill in England. His contribution to cotton textile technology was that he decided to move to America. He memorized a complete working plan for the Arkwright power spinner. This was banned at that time. Slater pioneered America's industrialization in a significant way. Later he became one of the biggest producers of cotton processing machinery in America.

5. *Samuel Crompton's Spinning Mule.* A musician as well as a mechanic, Crompton was a dreamer. He dreamed of improving the society he lived in. He developed his spinning mule in 1779, which helped perfect the spinning of fine cotton yarn. He combined and improved the key features of Arkwright and Hargreaves machines. He put the mule on a movable carriage for stretching and twisting the cotton yarn. The machines were getting increasingly more complicated and were increasing the number of spindles. A skilled spinner was required to supervise and intervene a rather complicated cycle, and therefore at first the spinning mule was not very well suited to use a power drive. For many decades, the same principle was used in most of the world's textile spinning industry.

Crompton was too poor to patent his invention, and he sold it to others for a small amount of money. Many other people made large fortunes from Crompton's invention, or by the production and selling of cotton fabrics using his invention. Many years later, the English House of Commons acknowledged his contributions, and granted him 5000 British pounds for his pioneering contribution.

So far the textile machines were developed and designed by common sense. These machines were designed by amateur handymen who were not trained mechanics. Hargreaves was a carpenter and weaver. Arkwright was a barber and hair trader. Crompton was a farmer and weaver. They all had learned to make their machines themselves. They worked with or were trained by someone who had worked under trained mechanics, such as Henry Maudslay, who invented and improved a lathe.

The textile yarn spinning had gained phenomenal increases in productivity. Thus weaving became the slowest link to making the final cloth. Weaving looms were waiting for the power drive.

6. *Edmund Cartwright's Automatic Powered Loom.* Cartwright, a clergyman turned inventor, invented the world's first automatic power loom in 1785. He leveraged many of the inventions that were invented by others before him. Initially the power loom was a crude machine to work with. The fabric produced was of coarse weave. It took substantial time before more people switched from hand loom to power loom. Cartwright's contributions helped improve the performance and efficiency of textile processing. Until then, many of the weavers and spinners were still using the hand loom as improved by John Kay and his son. The power loom also replaced the privileged powerful position of the hand-loom weaver. Over three decades later, in 1822, Roberts designed a true self-acting mule.

In the fourth quarter of the 18th century, the finishing processes for cotton textiles were also significantly improved. Berthollet introduced chlorine-bleaching, and new synthetic dyes were discovered. Thomas Bell invented cylinder printing.

Until 1800, cotton-spinning and weaving innovations were combinations of using various parts of the traditional spinning wheel. Except for the use of rollers, these technological inventions did not understand the sophisticated principles of science. The inventors and innovators merely responded to the market pressure and higher demands.

7. *Eli Whitney's Cotton Engine or "Gin".* A fresh Yale graduate visited some friends in Savannah in the American South. He was impressed by the plantation life of his well-to-do friends with moonlight parties, and he decided to stay on for a while. During the parties his Southerner friends often complained of the labor, time, and effort required to separate the cotton fibers from the seed in a cotton flower. The work had to be done manually or by "churgha ginning," usually by slaves and other hired help. A worker could thus clean only about one pound of raw cotton per day.

Young Whitney, with some mechanical skills, saw this and worked for a few months to develop a cotton saw engine or gin to do the separation by machine. A hand crank turned the "saw-gin," whereby a mass of raw cotton was plucked apart by revolving spikes. The lint was carried away while the seed dropped down. Using Whitney's cotton gin, a worker could easily clean as much as 50 pounds of raw cotton in a day. With time, larger cotton gins were designed and produced, which could clean even larger quantities of raw cotton every day. But the principle remained the same.

Impact of Cotton Gin on the American Economy. Prior to the invention of the cotton gin, in 1791 America shipped only 400 bales (500 pounds per bale) of cotton to Europe. In 1800 the cotton shipments increased 75 times to 30,000 bales, or about 1000% growth per year. By the end of the first decade of the 19th century, about 180,000 bales of cotton were exported by America to Europe in 1810.

Whitney's Invention and Slavery. In no time, cotton became the economic "King of cash crops in the South." To plant, cultivate, pick, and process increasingly larger amounts of cotton, more and more farm help was needed. Just 6 years prior to the invention of Whitney's gin in 1792, the entire South had unanimously agreed to abolish the importation of slaves. But with the improvement in the processing of cotton with Whitney's gin, the situation changed dramatically. There was money to be made by the cotton farmers. And the increasing work could not be done by the British colonists themselves. More and more slave labor was imported. In the first decade after the invention of Whitney's gin, the slave population of the Southern states was increased by 33%. Within one decade, more than one million slaves were working on cotton plantations in 1810.

Cotton had a different impact on the American North also. Until the phenomenal growth of the popularity of cotton, the American people primarily used wool and linen (linsey-woolsey). Cotton was until then processed primarily in homes. As the amount of cotton available grew rapidly, entrepreneurs began setting up machines to process and weave cotton fabric. Quaker merchant Moses Brown hired Samuel Slater to build a cotton mill in 1793 in Pawtucket, Rhode Island. The mill employed Slater, a widow, and five children. The children were paid 60 cents per week, and they worked 14 hours a day in the summer and 12 hours a day in the winter. The widow received $1.60 per week. For the pioneering development and modification of a spinning frame machinery, Slater is called the father of the American hydromechanical spinning.

With the abundance of rapid waterways in New England and the development of hydraulic power, textile processing mills grew swiftly in the North in the early 19th century. In 1822 the Merrimack Manufacturing Company commissioned its historical first mill at Lowell, Massachusetts. In no time, New England became a rival to England in the production and manufacture of cotton cloth. New Bedford became the "Bolton of America." Cotton thus also became the "King of the North."

Initially the textile mills did not grow equally rapidly in the Southern states. During 1817–1822, the South built mills in North Carolina. Savannah, New Orleans, and other seaports in the South traded cotton goods to worldwide markets. By 1860, before the beginning of the Civil War, the South had increased its slave population to over 4 million. The American South was by now a strategic source of raw material to the textile industry of Britain. The United States exported 2.5 billion bales of cotton to England in 1859–1860, compared to only 563,000 bales imported from the East Indies.

The Civil War and Cotton-Picking in the South. The 4 years of Civil War, combined with the liberation of slaves, caused havoc on the industry and commerce of the South. In 1881 the International Cotton Exposition met in Atlanta, showcasing improved methods for cotton planting and ginning. The cotton industry started reviving in the New South.

Southern textile industry saw a rapid growth in the years between the two World Wars. Soon the New South was overtaking the North in textile production. In the last year of World War I the North had over 18.1 million active spindles, whereas the New South had 14.5 million active spindles. By the beginning of World War II, the South increased its textile processing capacity to 17.9 million active spindles whereas the spindles in New England dwindled to 5.3 million spindles only. By 1960, the United States had 23.8 million spindles; 18.9 million of these were located in the South, whereas 4.9 million were located in New England.

With the industrialization brought about by the two World Wars, relatively lower wages, and the development and spread of electric power, the industrial textile mills grew in the American South. By the end of the 1950s, about 80% of American textile technology was south of the Mason–Dixon line. The pattern did not change thereafter.

Source: Adapted from various sources listed at the end of Chapter 2.

CASE 2B: TECHNOLOGY MANAGEMENT PRACTICE

HENRY FORD: ASSEMBLY MOVER OF AMERICA'S AUTOMOBILE AGE

Ford's Farmer Father

Henry Ford was born on July 30, 1863, in a farmer's family in Dearborn, Michigan. When Henry was 13 years old, he had already lost his mother and three siblings. His widower farmer father with five children noticed that Henry hated the farm life, and he disliked the chores of chicken and cow that came with it. Henry, however, liked tinkering and mechanical repairing of farm implements. He fled from home at the age of 16, and he took an apprenticeship job in Detroit to repair steam engines. He also moonlighted fixing clocks and watches. Later, he even thought of going into the business of making a watch for 30 cents. But he rejected the idea because it was not a necessity for all Americans. At 17, Henry was a certified journeyman machinist. (Sources are provided at the end.)

Mechanic Instincts

In the words of Thomas Edison, Henry Ford was a "natural-born mechanic." Henry learned the logic of a machine by looking at it once. He might not have been able to create a new machine, but he was good at improving an already existing one using the known ideas already invented by others.

At a Palm Beach automobile race crash, Henry Ford picked up a smashed part of a French car and learned that vanadium alloy could give him three times the strength per weight of steel. Later he used this knowledge to make his Model T lighter. Because of his single-minded search to succeed, he also gained a diverse range of mechanical experience in the American industry of the period. From repairing steam engines for sawmills at Westinghouse, he began repairing an internal combustion engine at Eagle Iron Works.

In 1889, at the age of 26 years, Henry Ford married a girl next-door, Clara Bryant. He was a devoted husband. Two years later, in 1891, they moved because Henry joined the Detroit Edison Company. Henry Ford, with his mechanic instincts steadily progressed through the organization and became the chief engineer of the company.

There, Ford met the great inventor Edison in August 1896. Henry shared his vision of tinkering with a self-propelled gas-driven buggy. Edison encouraged Henry and agreed with him that there was a significant future with a light but powerful engine for general use. Earlier in June 1896, Henry Ford had already mounted a gas engine on a buggy car. He had driven the car in late hours of the night on a rainy day. As the car rolled out of a broken wall, his wife Clara held an umbrella over his head. Another friend piloted ahead of him on a bicycle, to shoo away any horse-driven carriages clear of the still unpredictable contraption. However, it required 3 years more before Henry Ford was convinced about this radical idea for mass transportation.

Birth of a Motor Company. As America edged toward the 20th century, in 1899 Henry Ford left the Detroit Edison Company. Henry Ford started the *Detroit Automobile Company.* He received the blessings and capital from the mayor and other local capitalists. According to the prevailing business practice, Ford contributed no cash but gave his designs and technical expertise for his share. This company built 25 cars and went belly up.

In 1901, Henry Ford revived his dream and started the *Henry Ford Motor Company.* He now owned one-sixth share in the company, with the rest held by other financiers. They differed on many things. Henry Ford dreamed of building a cheap car accessible and affordable to all. The financiers believed in making expensive cars for the rich. Henry Ford believed in publicizing his automobile performance by building racing models. The financiers felt this was a distraction and drain of their resources. Finally, they brought in Henry Leland as a master machinist to counter Henry Ford.

Ford was confident of his ability to build racing cars. In October 1901 he challenged and defeated the current speed-car champion Alexander Winton. The next year, Barney Oldfield set a new American speed record in Ford's 999. This for the first time confirmed Ford as a hero in the eyes of the native Detroit residents.

Interchangeable Parts

At that time, Ford still produced his automobile one at a time. He had a limited idea of how to make cars in mass numbers. Henry Leland, on the other hand, had some knowledge and inclination toward making cars with interchangeable parts. In 1908 he demonstrated vividly in front of the British Automobile Club, and the world automobile community by disassembling three Cadillacs into component parts, mixing them, reassembling them, and then driving them for a few hundred miles without a glitch.

The two Henrys, Leland and Ford, often clashed. Even though Leland had a few technical tricks to teach Ford, Ford was not the type to listen docilely to what could or could not be done. Many of Henry Ford's subordinates learned to leave out such conclusions for Henry Ford. The pioneer in him always propelled him to prove the impossible. Ford became frustrated and left. Leland, on the other hand, stayed on and later founded Cadillac Motor Company. And then Lincoln too.

Third Time Around: Ford Motor Company

In 1903, Henry Ford formed the *Ford Motor Company.* This was a 250 × 50-foot plant employing a dozen employees. This time he owned 25.5% of a total of $28,000 in stock split into 1000 shares. A matching 25.5% stock share, or 255 shares, was held by a coal dealer, A. Y. Malcolmson. The remaining 49% was owned by 10 other investors, with 49 shares each. This included the Dodge Brothers, whose machine shop was to supply parts, and Albert Strelow, who provided his own shop as the factory space for the new company.

Soon the original shareholders lost faith and disposed of their holdings. Malcomson got $175,000. Strelow sold off his shares for a mere $25,000. On the other hand, a Ford share worth $100 in 1903 was bought back by Henry Ford in 1920 for $262,000, a 2620-fold increase of the original investment in 17 years.

Product and Production Revolution. In 1908, the Ford Motor Company launched its Model T car in two versions, and it priced them at $825 and $850.

Some historians consider the design of the Ford Model T to be the most successful single technical design in the American history. The design perfectly fit the context in which Model T was introduced. (Sources are provided at the end.)

In 1904, only one in 14 miles of roads was paved in the United States. The rest of the roads were washed out by rain, and they got worse when they dried. To meet the challenge of these road conditions, the Ford Model T was designed to be a sturdy car to take the beating. Of course, such country roads had no service or gas stations. Thus Model T was designed simple so that anybody could repair them.

Ford fought frequently with his stockholders over the type of car he ought to be making, as well as over the target markets. The financiers wanted the status-quo design, a few expensive and luxurious cars priced high with a fat profit margin for the few rich professionals. Henry Ford wanted to design and make lots of cheap cars and sell them at a small profit margin, to a very broad segment of people. And he proved right. The Ford Model T cars sold in such large numbers that after the first three years, Ford production supervisors could not forecast their annual sales. All they had to do was make them as fast as they could—hence Ford's rush to speed the production operations as fast as he could.

The two models of Ford Model T enjoyed a dominant market share of 50% or more for many years. To meet such high customer demand, Ford employed an innovative production process, namely a continuous assembly line. This was a modification of the production process first introduced by rival Ransom Olds. Henry Ford also innovated some new labor practices. He offered an unheard of daily wage of $5 a day, and he introduced an 8-hour work day in 1913.

Ford's early factories, such as the Mack Avenue plant and the Piquette Avenue plant, opened in 1903 and 1904, respectively. They made cars like everyone else was making them at that time, using stationary workstations where parts, supplies, and subassemblies produced were assembled together. But between 1906 and 1914 Ford radically transformed the production process from the stationary way to the moving assembly line. In the new production process workers did a few specialized tasks while a slowly "birthing" car moved past their work stations.

The biggest challenge for the moving assembly line was on the material handling side. As the assembly line was moving at a constant brisk pace, all the needed parts, supplies, and subassemblies had to be available at the workstation ahead of time. Ford built an unprecedented massive plant at Highland Park in 1910. It included large four-story-high manufacturing buildings, one-story-high machining buildings, an elegant electric power plant using gas turbines, and railroad yards to bring in materials and other supplies right inside the plant.

The workers' tasks at the Ford plant were designed in such a way that even an untrained immigrant worker could easily do the simplified repetitive tasks. America, with a population of 80 million in the first decade of the 19th century, was attracting over 800,000 people every year. They mostly came from the rural regions of Southern and Eastern Europe. Most of them did not know much English. Ford and many other companies ran English-language schools for their immigrant workers. Some immigrant workers, including the American workers, spread the word that the mind-numbing repetitive work caused "Forditis." This was a set of different physical and mental conditions in workers working on the assembly line. The fear spread, and in 1913 worker turnover at Ford plants reached 370% a year (see sources at the end).

Ford's 8-Hour $5 Workday

In January 1914 Ford announced that a work shift at Ford plants would be 8 hours instead of the 9-hour workday common in all factories at that time. Also announced was that Ford would double the daily wages to $5 a day, instead of the $2.50 they and every other plant paid at that time. The announcements attracted millions of workers from all around the country. Everyday it created riots in front of the Ford plants. The Ford Way shocked the American industry.

But the 8-hour $5 workday was offered to immigrant workers with some strings attached. Their homes were inspected for cleanliness by inspectors from Ford Sociological Department. Workers were also asked and checked if they were legally married. Some historians would argue that many other American employers put similar demands on their immigrant workers.

Ford's growth seemed unstoppable. During World War I, Ford started the River Rouge plant. This was a 2000-acre factory with more than 100 buildings. Ford had envisioned a fully vertically integrated enterprise system. Boats brought basic bulk raw materials to the docks of the plant. Ford owned hundreds of thousands of acres of hardwood forests in upper Michigan, and the company processed its lumber there. Ford also bought iron mines in Michigan to make his own steel. Rubber plantations in Brazil were purchased to produce their own rubber. Glass was also produced in-house. Ford controlled the entire production process from the raw materials to the finished cars.

Ford Paradox

Henry Ford hated the constraints of a farming life, and he preferred the urban settings to the rural settings. To many Americans of his time, Henry Ford was "a simple man, an honest man, whose virtues were manifested in the products he made: plain, cheap, durable, and simple enough to be repaired by any man with a simple set of tools and an ample supply of elbow grease" (see Source at the end).

Ford perceived that an American was relentless and that he "wanted to be someplace he ain't. [And] as soon as he gets there, he wants to go right back."

Ford didn't care about the prevailing paradigm that the auto-driven vehicle was a luxury to be built and owned by only the rich. The Ford car helped average Americans stretch their reach into the continental American Frontier. It was faster than ever and more reliable. Ford succeeded in achieving his mammoth mission—that is, the production of:

a motor car for the great multitude [that is] large enough for the family but small enough for the individual to run and care for. [It is] constructed of the best materials, by the best men to be hired, after the simplest designs modern engineering can devise. [Yet, it be] so that [one] making a good salary will . . . own one [and be able to] enjoy with his family the blessings of hours of pleasure in God's great open spaces.

Ford championed American individualism, and he hated bureaucratic interference by others. This ranged from statisticians to stockholders, from federal regulators to union rioters. He bought over the meddling stockholders, and he couldn't understand what more "a union could give" beside levying their tax, when he was willing to pay workers $5 a day while the going rate was $2.50.

Wary of Wall Street Bankers

Ford, like his friend Edison, very reluctantly dealt with Wall Street, and he hated bankers the most. He believed Andrew Jackson, who stated that bankers try to conspire to confuse the average Americans and grab their hard-earned savings. He was glad that he had risen beyond their reach. He subscribed to the mood of the day that Ford produced cars that helped people run, while the bankers created the Great Depression. He hated Bernard Baruch, a prominent American in his time.

Ford first ran as a senator and seriously considered running for President in 1924, to counter such financiers' influences. He fought hard against patient cartels, but he donated his personal wealth to continue patient care in the Depression years.

Ford had other character flaws, which some believe came after he became very rich and famous. Some considered him racially prejudiced (Livesay, 1979, p. 161).

Henry Ford cared little for his only son, Edsel Ford. Henry relied more on a bodyguard and his own tactics to run his worker force. Henry Ford then compensated the bodyguard by making him his right-hand man and the second most powerful person in his company.

But, paradox or not, Henry Ford's achievements, more than those of anybody else, crafted the America's 20th century. He built, owned, and completely controlled a billion-dollar corporation. But, Henry Ford's values were deeply rooted in the 19th-century daring as well as in dogmas. He was driven by delivering results for the future, and he did not care much about the past or what others thought of him in the present.

Terrible Twenties: Ford's Fall from Grace

The sales of the Ford Model T started sliding during the 1920s. Ford tried a number of marketing approaches. He reduced the price a number of times and increased the dealers. But these efforts produced little favorable results. Finally, Ford agreed with others and decided to introduce a new Ford Model A car. In 1927, Ford plants were closed for 9 months to retool for the model change. In the meantime, General Motor's Chevrolet overtook Ford for the market leadership, and it never let go of its position since then.

Sloan's Strategy for General Motors

Alfred P. Sloan, the General Manager of General Motors, changed the norms of the auto industry. He offered a different "customized" model to each customer. Henry Ford on the other hand offered a car in any color as long as it was black. Furthermore, Henry Ford's Model T was built to last—rugged, reliable, and economical. The style remained unchanged from one year to the next year. In 1923, Sloan of GM introduced annual changes in auto styles to create a certain amount of customer dissatisfaction with their older models.

General Motors thus became the new king of the auto world.

Source: Adapted from Staudenmaier, John M. 1994. *Invention and Technology,* Fall:34–44; and Livesay, Harold C. 1979. *Men Who Shaped the American Economy: American Made,* New York: Harper Collins. See Chapter 6, "The Insolent Charioteer, Henry Ford," pp. 139–181; and Chapter 9, "Superstar of the Worldwide Car Henry Ford II," pp. 241–266.

CHAPTER 3

Proprietary Know-How, Intellectual Property, and Research & Development: The Sources of Scientific Industrial Revolution

PREVIEW: WHAT IS IN THIS CHAPTER FOR YOU?

With the dawn of the 20th century, systematic experimentation and the scientific research ushered in the Scientific Industrial Revolution in the United States. With the development of proprietary know-how for a sustainable competitive advantage, America became the new world leader, in what is often referred to as the American Century.

We will review five alternate routes for gaining proprietary intellectual know-how. These include:

1. Charles Goodyear's accidental discovery
2. Thomas Edison and Baekeland's systematic experimentation
3. Wallace Carothers' scientific research
4. Sony's acquisition
5. Kodak's Advanced Photo System alliance

Working definitions of basic research, applied research, development, and other terms will be discussed. Empirical models such as the pattern of innovation and S curves will be reviewed for analytical guidance. Once we understand the nature of proprietary know-how, we will look at ways to protect our intellectual property. Patenting will be explored in detail.

At the end of this chapter we have a case on Wallace Carothers, the inventor of a family of popular polymers such as nylon, polyester, neoprene, and others. at E. I. Du Pont's Experimental Station. Carothers also laid firm foundations for the two fields of polymer science and fiber science. The second case lists the milestones of technological achievements by researchers at Bell Telephone Laboratories.

THE SCIENTIFIC INDUSTRIAL REVOLUTION
AND THE AMERICAN CENTURY

The Industrial (Machine) Revolution in England was fueled by the development of production and operations technologies, such as the steam engine and textile machines. In the late 18th century, but mostly in the 19th century, England leaped into the position of the world's new industrial leader. By the end of the 19th century and the beginning of the 20th century, however, the center of economic and industrial gravity started shifting to North America. Why did this shift take place from a big empire to this young nation?

There were some interesting technology-related factors contributing to this. The technological projects such as the ground-breaking construction of Erie Canal and other waterways in the 1820s, the construction of coast-to-coast railroad connections, and other improvements in the infrastructure of the Panama Canal played a big role. The end of the Civil War in the 1860s ended the North–South fight over slavery. But the major seeds of economic growth and wealth creation came from the birth and growth of new technologies which gave birth to new industries. The electrical, communication, and chemical industries were born in America. The technologies for all these new industries had one thing in common. They were based on a systematic or scientific research. These were supported either by pioneering individuals or by visionary corporations.

The first few technological discoveries were made by the development efforts of individual researchers such as Charles Goodyear (who invented vulcanization of rubber), George Eastman (who developed a portable Kodak camera for amateurs), and Leo Baekeland (who discovered the route to plastics). Subsequently, the focus shifted to corporate research in laboratories, such as at the General Electric Research and Development Center, Bell Telephone Laboratories, and the Du Pont Experimental Station. There is very little doubt that this Scientific Industrial Revolution launched America into becoming a world leader in the 20th century. Because of America's rise to stardom, the new century has often been referred to as the American Century.

AMERICA'S TANGIBLE AND INTANGIBLE RESOURCES

Tangible Natural Endowment

The New World, with a large continental land mass, was no doubt endowed with rich tangible resources. This included an enormous stretch of fertile land for agriculture, easy access to big rivers, and rich deposits of coal, iron ore, minerals and fossil fuels.

However, unlike in the case of England and much of Europe, there was not enough supply of manpower in America to exploit the natural bounties.

America therefore often favored mechanization of work over manual operations. This gave birth to a love and fondness for new technology. Technology helped America to overcome its limited human resource and it enabled America to tackle its enormous challenges against nature.

A Valuable Intangible Resource: A Positive Attitude to Change

Another resource, and perhaps a more valuable resource than tangible resources, was America's intangible but extremely valuable attitude toward change. This progressive outlook was a major reason for the rapid rise of this young nation as the new industrial leader of the world. On the other hand, in Europe in general, and in England in particular, even in the 20th century, there was a tendency to frequently slide into a tradition-bound conservatism. There was reluctance to disturb the age-old order whereby the rich get richer by cornering all the new opportunities while the poor remained dependent on the mercy of the rich.

In America, on the other hand, the rules were often very different. Most of the early American immigrants left their far-off home countries in Europe because of persecution and exploitation in their home countries. They were therefore excited by the new freedom, which included the liberty to pursue happiness. Many of them were eager to try something new, almost anything new. Further, while doing so they felt there was little shame when they failed at first or did not succeed.

Soon, an "I Can" attitude of Americans gave birth to many new technologies. America thus became the natural home and center of gravity for (a) the new technologies based on human efforts and (b) the subsequent new industries resulting from these technologies.

New Indigenous Trades and Technologies

The early waves of America's immigrants came from the northern and southern parts of Europe. These Europeans brought with them the traditional European skills such as printing, basket-weaving, and candle-stick making. They were initially inclined to make their living from these crafting skills. However, very soon they either faced too much competition in these crafts or (b) exhausted full potential by saturating their markets. To survive and grow in this competitive arena they had to reinvent their futures. And unlike in their native home countries in Europe, their new homeland in America gave them freedom to dream and try new activities. With this freedom, the new Americans ventured into inventing new technologies to pursue their path to prosperity and more economic freedom.

This is the main reason why America rapidly became the home of the Scientific Industrial Revolution in the 20th century. This can be illustrated with the birth of the scientific-research-based chemical, electrical, and telecommunication technologies. These were the high technologies at the dawn of America's 20th century. The discussion below illustrates how American immigrants used two of their newly found freedoms—namely, to dream and to pursue happiness—to invent one new technology after another.

ALTERNATE ROUTES TO CONCEPTION OF NEW TECHNOLOGY

Given the freedom to pursue happiness and prosperity, a new technology can be conceived in one of the following five ways.

1. By accident—the Charles Goodyear way
2. By systematic trial and error—the Edisonian way
3. By scientific research—the Bell Laboratories way
4. By acquisition—the Sony way
5. By alliances—Kodak's Advanced Photo System way

1. Accidental Conception of Technology

The luckiest way to develop a new technology is by accident. History of technology is full of instances where people were trying to do one thing but serendipitiously hit upon something else, which led them to a new technology. And a pot of gold.

Potato Chip Technology. One famous accident is about the invention of potato chips—the staple food for most teenagers and college students around the world. According to the business legend, at a restaurant in Upstate New York, there was a particularly picky customer. He kept complaining about how soggy and thick his fried potato order was, and he kept sending the dish back to the kitchen to have the potatoes cut thinner and fried more. The cook got real upset with this pesky request. To spite the customer, he sliced a potato as thin as a paper. Then fried them so long that they curled. He sprinkled a lot of salt on them before sending them to the picky customer. He waited in the kitchen expecting the customer to get really wild. He waited anxiously for some time, but no complaints came in. Instead the owner of the restaurant came in all excited, telling the cook that the customer loved those paper-thin fries so much that he was asking everybody else to try them too. Soon some more orders poured in for the paper-thin fried potato chips.

The fried potato chips became a standard novelty of the restaurant's menu. The cook and the restaurant's reputation spread quickly, and many customers started pouring in from far-off places. Other restaurants imitated the deep-fried potato chips. As they say, the rest is history. Soon we all got hooked on potato chips, and one of the chip manufacturers challenges us to try just one. Potato chips have become a staple food of the American diet.

Another industrial example is the accidental discovery of the thermal vulcanization of rubber by Charles Goodyear, the father of America's rubber industry.

Charles Goodyear's Vulcanization of Rubber. Charles Goodyear (1800–1860) once said that he was encouraged in his long pursuit of his goal to vulcanize rubber and give rubber a stable shape by his belief that "what is hidden and unknown and cannot be discovered by scientific research will most likely be discovered by accident, . . . by a man who applies himself most perseveringly to the subject and [who] is most observing of everything related thereto."

The story of rubber is full of accidents. In 1735 a group of European astronomers, on an expedition to Peru, accidentally discovered that the native Indians collected the sap of a particular tree. In the heat of the sun or fire, the gummy substance was hardened. The native Indians often used that to cover their feet. The Europeans brought the sticky stuff with them as a curiosity with no practical application.

Many years later, the famous English scientist Joseph Priestley accidentally discovered that he could use the hardened gum to erase his mistakes in manuscript by rubbing the gummy substance over them. He therefore gave it the name "rubber." The French mixed some rubber strands with cotton threads to make "elastic" garters for women and suspenders for men.

In 1823, Charles Macintosh (1766–1843) accidentally discovered in England that rubber dissolved in naphtha at certain temperatures. He used this rubber solution to make waterproof fabric for mackintosh raincoats.

In the 1830s, a European ship brought some pairs of crude Indian rubber shoes to America. Surprisingly, the rubber shoes became instantly popular as shoe covers for the rainy days. The goods made of India rubber sold so well that it led to the establishment of the Roxbury India Rubber Company for making rubber laminated fabrics in America. Many other companies started similar businesses. But they soon discovered that the heat in summer melted the rubber goods and produced a very bad odor.

Charles Goodyear bought a rubber product (a lifesaver) made by Roxbury, and he worked on improving its air valve. He was told by a Roxbury salesman, again by accident, that the rubber itself needed far more improving. Goodyear took that advice seriously and searched for ways to improve rubber.

Goodyear, without knowing much about rubber, started doing experiments by mixing gum with all kinds of household substances. He used salt, sugar, pepper, oil, sand, and even cheese and soup. He then used a kitchen rolling pin to make thin rubber films. These films were made into rubber shoes. But the heat in summer melted them fast.

He tried many other things, but nothing helped to stabilize the unstable form of rubber under heat. He was left with little money and had to live off the charity of his brother-in-law. One day a rubber specimen was carelessly dropped over a hot stove. This charred it like a leather. That was nothing unusual, except that at the fringes the rubber sample had cured perfectly well. Goodyear noticed the thin strip of fringe, and knew that he had accidentally discovered what he had been looking for so many years. He soon found out that carefully controlled heat would cure rubber to the desired form. He called the process *vulcanization,* after the Vulcan god.

These examples indicate that accidents can help a keen mind discover new technologies for product innovations as well as process innovations.

In the 20th century, the accidental conception of new technologies could still happen to anyone of us. But, as more and more sophisticated products and processes were discovered by a systematic trial and error or by scientific research and development effort, the accidental conception of new technology became increasingly rare. With the scientific industrialization in the 20th century, the discovery of new products and processes evolved to become so complex that a simple accident seemed unlikely to produce a major new technology with significant commercial opportunities.

2. Technology by Systematic Trial and Error

New technology can also be developed by systematically trying all possible alternate approaches—until a favorable accident takes place. Very often the invention of technology is a numbers game. The more trials one makes, the higher the probability of a successful outcome. The economic returns from the successful trials justify the frustrating journey and the efforts spent on the not-so-successful other trials. Learning took place all the time. Great inventors, such as Thomas Edison, did not consider these "other" unsuccessful trials as failures, but as steps to get closer to their success.

Thomas Edison: The World's Champion Inventor. Thomas Alva Edison (1847–1931) is often considered modern history's most prolific inventor. In his lifetime he obtained over 1000 patents. His approach to developing technology was to systematically try all possible approaches to invent commercially profitable products and processes. He did so without searching for an in-depth understanding of the scientific principles involved in a technological innovation. By trial and error the inventor used what worked, and he learned from what did not work.

For a suitable filament for his world-famous incandescent lamp, Edison and his associates tried hundreds of different metals and a wide range of substances including alloys. Most of them failed. But Edison and his associates did not give up. Each failed trial taught them what did not work, and it brought them closer to success. Finally, they discovered that not metal but a carbonized sewing thread produced the long-lasting electric lamp. Their first light bulb used a filament produced from burnt thread. It lasted for 40 long hours. It may have produced light for a longer time, but Edison wanted to see the effect of changing voltage and amperage of electric current on the life of the filament.

Once someone addressed Edison as a scientist. Edison vehemently denied and insisted that he was not a scientist. He said that he preferred to be called an inventor. He was interested in producing inventions that had practical utility and that made him some profit. He referred to his Menlo Park facility, where he did his trials and errors, as his "invention factory" and not a laboratory. Historians claim that the only "scientific" contribution Edison ever made was what was known as the Edison Effect. It is the phenomenon of electricity passing through metals. But he had more than 1000 patents to his credit, more than anybody else.

Edison's commercial success "motivated" many others to set up Edison-like invention factories. The Edisonian approach to the conception of technology involved a trial-and-error method which did not invest too much effort in the development or understanding of the underlying principles of nature. For example, in selecting a burnt sewing thread as their filament for the incandescent bulb, it is unlikely that Edison or his associates knew that the burnt filament conducted electricity or lasted a long time because of the formation of a graphite carbon-like structure in the burnt filament. Edison did not have the luxury of using an X-ray machine, which was yet to be discovered, to identify the molecular packing in the burnt filament. Nor was he interested in discovering such knowledge.

Baekeland's Path to the First Plastic. In 1889, Leo H. Baekeland (1863–1944), a Belgian chemistry professor, was on his honeymoon in New York. There he met Richard Anthony of the largest photographic company of New York City. The Anthony Company, later changed to Ansco, produced photographic products. The company often faced a hard time in the photodevelopment process. In the late 19th century, the chemical industry in America was yet to take the form of a well-defined profession. When faced with a production problem, the chemical producers looked for a chemist willing to study and fix their problem. The manufacturer could only hope that in the meantime their losses were minimum and that nobody was hurt.

The 26-year-old Belgian chemist had already invented a process in his country to develop a dry photographic plate in water—instead of using chemicals. Baekeland was interested in coming to America because of the freedom to try new things. Anthony saw potential in the young chemist, and he offered him an appointment.

Baekeland worked with the Anthony Company for 2 years. He solved many problems for them. But then he was ready to move on and be free to try new ideas of his own. In 1891 he opened a practice as a chemist, just like a physician or a lawyer would do. But he soon discovered that to make ends meet he had to take many simple projects in widely different areas. Professionally, he was not going anywhere.

A short spell of sickness and fast disappearing funds forced Baekeland to think carefully about what he was doing. He decided to go back to his old interest of developing a photosensitive paper to help even amateur photographers take good pictures in artificial light. At that time the photographs were primarily taken by professional photographers who made their subjects sit still for hours in the sunlight.

After 2 years of persistent effort, in 1893 Baekeland was able to perfect a process to make such a photosensitive paper. He called it *Velox*. However, the American economy was going through a serious recession and a deep depression. There were, therefore, no industrial investors or buyers for the proposed new product. To manufacture Velox paper, Leo Baekeland joined Leonardi Jacobi in a partnership to start Nepera Chemical Company in Nepera Park, Yonkers.

George Eastman's "Kodak Moments". At that time, George Eastman (1854–1932) of Rochester, New York was also trying to change the old ways of photography by introducing a number of innovations. While growing up, George became interested in photography, but he hated carrying the hard-to-use bulky picture-taking equipment he needed. Until the late 19th century, it took everyone a lot of time to make the photosensitive glass plates used in a big bulky camera.

In 1880, while working as an accountant, in his free time George Eastman developed a faster and more efficient process for coating dry photographic plates. But the glass plates were still bulky, heavy, and hard to handle. George Eastman was not completely satisfied. With W. H. Walker, George Eastman then pioneered the use of a transparent roll of photographic film. Together they founded the Eastman Dry Plate and Film Company in 1884.

Four years later George Eastman introduced the first Kodak camera in 1888. The portable new camera was loaded with a roll of sensitized paper on a holder. After a few market trials, in 1892 George Eastman launched a Kodak camera loaded with a roll of photo-sensitive film for 100 exposures. He priced the camera at $25.

To market his camera, George Eastman innovated one of the most famous advertisement lines, "You press the button, we do the rest." The amateur photographers were asked to take their pictures and then bring the camera back to Eastman Kodak. Kodak would develop the pictures for them. And for another $10 Kodak promised to load their next roll of film. Later, George Eastman simplified the camera so much that he launched it into the market as the *Brownie* for kids.

These efforts turned Kodak cameras and photographic picture-taking into a common household thing. George Eastman, with his technical inventions and marketing innovations, successfully shifted the focus of photography from the professional photographer to the amateur family man and woman. He did so by removing most of the hassles of taking photographs. A large number of amateur photographers could thus focus on the excitement of capturing their memorable moments—promoted by the company as "the Kodak moments."

Eastman Invites Baekeland. In the last year of the 19th century, Leo Baekeland was still struggling with his Velox photosensitive manufacturing business. One day he received an invitation letter from George Eastman in Upstate New York. George suggested that if Baekeland was willing to sell Kodak his Velox manufacturing company, he was welcome to visit Rochester, New York for a talk. During the long carriage ride from Yonkers, New York City to Rochester in Upstate New York near Niagara Falls and the Canadian border, Baekeland kept wondering if he would get even $25,000 for his manufacturing process. George Eastman saw Leo Baekeland, invited him inside his office, and right away offered him one million dollars.

Baekeland immediately took the offer. At the young age of 37 and in less than a decade since coming to America, this new immigrant became a millionaire. Having seen some hard times earlier, Baekeland could now afford two things he had always wanted to do. These were to travel to Europe and to do his research in a well-equipped laboratory.

After returning from a long trip to France, England, and Italy with his children Baekeland got the idea to synthesize glass found only in nature. He set his goal to synthesize a glass-like material that could be shaped easily. By that time, chemists had already learned to break down substances into their components. They had learned to combine components to synthesize altogether different substances. Ammonia was one of such early substances synthesized.

In a similar manner, Baekeland started with two of the simplest chemical substances known at that time, phenol and formaldehyde. He had known from his past experience that when the pungent-smelling liquid phenol was reacted with pungent formaldehyde gas, they produced a brown tar-like substance with many bubbles and no practical use.

Bakelite, the First Plastic, Is Born. Baekeland systematically experimented with the various stages of this reaction. He found that adding air pressure reduced bubbles produced by the reaction. On the other hand, increasing temperature seemed to speed the reaction. He also noted that adding acid or alkali controlled the hardness of the final product. He perfected his process for 2 years and received 400 patents. Finally in 1909, Baekeland produced a number of useful articles from this new synthetic material, and he demonstrated them to a group of chemists in New York City. These were pipe handles and cigarette holders to substitute the current use of amber. He called his new substance *Bakelite*.

Bakelite was harder than hard natural rubber, yet it could not be easily scratched. Unlike rubber it retained its shape well. It was an excellent insulator for heat, electricity, chemicals, and so on. He also developed a process to use Bakelite to reinforce cheap wood into a hard wood.

During World War I, Bakelite was soaked in a fabric to make airplane propellers and automobile parts. After the war was over, the use of Bakelite grew with the rising popularity of radios. A lot of Bakelite was used in wireless equipment as an insulator.

Bakelite was the first of many synthetic plastics that boosted the economy of America enormously. With systematic experimentation and careful observation, Baekeland became very good at synthesizing the right quality of Bakelite resin. He made many useful items from his plastic resin. Yet, he had no idea why the material behaved the way it did. Nobody knew about the molecular structure of these new types of materials. They were so different from the common inorganic or organic substances commonly used at that time.

3. The Scientific-Research-Based Conception of Technology

America's Late Start for the Nobel Prize Run. In the beginning of the 20th century, what George Eastman and Leo Baekeland did was the common practice. They invented new products and processes by a systematic trial and error. The inventors often did not know the scientific basis behind their discoveries.

For the first six years of the American century, no American received a single Nobel Prize. In 1901, the Nobel Prizes were started in Sweden to recognize the scientific discoveries around the world. The first American Nobel Prize was won by A. A. Michelson for Physics in 1907. He discovered an innovative new way to measure the velocity of light.

Since then, American scientists and engineers have done very well, and they are way ahead of their counterparts in other parts of the world in acquiring Nobel Prizes. Much of this success can be attributed to the importance that growing corporations gave to scientific research to stay ahead of their competition.

Impact of Edison's Invention Factory. The first industrial laboratory in America was pioneered by **Thomas Alva Edison (1847–1931).** In 1876 Edison started his Menlo Park laboratory in Edison, New Jersey as a central place to invent new things. Edison called it his "Invention Factory." When somebody called him a scientist researcher, he vehemently denied it and said that he was not interested in doing scientific research for finding some fundamental principles of nature. His goal was to invent new, useful, and profitable things. He was bitter from a previous failed experience.

In 1868 Edison invented and patented his first of many inventions. This was a device to speed up the recording of votes in a legislative assembly. Congress refused to buy Edison's invention, preferring the slower voting process by show of hands. Edison vowed to never again invent something which nobody wanted. Notably among his successful inventions were the electric light, the phonograph, and the motion picture projector.

Corporate Research: Bell Style. In 1876 in a Boston attic, **Alexander Graham Bell (1847–1922)** invented his telephone with identical microphone and receiver. Human speech was converted into vibrations in a membrane which produced electric current pulses in a solenoid. These pulses were transmitted and produced vibrations in a membrane at the receiver end.

Barely a year later in 1877, Alexander Graham Bell predicted that "wires will unite the head offices of the Telephone Company in different cities, and a man in one part of the country may communicate by word of mouth with another in a distant place." This prophecy was truly amazing, considering that during this time the profession of electrical engineering was yet to emerge. The distribution of electricity and the electric light bulb were yet to brighten our night lives.

The General Electric Research Laboratory. The first U.S. corporation to start a laboratory was General Electric (GE), which grew from the Edison Electric Light Company. In 1900 GE started a research facility outside Schenectady, New York to figure out ways to improve equipment for generation and safe distribution of electricity. Some research efforts were directed at applying electricity safely to run daily appliances. At first, **Willis R. Whitney,** an assistant professor of chemistry at the Massachusetts Institute of Technology, was asked to spend 2 days a week to guide research at GE's manufacturing facility. Later, he resigned from MIT to become the Director of the world's first industrial research laboratory. GE at that time was facing severe competition from lamps sponsored by competitors such as Westinghouse (Note 3.1).

In 1909, GE attracted **Irving Langmuir (1881–1957),** a physical chemist, to do research on extending the life of a light bulb. Thus in 1913 Langmuir improved the life of the tungsten lamp by filling an inert gas in the bulb. This was studied and was found to reduce the evaporation of tungsten particles. Later he branched out into fundamental scientific research. His new goal was to understand the underlying scientific principles and generate new knowledge rather than to develop a specific practical application. The underlying assumption was that such fundamental research could lead to discoveries, technologies, and products which are totally new and radically different from the existing ones.

The GE Research and Development Center helped the parent company GE lead in many technology-intensive sectors. For many years, GE was the undisputed leader among companies in getting the maximum number of U.S. patents.

The Birth of Bell Telephone Laboratories. American Bell Company, the predecessor to American Telephone & Telegraph Company (AT&T), started the world's most famous Bell Telephone Laboratories in New Jersey. The young telephone company was facing many technological challenges. The transmission of the human voice through wires generated energy signals which were much smaller than the signals produced by pounding a key in a Morse code telegraph. The human voice also had a very broad spectrum of frequencies. On top of all this, even the telegraph pulses were known to weaken and get distorted when transmitted along a cable wire over long distances. There was still a lot to learn, technologically speaking.

In the 1880s the American Bell Company was doing some research and a lot of development work for improving telephone connections across long distances. This research work was done in Western Electric's Engineering Department, which was the hub for AT&T's manufacturing subsidiary. As a result of these efforts, amplifiers were developed. Such successes reinforced the management's faith in the central laboratory's mission to primarily do research.

AT&T considered basic research as the study of principles governing the nature. The management's commitment was to support research and exploratory development "by all means . . . if at all possible" (Note 3.2). Thus increasingly larger number of scientists and engineers were hired. In 1914, there were 550 technical people working in the Bell companies. By 1924, they grew to 3000.

In September 1924, a notice was posted on the bulletin boards of AT&T and Western Electric; it stated that a new research organization was about to be formed and would be known "by some such name as Bell Telephone Laboratories" (Note 3.3). In January 1925 the new company started its pioneering work. This was later jointly owned by both AT&T and Western Electric. The Lab's mission included meeting the needs of the operating telephone companies as well as supporting the manufacture of equipment to communicate effectively.

This joint organization structure helped. For example, when electronic switching systems were being developed, while the solid-state devices were still not fully developed, the two groups of researchers could learn from one another. Bell Labs integrated the horizontal flow of information across all stages of a technological transformation process, from the development and design stage to the manufacture and installation of equipment.

When the Bell Telephone Laboratories were born, telephony was already 50 years old. At that time the average monthly charge for home telephone was about $3.60. A 3 minute coast-to-coast call was billed at $16.50, and it took on average 8 minutes to get the connection. The average daily wage for a manufacturing job was $5. The head count of about 3000 employees of Bell Laboratories in 1924 expanded to more than 30,000 by 1994.

The funding support for research at Bell Telephone Laboratories continued even during the Great Depression of the 1930s. For this persistence, AT&T was rewarded with rich returns. During this period, Bell Laboratories developed crystal filters which enhanced the technology of sending many calls over the same transmission channel. Coaxial cable and more efficient switching systems were also developed. A new understanding of acoustics was achieved.

During World War II, all research not directly related to America's war effort had to stop. Bell Laboratories redirected its efforts to military applications. Waveguide systems and high performance electron tubes, as well as improvements in radar, were developed as a result of these efforts. These inventions helped America and the Allies win World War II. During the postwar years the focus was changed to innovations in electronics and the application of sophisticated material properties.

Over the years, researchers at Bell Laboratories invented and discovered an impressive list of technologies (see Case 4B for a list of Bell Laboratories" achievements in the first 50 years). This included the development of information theory, early computers, and fiber-optic technology. The birthplace of the transistor and certain types of radars was also Bell Laboratories. The mission of the pioneering research laboratory was affected when the U.S. courts ordered the breakup of AT&T into "Baby Bells." Some of the research effort went to Bellcore laboratory.

The Birth of America's "High-Tech" Chemical Industry. In the 19th century, the accidental discovery of synthetic chemicals for dyes and other substances started new industries in England and Germany. For example, **William Henry Perkin** successfully produced the first synthetic dye from coal tar. A new and enlarged German nation aided German scientists and encouraged businessmen to produce chemicals that would make Germany self-sufficient and independent of imports. Thus low-priced German dyes and drugs were exported in large quantities from Germany to England and America. The importing countries received these cheaper goods often at the expense of sacrificing support for the hard-pressed indigenous manufacturers or developers of these essential substances.

Impact of World War I. The sudden outbreak of World War I changed the chemical scene in 1914. The English industrialists rushed to start their own chemical industry from scratch. When America entered the war in 1917, it seized control of the German patents under an Alien Property Custodian Act. A Chemical Foundation was created to license the German chemical processes to American firms for a small fee. This gave birth to the giant industrial chemical corporations such as American Cyanamid, Dow, Du Pont, Eastman, Union Carbide, and so on. Du Pont took an interest in research on dyes and synthetic rubber, while Eastman focused on producing photographic chemicals. During the war years a Celanese plant was quickly built to produce artificial silk or cellulose rayon. After World War I, the college graduates with chemical engineering and chemistry backgrounds became professionals with great demand in America's fast-growing chemical industry.

Du Pont's Explosive Rise. E. I. Du Pont played a very strategic role in World War I. England was caught by surprise when the war broke out in 1914. The English were still producing trinitrophenol (TNP) explosives used by them during the Boar War. In the meantime, Germans built enormous competitive and military advantage by producing far superior trinitrotoluene (TNT) explosives. With the onset of the war, the Allied Armies were forced to turn to Du Pont. This explosives company had earlier boosted its TNT production to meet the high demand for explosives for constructing the Panama Canal project supported by the American government. Besides the English, the Russians, French, and Italians also placed orders for large quantities of TNT from Du Pont.

By 1918, Du Pont had expanded its explosives manufacturing capacity and was producing more than 50 times its production in 1914. But then the war ended suddenly. All the orders for explosives from the Europeans were canceled. Du Pont was afraid of collapsing under its own huge organization and production capacity, and it had to rush to search for new products and technologies. In 1920 Du Pont acquired the French and English patents to produce cellulose rayon, and it developed its textile and tire applications. In 1923 Du Pont licensed the Swiss patents to produce cellophane cellulosic film. In 1928 Du Pont acquired acetate rayon process too.

Du Pont's Experimental Station. While still searching for new technological areas to diversify into, in 1928 Du Pont decided to set up a new laboratory to do fundamental research. This was a very different mission compared to Du Pont's past goals. In the past, Du Pont laboratories did only the development work. These efforts were always directed to very specific practical applications. The new goal for fundamental research was to understand the principles of nature.

To guide the new mission for fundamental research, a 32-year-old Harvard university professor of organic chemistry, Dr. **Wallace H. Carothers (1896–1937),** was hired. Carothers' research interest was in studying materials with giant macromolecules. These materials were very poorly understood at that time.

Most chemists did not know then, but many natural and man-made materials known at that time were made of large molecules. For example, starch, carbohydrates, and proteins have giant molecules. So do fibrous substances such as cotton, wool, and keratin. Even Bakelite was made of macromolecules—even though Leo Baekeland did not understand its molecular structure. During the 19th century most macromolecular substances were produced either by systematic experimentation or by accident.

It was under such conditions that Du Pont decided to start an Experimental Station at Wilmington in Delaware. The corporate mission was to do fundamental research, without any specific commercial application in mind.

Wallace Hume Carothers was approached by Du Pont in 1927 to head this new initiative. The financial backing of a chemical giant company was attractive for a 31-year-old Carothers to leave teaching at Harvard. (For a detailed profile of Carothers' career see Case 3A.)

From the start, Carothers clearly set a challenging objective for his fundamental research. He wanted to prepare large molecules of known structure, through the use of established reactions of organic chemistry. And then he would investigate how the properties of these substances depended on their molecular constitution. In other words, he would develop the theoretical basis behind the polymerization reactions.

Synthesis of Nylon 66 Polymer. Carothers focused his research on the synthesis of a polyamide polymer. He had selected a simple polymer with molecular linkages similar to the molecular linkages in natural wool and silk. According to his polymerization theory, the polymer was expected to produce a fiber with good elasticity. On May 23, 1934, such a polymer was synthesized, and its hot solution was filled in a hypodermic needle. As the polymer substance was squirted through the tiny opening, world's first completely synthetic fiber was formed. This fiber was stretched several times under cold conditions. The fully drawn fiber was rigorously tested by Carothers' team. It showed outstanding properties, superior to those of the natural fibers used at that time. Carothers' polymerization theory predicted that. Carothers had already spent about one million dollars to develop polymer 66. The development engineers and others took over from there, to scale up the production process to a commercial scale.

In 1938, Du Pont publicly announced the sale of the new synthetic fiber in the form of toothbrush bristles. The fiber was called "nylon 66." Some historians believe that the name was derived by combining the names of the world's two major cities, New York and London. The number 66 represented the numbers of carbon atoms in the raw materials used to produce the new synthetic polymer.

In 1938 nylon synthesized by Du Pont changed the clothing and fashion industries. Nylon was offered to public in limited quantities as yarn for stockings. For the first time a synthetic fiber was offered to customers as a premium product with superior properties, rather than as a cheap artificial substitute imitating the real natural stuff. During the World War II years, when the supply of silk from the Far East was blocked by the Japanese rule in the region, nylon was extensively used in leggings to substitute silk. Within 1 year of synthesis of first nylon by Carothers, more than 64 million pairs of nylon stockings were sold in 1939. Nylon goods sold like hot cakes, and Du Pont was happily on its way to prosperity and profitability. Carothers, however, was not alive to see the day. He had taken his own life in 1937, before the age of 42.

The Technology Journey from Baekeland to Carothers: A Comparison.
The chemical industry in America evolved a long way from Baekeland's Velox in 1893 and Bakelite in 1907, to Carothers' nylon 66 in 1934. It did not matter to Baekeland that he did not fully understand the theoretical principles behind

the production of Bakelite. Carothers on the other hand was a scientist with the primary mission of understanding the principles of nature. He was specifically doing research looking for proofs to support his polymerization theory. It did not matter to him whether a commercial product emerged as a result of that search or not.

In the 1890s, Baekeland did research on his own and was forced to become an entrepreneur to commercialize his Bakelite technology. Carothers, on the other hand, was funded by Du Pont, a large future-looking corporation, and given considerable freedom and funds to unfold the secrets of nature. By doing so, the sponsoring company was able to commercialize radically new products in the market with limited competition for a long time.

Baekeland was forced to take a great risk in commercializing his technological discovery himself. Carothers, on the other hand, had the freedom to do research without risking his own funds. But the knowledge he created became the property of the sponsoring corporation. Carothers benefited from the support provided by the development engineers and many other Du Pont employees in turning Nylon 66 into a commercial success. What he and his associates achieved in 4 years might have taken four times as long if he worked by himself. And yet, on his own, he might not have ever invented what he did with the help of others in a large corporation like Du Pont.

By the 1930s, many cowboy-like Lone Ranger entrepreneurs of the 19th century were replaced by the "corporate suits" who developed new technologies in teams, using other people's money.

Over the years, synthetic plastics and polymers made an enormous impact on the economic growth of the United States. From 1920 to 1930 the use of synthetic plastics increased 10-fold from less than five million pounds a year to more than 50 million pounds a year. The growth was mainly because of new applications of plastics in radios and automobiles. Similarly, the synthetic fabric reduced the demand for leather for use in automobile upholstery. Otherwise the high demand and price of leather would have driven the price of beefsteaks sky high.

The above-mentioned historical events have illustrated how scientific research at laboratories, such as those started by General Electric, AT&T, and Du Pont, reaped rich dividends for their corporate sponsors. All these companies have retained their leadership positions in their respective industries.

In the 1990s, AT&T was still the oldest and the biggest telecommunications company in the world, and Du Pont was still the highly respected chemical giant in the world.

Management of Research Organization. What were the experiences gained from running these early research laboratories? Were there any lessons that organizations could apply in the 1990s to improve the productivity of their research and development centers? Given below are some observations.

Two Continuities of a Research Organization. In 1971, Bell Laboratories published a book entitled *Synergy.* This was written by H. W. Bode, a former vice president of Bell Laboratories, who later took the position of Gordon McKay Professor of System Engineering at Harvard University. Professor Bode discussed the role of Bell Laboratories in developing synergy by two types of continuities involved in developing a technology.

First, continuity was the synergy over time. This could be referred to as the *historical "learning" synergy.* This is the synergy gained from one research project to the next project. The Bell Lab researchers could build on the experience and expertise gained from developing the preceding level of technology, and they could transfer that to the next improved level of technology. In Professor Bode's perspective, this synergy in time was very important, almost as important as the second type of synergy.

The second type of synergy was the *horizontal cross-functional synergy* across the different departments of research, development, design, manufacture, installation, and so on. The horizontal synergy helped a physicist to hand over a theoretical concept to an electrical design engineer to design an electric circuit based on their shared understanding.

The Third Synergy. In the organizational setting of the 1990s, there is a third level of synergy that is often missing and much more critically needed—namely, the hierarchical synergy of vision provided by the upper echelons. This includes the understanding of the technological challenges facing an organization by senior executives of the organization. The bigger the technological and competitive challenges facing an organization, the more synergistic the integration the organization requires along its hierarchical layers. The hierarchical synergy requires two-way integration. The senior executives must learn and fully grasp the technological challenges of the business they are in. On the other hand, the technical employees in the lower layers of the hierarchy of a large organization need to know the "business case" of the technologies and the operations they staff. They cannot afford to ignore the competitive and market challenges of their employer. They must learn what their stockholders and investors expect from them.

Together, the *historical temporal synergy, the horizontal team synergy, and the hierarchical tiered synergy* can provide a formidable source of sustainable competitive advantage for a technology driven, research supported enterprise.

The Bell Vision: People, Purpose, Environment. Research organizations also need priorities of values and vision for the future. There is often competition for resources. Priorities of values can help guide through alternatives. Bell Laboratories were fortunate in having a clear priority of values, along with vision for the future.

For many years, Bell Laboratories were guided by the vision provided by Dr. James B. Fisk. He was its head for a quarter century from 1959 to his retirement in 1974. For Bell Laboratories' success he prescribed "people, purpose, and environment." He strongly believed that the best science and technology can only emerge by hiring, attracting, and retaining the best minds. To challenge and stimulate them, such best minds need an environment of freedom and flexibility of purpose. At Bell Laboratories, the vision for the future was to further the frontiers of communication technology.

The technological success of the Bell Laboratories inspired setting up of similar laboratories by other leading companies. IBM, Eastman Kodak, and much later Xerox followed in AT&T's footsteps—but not always with similar success. In the 1970s, Xerox's Palo Alto Research Center (PARC) facilities played a pioneering role in the development of many desk-top computer-related technologies. But these did not help Xerox generate the wealth the way other companies did.

Xerox: The Photocopying Pioneer. Consider Xerox, a company based on a breakthrough proprietary technology, which later failed to cash in on some of the equally radically technologies developed by its own research center.

Chester Carlson (1906–1968), the father of photocopying technology, worked as a patent analyst for Mallory Company, an electrical component producer in New York City. His work required him to draw and make multiple copies of different documents and drawings over and over again. He wished there was a better way, preferably a machine to do that tedious manual work. In his spare time, Carlson started tinkering with developing a technology to electrostatistically transfer the image from an original document to a blank paper. After much persistent effort, Carlson was able to successfully transfer a crude image of "10-22-28 Astoria" written on a sheet of paper. He received a patent for the invention of electrography in 1937. But commercializing the electrography process proved to be a much harder hurdle than inventing the photocopying technology.

For 20 years Carlson searched for a sponsor to fund the production and marketing of his revolutionary technology. But all he received was rejections from corporate leaders like GE, IBM, Kodak, and RCA. They all thought that the unproven photocopying technology was an expensive replacement for the carbon copy paper used at that time. Finally, in 1960 a small paper supplier, Haloid Company of Rochester, agreed to fund the "unwanted" technology. The company, on the advice of a university professor, changed its name to Xerox Corporation.

Xerox's Palo Alto Research Center. Xerox Corporation launched its first "914" photocopier, a machine that made photocopies on a 9-inch × 14-inch paper. After some initial gestation period, Xerox hit a runaway success. The customers made more and more photocopies using the clean and odorless Xerox machine. And quickly they started discarding the smudgy carbon papers and smelly ammonia-based photographic process used earlier. The photocopying technology changed the entire atmosphere in the offices and other workplaces around the world.

Until the early 1970s, Xerox's research efforts were primarily directed at improving its photocopying technology based on Chester Carlson's patents. This required understanding of photoreceptors, laser scanners, image storage, and so on. Much of the needed research efforts were carried out at Xerox's Webster Research Center, near Rochester, New York. However, Xerox soon saw the early signs of increasing intensity of competition in the photocopier market. As a result, Xerox became actively interested in diversifying into other related businesses, particularly in the area of office automation.

In this new vision for Xerox, the photocopiers were expected to be a terminal part of a bigger document processing system. This added new items to Xerox's technological needs and research agenda. For example, new technologies were required for a user-friendly human interface with a digital system. Office equipment was expected to be miniaturized with microelectronics. The research for office automation and digital systems was done at Xerox's Palo Alto Research Center (PARC) in the Silicon Valley near Stanford University in Northern California.

The R&D efforts at PARC soon produced some impressive new technologies for the fast emerging Information Technology Age. For example, the PARC researchers developed technologies for the Graphic User Interface (GUI), use of a "mouse" as a system input device, and a "fax" machine for document transfer over long distances, thereby fulfilling Xerox's future requirements for a user-friendly automated office.

Many of these technologies, however, did not appear in the marketplace in Xerox's products. Instead similar technological innovations showed up in Apple Corporation's Lisa computer (1983) and Macintosh computers (1984). In 1981, Xerox management decided to pursue distributed network approach to computing and documenting systems, rather than developing a dedicated desk-top computer system.

The Xerox researchers had developed a stand-alone computer, Alto system, which used the SmalTalk™ programming language. But Xerox management decided to support an interlinked network system consisting of a series of peripheral equipments including an intelligent typewriter, Ethernet local-area network, and Star workstation. The system was priced too high for many customers to get excited about it. And, it failed to make much money for the photocopying pioneer.

On the other hand, Apple Corporation's co-founder Steve Jobs targeted Lisa and Macintosh computing systems, inspired by Xerox PARC innovations. He targeted these computing systems at the personal computer market segment. Using similar technological research, Apple was able to cash its new technologies in the market in a big way, while Xerox missed the opportunity to do so with its proprietary pioneering technologies. Soon Apple's Macintosh became a big player in the personal computer technology, whereas very few people ever heard of Xerox's Alto (Note 3.3).

Summary So Far. In conclusion, in this section on conception of propri-etary technology by scientific research, we reviewed how in the 20th cen-tury the United States became the birthplace of new scientifically created technology-based industries. The freedom to pursue prosperity first tempted many individual inventors such as Baekeland, George Eastman and Chester Carlson to create technologies for plastics, Kodak cameras, and photo-copiers respectively. Such liberty to create later helped corporate innovators such as General Electric, AT&T, and Du Pont to rise in their fields by in-venting new technologies and changing the rules of competition in the mar-ket place. Du Pont's nylon polymer had a devastating effect on the natural-fiber-based textile companies. GE's electric energy edged out the water-based motive power with electricity as the prime mover. AT&T's tele-phones, transistors and other electronic devices changed the way we com-municated with each other for decades.

Next, we must continue our investigation of other alternate paths to the conception of new technologies.

4. New Technology by Acquisition

Not all companies can and should develop all the technologies they need on their own. Development of new technologies often require large investments of financial and human resources in highly risky ventures. Different compa-nies have different core-competencies. In other words, they have unique skills in which they are the best in their business. These rare, hard to imi-tate, and value-adding competencies provide sustainable sources of compet-itive advantage. Acquisition of a technology could be a viable strategy un-der certain circumstances.

In Japan during the reconstruction years in the 1950s and 1960s, soon af-ter a complete devastation in World War II, many Japanese companies ac-tively acquired technologies from their American counterparts. For exam-ple, Sony Corporation licensed transistor technology from the Bell Laboratories for its application in their miniaturized electronic radios. The transistor radio's success, based on an acquired technology, helped Sony be-come a global leader in audio and video technology.

In 1968, Busicom, a Japanese company specializing in scientific calcula-tors, contracted a newly born Intel to leverage its chip-making technology for a new electronic calculator. The Japanese designers had designed their calculator with a set of 12 integrated circuits on a printed circuit board. The Intel designers proposed a better solution—that is, to build a single semi-conductor chip that could be programmed to act as a calculator. Thus the first microprocessor chip Intel 4004 was designed and introduced to the world.

The irony of history is that later Intel saw a potential for wider applica-tions of the microprocessor chip they had developed for Busicom. To apply microprocessor technology in other appliances, Intel licensed back the microprocessor chip technology from Busicom for a one-time payment of $60,000 (Note 3.4).

NIMBY and NIH Syndromes. Corporate allergy and resistance to learn from others is called the "Not Invented in My Back Yard" (NIMBY) or "Not Invented Here" (NIH) syndrome. These are caused by the "seasonal" shifts—that is, sudden changes in one's familiar competitive environment. Instead of adapting and responding to the uncontrollable external changes in the markets, there is a tendency to drag one's feet with the business as usual. Because there is not enough time to develop one's own response, learning from other successful adaptors could help. But, because those solutions were not developed in a company's own back yard, they are quickly rejected. Managers should look out for such techno-allergic behavior. Acquisition of relevant and value-adding technologies and lessons could help catch up fast with the changing season.

5. New Technology by Alliances

Alliances with Universities. In the 20th century, the universities became one of the most common sources of publicly funded (and generally still unproven) technologies based on their basic research. They have a large pool of knowledge workers, dedicated to generation and exchange of knowledge. But they often have strings attached to their participation in collaborations with industry. Their primary goal being teaching and/or the pursuit of public knowledge, their time frames are defined by their teaching and academic schedules. The tangible measures used to evaluate the outcomes of a university are number of graduated students, research publications, grants acquired, and so on. Such a list often does not include participation or contributions to inventions or innovations. Furthermore, the incentives and rewards of university faculty are often tied to the "publish or perish" principle. The university faculty members are therefore very keen to publish their findings as early as possible. The universities are sometimes accused of being ignorant about the business side of technologies.

Industrial enterprises, on the other hand, often have more technical resources. They also do not mind doing trial-and-error type of development work to arrive at the best solution. They are generally short of head counts and almost always rushed for time. Often company employees are very paranoid about disclosing their secrets or findings to their current or potential competitors.

Alliances with Government. Governments in most countries are often a major source of research-and-development-based technologies. See Exhibit 3.1 for the major technological achievements supported by federal funds in the United States during the first 50 years since the end of World War II. It is an impressive list of technologies, consisting of such breakthrough

EXHIBIT 3.1. First Fifty Years of Federally Supported Technology-Related Achievements

World War II

1945	Manhattan Project to develop and produce the first two atomic bombs.
1946	ENIAC—the first large-scale U.S. electronic computer.
1951	Three-dimensional spiral structure of proteins discovered by Linus Pauling.
1951	Cortisone and cholesterol synthesized for the first time.
1951	Salk and Sabin develop polio vaccines.
1957	Bardeen, Cooper, and Schrieffer develop superconductivity theory.
1958	The first U.S. satellite, Explorer I, launched into space.
1960	First successful laser demonstrated by Theodore Maiman.
1967	First coronary bypass operation developed.
1969	Communications network, ARPANET, developed. later becomes Internet.
1973	DNA splicing in *E. coli* bacteria demonstrated by Stanley Cohen and Herbert Boyer.
1976	Soft landings made on Mars surface by Viking I and II.
1976	Ozone layer depletion by chlorofluorocarbons reported by National Academy of Sciences.
1970s	Development of nuclear magnetic resonance imaging.
1981	The space shuttle flies to space and back for the first time.
1985	Third form of carbon, Buckminsterfullerene, discovered.
1989	Human Genome project commissioned.
1990	Hubble Telescope launched into space.
1994	Fermi National Accelerator Laboratory discovered Top Quark.

Source: Langreth, Robert. 1995. Dark days for science? *Popular Science,* October:74–80.

technologies as the first large-scale electronic computers, cortisone and other life-saving pharmaceutical drugs, lasers, communication networks including Internet and its earlier ARPANET versions, nuclear magnetic resonance imaging, and so on.

Since the government-funded research and development is paid by public and tax dollars, the governments often are keen to share the nonmilitary and national security-related technologies with the civilian corporations.

Since the end of the Cold War and the breakup of the former Soviet Union, however, the government-funded research and development in the United States have been under a severe attack. In 1995 the federal funding for government research and development received a sharp cut by a very cost conscious U.S. Congress. The Clinton government proposed about $73 billion on research and development during the fiscal year 1996 (starting October 1). In current dollars, this was $169 million, or about 0.2%, higher than the amount spent by the U.S. federal government on R&D in the previous fiscal year. (See Exhibit 3.2.) But this increase turned into an annual 2.6% reduction in constant dollars if the 2.8% inflation rate for the year (according to the U.S. Office of Management and Budget) was factored in.

In terms of allocation of R&D resources to the different federal agencies, the Department of Defense still spent 52% of the total federal R&D budget in the United States. The civilian research and development accounted for $34.9 billion, with a meager 0.4% increase over the previous year.

The Clinton government also sought $1.1 billion for the construction of information superhighway, including the high-performance computing and communications infrastructure. Environmental and global change research, as well as research conducted by 12 federal agencies, was allocated $2.2 billion in fiscal year 1996—a 1.0% decrease over the budget in the previous fiscal year. Environmental and global change research was allocated $356 million, a rise of 5.1%. The worst news for the year was that the National Science Foundation's funds for academic research were reduced by 18% to $100 million only. The National Science Foundation explained this fall due to the recently completed launches of its twin Gemini telescopes.

EXHIBIT 3.2. U.S. Federal Research and Development Budgets

Federal Agency	Fiscal Year 1993 Actual	Fiscal Year 1995 Projected	Fiscal Year 1996 Proposed	1995/1996 Percent Change
1. Defense—Military	38,898	36,272	35,161	−5.9
2. Health and Human Services	10,472	11,676	12,123	+1.0
3. National Institutes of Health[a]	10,326	11,321	11,789	+1.3
4. NASA	8,873	9,455	9,517	−2.1
5. National Science Foundation	2,012	2,450	2,540	+0.9
6. Agriculture	1,467	1,554	1,499	−6.4
7. Commerce	793	1,284	1,404	+6.5
8. Transportation	613	687	755	+7.2
9. Environment Protection Agency	511	589	682	+13.0
10. All others	1.957	2,110	2,077	−4.4
R&D facilities only	2,727	2,063	1,962	−7.7
R&D Total	72,493	72,714	72,883	−2.6

[a]NIH numbers are separated from HHS totals; and from Cowen R., and Adler T. 1995 February 11. Federal budget 1996: R&D would fall, *Science News,* 147:86–87.

Source: Adapted from figures provided by the U.S. Office of Management and Budget (OMB). Rates adjusted by deducting OMB's projected Fiscal Year 1995 inflation rate of 2.8%.

Was this an exceptionally bad year for the government funding for R&D? Not really. The news for the near future seemed even bleaker. It was projected that by the year 2002, the federal R&D budgets would be slashed to about $60 billion, or about 20% less than the $72 billion budget in 1995. (See Exhibit 3.3.) Whereas the Department of Defense budget would re-

EXHIBIT 3.3. The U.S. Federal Research Agencies

	Annual R&D Budgets ($ million)		
The U.S. Federal Agencies	1995	1996	2002
1. Department of Agriculture 10% cut in agro-research.	1,745	1,388	1,140
2. Department of Defense Same or slightly more funds.	35,448	NC[a]	NC

3. Department of Energy
Severely cut research on energy
  efficiency, fossil energy, and renewable
  energy.
	4,528	3,142;	2,108

4. Department of the Interior
National Biological Service dissolved,
  and U.S. Geological Survey received
  20% less.
	864	668	474

5. NASA
An environmental satellite program.
  Earth Observing System cut drastically.
  Privatization of Space Shuttle
  proposed.
	12,828	12,211	8,672

6. National Institutes of Health (NIH) Reduction in 1996, then budget freeze for 6 years.	11,330	10,451	8,752

7. National Institute of Standards and
Technology
Eliminated Advanced Technology
  Program, which funded study of
  cutting edge technology by
  corporations. Future of Baldrige
  National Quality Award Program
  unclear.
	764	335	NC

8. National Oceanic and Atmospheric Administration A number of national weather service stations likely to be closed.	1,980	1,655	1,264
9. National Science Foundation (NSF) Social and economic research affected.	2,691	2,393	2,147
10. Office of Technology Assessment (OTA) Stopped the advising to Congress on science and technology issues.	22		
	72,200	67,691	60,013

[a]NC, no change.

main unchanged, most of the other federal agencies would probably have drastic reductions in their funding. The irony was that the Advanced Technology Program under National Institute of Standards and Technology (NIST), which funded the study of cutting-edge technology by corporations, was drastically hit. The Office of Technology Assessment, which had a 1995 budget of only $22 million and which was assigned to advise the Congress on science-and-technology-related issues, was altogether stopped.

It is amazing that many in America complained about government support, and at the same time attributed the success of the Japanese competitors to the support the Japanese companies received from the Ministry of International Trade and Industries (MITI) in the Japanese government. However, at the same time, the federal programs which could play a similar role of guiding the U.S. corporations are given minimum financial support.

In summary, in the years ahead, the potential alliances with the government agencies for gaining publicly funded new technology are expected to decrease drastically.

Alliances with Noncompetitors. Sometimes companies scan and gather competitive intelligence to find out other companies who may have already developed certain technologies that they could acquire or ally with. Such alliances save significant amounts of time and efforts required for "reinventing the same wheel." As mentioned earlier under "acquisition of technology from others," the Japanese companies like Sony, Toshiba, Canon, and others have benefited enormously by building synergistic alliances with other noncompetitors. For example, Eastman Kodak had alliances with Canon to leverage their photocopying technology.

Some other companies, on the other hand, are biased by their "not invented here" (NIH) bias. They cannot accept the fact that other companies or their competitors could develop superior technological competencies. Such companies often are forced to regret their "denial" later as they lose their market power.

Technology-Based Alliances with Competitors. This may seem like a strange way of acquiring new technology, but it is happening. The primary reason for "dancing with the enemy" is that such alliance could reduce the cost of developing a sophisticated new technology. By joining hands, the competitors reduce their risk of failure.

The Advanced Photo System Co-Development. Eastman Kodak, the pioneer of photography technology, made an unprecedented move in the 1990s. Previously, Kodak introduced its world famous photographic products—from the Brownie camera in 1900 to the disc camera and the photo-CD many decades later—launched solo by the company. However, in 1995 a revolutionary new photographic technology system, called the Advanced

Photo System (APS), was announced. Kodak CEO George Fisher considered APS to be as revolutionary as George Eastman's launch of the Brownie camera in 1900.

APS consisted of a radical new 25 millimeter film format, a new smart camera which could take pictures in three different formats, and a new photo-finishing system. What was more radical was that Kodak developed the new technology standard in collaboration with its four Japanese partners, including rival Fuji Photo, Canon, Nikon, and Minolta. With a mission to develop a new world standard for photography, Kodak and the co-developers of new technology also licensed it to 40 other companies to develop APS products and services.

Why did Kodak dance with the enemy to develop APS? One can speculate. During the 1980s, senior managers at Kodak realized that picture-taking in America in particular, and in many other industrial countries in general, was a maturing activity. With a heavy dependence on revenues from silver-halide photographic film, Kodak faced a critical assault from the emerging digital imaging technology. The rival Fuji Photo and other camera producers in Japan also concurred.

All of these companies realized that they faced a far bigger threat from digital imaging technology than they did from each other. They also knew that to revitalize the photograph-taking habit in customers, they would have to make the photo-taking an extremely user-friendly event. Many amateur photographers were still having a hard time loading films in their cameras. They wanted different formats. They wanted a more compact, better picture-taking, user-friendly camera.

Kodak and others also knew that one company alone could not revitalize the world-wide photography market. In April 1991 Kodak asked the general manager of its new Advanced Photo System business to form a cross-functional technology development team. High-performing and multi-skilled members were drawn from various groups, including marketing and market research, R&D, film manufacturing, camera design group, and photo-finish equipment service group.

A few months later in late 1991 Kodak signed a joint technology development agreement with Fuji, Canon, Minolta, and Nikon. They were called G-5. Their goal was to give the photographic industry an altogether new technology platform, from film to photo-finishing. Many patents for photography technology that were needed to develop the new technology were swapped or pooled together. The G-5 global partners worked together for the next five years, organizing alternate meetings in America and Japan. Finally, after billions of dollars invested by the five co-developers, the Advanced Photo System (APS) was successfully rolled out worldwide in February 1996. Kodak launched its APS products under the Advantix brand and the "Take Pictures Further" slogan. Some observers felt that the product rollout was a less than perfect "Kodak Moment," with some shortages here and there and inadequate information.

Kodak's researchers and technicians who co-owned the patents for the digital magnetic incorporated in APS received the "Inventor of the Year" Award from the Intellectual Property Owners, an association of companies and individuals holding patents (Note 3.5).

With the collaborative development for the Advanced Photo System, Eastman Kodak pioneered a new paradigm for revitalizing technology when the cost of development is too high for one company to bear.

DEFINING THE DOMAINS OF PROPRIETARY KNOW-HOW

In the previous section, we reviewed the five alternate ways to develop, acquire, or share new technologies. Before proceeding further, let us organize and define the different technology-related activities covered in this chapter.

Scientific Research

Scientific research is often defined as a systematic study of the relationship between the root causes and effects, or identifying the statistical correlations underlying the different natural and man-made phenomena.

The outcome of scientific research is usually generation of new scientific knowledge in the form of better understanding. For example, the description of the missions of Bell Telephone Laboratories and Du Pont's Experimental Stations, discussed earlier, indicated that the purpose of their scientific research was not limited to the development of a specific application. The input for research function (treated as a system) is some preconceived set of assumptions used as a hypothesis, and the ill-defined information about an observed phenomenon. The output of research is more organized knowledge about the phenomenon or the transformation process under study.

Development

Development, on the other hand, is generally concerned with the application of scientific, empirical, or experiential knowledge to either produce a new product, conceive a new process, or identify a practical application.

In relation to research, sometimes modified terms such as basic research, fundamental research, exploratory research, and applied research, are also used. These terms qualify the goal of a research endeavor. On the other hand, in relation to development, the practicing managers and policy makers often use expressions such as product or process development.

Do these terms seem to overlap in their scope? Perhaps they do. To understand the fine differences between these terms, some clear and precise definitions help guide us.

Basic and Applied Research

The National Science Foundation (NSF) has provided definitions to understand these differences. In 1967, the NSF defined basic research as the "original investigation for the advancement of scientific knowledge that do not have specific commercial objectives, although such investigations may be in fields of present or potential interests to the [researching] company."

To the NSF, applied research involves experimental studies that lead, "to the discovery of new scientific knowledge, and that have specific commercial objectives with respect to products or processes."

A major criticism against these definitions has been that they do not relate very well with the profit perspective of business organizations. Other industrial organizations and their associations have also tried to offer the definitions more appropriate for their industrial sectors.

A DYNAMIC MODEL OF PATTERN OF INNOVATIONS

The history of civilization is full of various kinds of innovations as different technologies evolved from introduction to growth and maturity stages. Is there a way to bring some order to this seemingly utterly chaotic world of technological innovations. Are innovations spontaneous sproutings of ideas, concepts, or procedures? Or do they follow a pattern, which can be preempted?

In the past, most studies of technological innovations were descriptive. These listed communication processes used and problem-solving methodologies employed during an innovation process. Professor James Utterback of the Massachusetts Institute of Technology, along with Professor William Abernathy of Harvard University Graduate School of Business Administration, searched for some systematic patterns of the innovations for a technology evolving over time. They reported their empirical findings in their ground-breaking 1975 article in the journal *Omega* (Note 3.6).

Product and Process Innovations

Utterback and Abernathy looked for any patterns of evolution for technological innovations, which they studied as product innovations, process innovations, and the innovations involving system integration.

They defined product innovations as ". . . a new technology or combination of technologies introduced commercially [and embodied in a commercial product or service] to meet a user or a market need. . . ." They noted that the product innovations changed over time, with a change in focus.

The production process innovations, on the other hand, were defined as "the system of process equipment, work force, task specifications, material inputs, work and information flows, etc. used to produce a product or service." They noted that over time the process innovation evolves and ". . .

becomes more capital intensive, [with] improved labor productivity through greater division of labor and specialization, . . . with rationalized flows, standardized product designs . . . and larger process scales. . . ."

Utterback and Abernathy cautioned that their unit of analysis was not necessarily the overall process of an entire firm, but it was the production process used to create its products or services. For a "simple firm" with a few related products or services, the production process could include all the operations of the firm. However, for a conglomerate, with diversified or integrated businesses, the processes are different in different business divisions of the firm. If an industry was highly fragmented, then the process for its technology traversed across many different firms. These guidelines helped the researchers compare process innovations across different firms, and over different time periods, in a consistent manner.

Patterns of Technological Evolution

It was noticed that the product innovations as well as the process innovations followed a pattern of evolution.

The product innovations evolved over time with a shift from their emphasis on product performance, changing to product variety and standardization, and later on to product costs. Similar variations were observed for the process innovations too. The researchers divided the pattern into three phases. In each phase, the focus of product innovations and the emphasis in process innovations differed.

Phase I: Introductory—Unconnected Stage. In this early stage of the development of a new technology, many unique and technically diverse product innovations were introduced rapidly. There was no consensus on the definition of a standard product. The innovators were not clear about the customers' preferences. The customers themselves often did not know what they wanted. These diverse products were produced by flexible, general-purpose, and disconnected production processes—so that innovators could respond to fast changes in the market requirements. There was low product and process standardization. Different producers tried to win the market by claiming that their products were with superior performance characteristics, better than those of other products. Product innovations were focused on performance maximization, while the process innovations were highly disconnected across the entire technology of transformation. Flexible and "organic" organizations innovated better than the rigid and "mechanistic" organizations.

Example: U.S. Automobile Technology. In the early stages of the U.S. automobile technology, there were over 200 auto makers. They offered a large variety of automobile products using a wide variety of engines driven by electric power, steam power, gasoline power, kerosene power, and other fuels available at that time. Chassis in some automobiles were covered, whereas in others they were not. Some of them were made of wood while others used shining chrome, brass, and other metal parts. During this stage of technology, the process efficiency and innovation played supportive roles in defining competitiveness. Product performance ruled.

Each automobile was crafted one car at a time. The production process operations for engine, transmission, or painting were completely disconnected and were done independently. This provided the flexibility when the market was not clearly defined in this introductory stage.

Phase II: Segmented Growth Stage. As the users of a product innovation gained familiarity while the producers learned new skills to produce their product, market uncertainty decreased. A few stable product designs evolved with sufficiently large demands. This motivated some firms to increase their production efficiency by using a more mechanistic production process. Different firms introduced their products with adaptive imitations of the successful features of other products in the market. As a result of this convergence of product innovations, a dominant product design emerged in the market. Thereafter fewer product innovations were introduced in the second phase compared to the first phase.

In this phase, the production process innovations increased over the previous phase. With pressure to increase production, the process innovations, such as automation and process control, took place in certain segments of the overall production process. The development of a dominant product design stimulated innovations in production segments which helped meet the fast growth in sales volumes. Firms that were able to rapidly respond to such growth in demand by suitable process innovations gained a higher share in the marketplace. Market shares and prices defined the competitiveness.

For the automobile technology in phase II, Henry Ford's Model T car became the dominant design of the automobile industry. The process innovation of the moving assembly line brought about integration in certain segments of the total automobile production process. Certain parts were fabricated on-line, and certain operations were carried out in a disconnected batch mode.

Phase III: Integrated Mature Stage. With further evolution in a proprietary technology, the basis of market competition shifted to prices, with lower profit margins. Markets were characterized by oligopolistic industrial structure. The product designs standardized, and few radical product innovations were introduced. Further product innovations involved incremental changes to less expensive standardized products. The key to competitive success involved production efficiency and economies of scale for the more capital-intensive competitors.

Eventually the product innovations and process innovations merged as an integrated interdependent system innovation. Every product innovation was accompanied with a related process innovation, and vice versa. An innovation in one part of the production process impacted changes in other parts of the production process. With shrinking profit margins, the total efficiency and total cost defined the competitiveness.

In Phase III, often further innovations in the production process became hard to fund—that is, until the shifts in market forces warranted development of an altogether new technology, with revolutionary changes resulting in a new series of product innovations and process innovations.

In the case of the evolution of automobile technology, first the Just-In-Time inventory management and eventually the lean production system corresponded to such integrated system innovation. Toyota and other auto makers integrated the entire supply chain. Ford Taurus was integrated with cross-functional teams.

The Implications of Patterns of Innovation on Management of Proprietary Technology

The foregoing discussion of a pattern of product and process innovations, as a technology evolved, has a number of implications on the management of technology in its different stages.

The Change Agents for Innovations. As a technology evolved from one phase to another, the change agents for innovations shifted from one type of individual to another.

In the first unconnected introductory phase of a technology, the change agents for innovations were those individuals (or organizational entities) who were most familiar with the specific needs of desirable customers and their market requirements. The innovations in this phase were not likely to take place in those individuals with more familiarity with the process technologies. Later, in the third systemic integration stage, the center of innovations moved to individuals with abilities to (a) integrate product and process innovations of a technology and (b) innovate with "system-like" technological solutions. In this stage, customer needs and market requirements were generally clearly defined, so individuals with those skills did not play an equally critical role. As mentioned earlier, the center of innovation would change drastically if the technology was to change radically to a different series of product and process innovations.

The Most-Likely-to-Succeed Type of Innovation. In the different phases of evolution of a proprietary technology, the type of innovation that was most likely to bring a bigger bang for the invested bucks also changed. In the first phase of the unconnected introductory stage of a technology, the most successful innovations were those which improved the performance of product features. In this phase, innovations in complex "system-wide" technological solutions were not accepted by the bewildered customers. The reverse was the case in the third integrated mature stage of a technology.

The Big Barriers to Innovations. In the unconnected introductory stage of a technology, the main resistance came from those who did not see the relevance of the perception of a customer need. In the 1970s, some systems engineers at a main-frame company could not appreciate the need for a desktop personal computer. They thought it would never grow beyond a few hundred hobbyists—until a few years later, when Apple cornered a big chunk of the new market. Their resistance and barriers to change were based on their perceptions of the potential need.

In a similar manner, some entrepreneurial-type marketers missed their cues for process efficiencies and system-wide integration as the desktop technology evolved to a later phase.

THE S CURVES OF TECHNOLOGY

Another ground-breaking model for managing research-driven technological enterprises was provided by Richard Foster, a Director in McKinsey & Company and the author of *Innovation: The Attacker's Advantage*. (Note 3.7). He proposed that technological leaders become losers because the productivity of research and development efforts varies. The productivity struggles first, increases fast next, and matures in the form of an S curve.

In the introductory stage of a technology, large investments of resources in research & development produce only limited technical advances. Beyond a critical mass of knowledge, when some of the key pieces of the technology puzzle are resolved, even small incremental increases in allocation of R&D resources produce large returns in the technical advances. This is the steep rising part in the S curve, with an accelerated growth stage. At some inflection point in the S curve, the rate of change of technical advancement per unit allocation of R&D resources starts declining. As the technology matures and reaches towards its "natural" limit, the marginal increases in allocation of resources to R&D could result in smaller improvements in technical advances for the relatively mature technology.

For example, as the semiconductor chip technology gets closer to natural physical limits, such as the minimum interatomic distance between particles, the additional resource allocations to R&D are not likely to overcome the physical laws of nature or provide a higher density of integrated circuits in a microprocessor or memory chip.

TECHNOLOGICAL DISCONTINUITIES

At some point in the evolution of a technology, the incremental cost of improving the old mature technology increases very high even to produce small improvements in the technical advances. Yet the leader in that technology must continue to invest more resources to protect its "sunk" cost and stay ahead of others.

In the meantime, other firms with either no or limited "sunk" costs invested in the old technology are tempted to allocate their R&D resources to develop radically new technology. Thus, often the substitution of technology takes place by the outsiders. They have a competitive disadvantage in competing with the old technology of the insiders. Their competitive playing field is either even or favorable in the case of an emerging new technology.

Foster illustrated the S-curve effect with the success of the Japanese digital watch attack on the Swiss mechanical watch technology. Michelin radial tire technology, with longer-lasting tires, quickly gained a double-digit high market share by substituting the bias-belted tires by Goodyear and other U.S. tire makers.

Exhibit 3.4 provides an interesting look into the market penetration of synthetic substances, substituting their natural counterparts. There is an amazing consistency in the penetration of discontinuous technologies (Note 3.7).

Joseph Bower and Clayton Christensen of Harvard Business School, in their 1995 article in the *Harvard Business Review,* reiterated a similar point. They noted that as technologies or markets change, the failure of leading companies to stay ahead as leaders of their industries was one of the most consistent patterns in business. They noted that Goodyear and Firestone, the leaders in the bias-belted tire industry, were late in entering radial-tire markets redefined by Michelin's radial tire technology. The leaders who invested aggressively in retaining their current customers often failed to make the needed technological investments for their future customers. The authors argued that the most fundamental reason was that the leaders stayed too close to their customers. There could be many other reasons, including rigid and arrogant bureaucracy, tired executive (human) resource, and short-

EXHIBIT 3.4. Technology Substitution for Synthetics

Technologies		Penetration Duration (years)		
Old	New	90 : 10%	50 : 50%	10 : 90%
1. Natural rubber	Synthetic rubber	1927	+29 yr	+27 yr
2. Butter	Margarine	1929	+28 yr	+28 yr
3. Leather	Plastics	1929	+26 yr	+28 yr
4. Natural fibers	Synthetic fibers	1940	+29 yr	+29 yr

term investments. They postulated that the new disruptive technologies demand a portfolio of competencies that is different than what the customers of the incumbent technologies demanded. There is a tension between the constancy and change (Note 3.7).

Silver-Halide Photography Versus Digital Imaging

Consider the silver-halide photography technology pioneered by Eastman Kodak over 100 years ago, as compared to the newer digital imaging technology emerging in the market in the 1990s. Eastman Kodak had huge investments "sunk" in the silver-halide photographic technology. Starting from the dawn of the 20th century, Kodak allocated a lot of resources in what was considered one of the most vertically integrated production process. George Eastman, with an eye for total quality, ventured to produce everything, ranging from the silver processing to film coating, camera manufacture, and photofinishing.

Robert Patton of the *Financial Times* postulated that just as video tape supplanted most of the home movies, digital imaging was on the verge of chipping away silver-halide film's dominance in still photography. The attackers were not only Hewlett-Packard in the United States, but Japan's Ricoh, Casio, and Sony companies, all of which were not previously active in the photographic film technology. In 1995, the resolution of digital imaging technology was not as high as that of a high-quality photograph. But, the digital imaging provided the advantage of easy manipulation and transmission with computers. Some of the digital cameras, such as Canon's EOS DC53 and Ricoh's DC-1 models, could record sound or short motions, along with still images (Note 3.8).

Hewlett-Packard, the king of printers for computers, was poised to take on Kodak and others in 1997 to compete for the $40 billion photography market. Hewlett-Packard's PhotoSmart digital photography kit, licensed from Konica Corp. of Japan, was a wedge to break into the Kodak turf. This could completely eliminate the photographers' trips to the neighborhood WalMart or drugstore photo centers. Instead the desktop would become the new digital darkroom for processing the captured images.

Thus far we discussed (a) the alternate ways to generate technologies and (b) the maturation and substitution of one technology with the next technology. Next, we turn our attention to the protection of the generated or acquired technology by patenting and other means.

PROTECTING INTELLECTUAL PROPERTY OF NEW TECHNOLOGY

The core wealth of a new technology or an innovative idea is in the intellectual knowledge that it embodies (Note 3.9). The societies and economies which recognize, reward, and protect ownership of intellectual property encourage new inventors to discover more and share with others what they have discovered. According to Abraham Lincoln, patents provided the fuel of interest to the fire of genius.

Owning an intellectual property is not always as noble as Abe Lincoln wanted us to believe. In 1997, a battle was brewing between the many independent inventors in America and the large corporations, over the ways patents for intellectual property were granted in the United States. The reason was that the U.S. Congress was considering changes in the U.S. patent law and was planning to reorganize the U.S. Patent & Trademark Office (PTO) into a semiautonomous government corporation. Changes were also proposed to publish the patent applications 18 months after filing, instead of when the patents were actually granted.

The individual inventors feared that the proposed changes would diminish their rights to protect their intellectual property. For years they had failed to receive support from many courts which did not uphold their patents. This changed in 1982 when the U.S. Court of Appeals was created for patent cases. The patent scene changed dramatically thereafter.

In 1997, an appeals court upheld $103 million in awards to Raymond V. Damadian and his company Fonar Corporation, who claimed to be the inventors of magnetic resonance imaging technology, in a lawsuit against General Electric Company (Note 3.10).

To fully grasp the issues surrounding the emotionally charged controversies related to patents, we must first explore some basic facts that govern the means for protecting the intellectual property of a technology or an idea.

Alternate Ways to Protect Intellectual Property and Technology

Patents are one of the many alternate ways by which the inventor(s) of the intellectual property of a new technology or an idea can protect and benefit from the efforts invested.

1. Patents. Patents are the technolegal documents filed with the federal government in the area of jurisdiction of the desired patent. In the United States, the U.S. Patent & Trademark Office processes the patent applications. Patents, when issued to an inventor restrict the manufacture, use, or sale of the patented product or process, except by the legal holders of the rights to the patent.

Patents cover an operating process, an equipment, a product, or its improvement. A patentable invention must be new, useful, not obvious, and not a variation of something already known. Inventors must also realize that their invention cannot be patented if it has been sold, used, or even made publicly known 1 year before filing for a patent.

Typically, the duration of a patent's protection is 17 years. It starts from the day of receiving the patent. (In 1997, the patent offices around the world were considering changing that duration to 20 years from the date of filing a patent). However, for a design it can be 14 years, and for a genetically engineered plant it is 18 years. The average cost for filing patent, patent search, and defending against objections could range between $5000 and $20,000 (or even more) depending on the scope of the patent.

2. Copyrights. Copyrights cover original works of authorship that are presented in a tangible form. Copyrights prohibit unauthorized copying of the copyrighted work, or publishing its imitation, by anyone other than the copyright holder. Duration of copyright is generally during the life of the author, plus an additional 50 years. The average cost of filing for copyright could vary from $150 to $300.

3. Trademarks. Trademarks cover words, symbols, and trade designs that represent the source or origin of the goods or services. Trademark protection restricts use of confusingly similar words or symbols to identify the sources of other goods. Duration of trademarks is for 10 years, and they are renewable thereafter. The average cost of filing a trademark, conducting a trademark search, and defending against objections could easily exceed $1000 or more.

4. Trade Secrets. An individual inventor or a firm could prefer to overlook the above-mentioned routes to legally protecting their intellectual properties. Instead they could prefer to simply hide it as a treasure or a trade secret.

Example: Coca-Cola's Trade Secret. Coca-Cola is a classic example of successfully using trade secret strategy in the highly competitive soft drink industry. The formula for Coca-Cola was developed in 1886 by John Styrth Pemberton, an Atlanta druggist. After more than 100 years, the composition of Coca-Cola was still a trade secret that was closely held by the successors of its inventors. There were a very limited number of individuals (nobody knew how many) who had access to the recipe of formulation of Coca-Cola.

If the inventor of Coca-Cola had patented the formula with the U.S. Patent Office, the composition of Coca-Cola would have become a public property after 17 years—that is, by the year 1903. In exchange, the inventor would have received a monopolistic use of the formula during the duration of the patent. Thereafter, everybody else could have used the recipe and the process to produce this "wonderful" drink.

By keeping the recipe of Coca-Cola a secret, its owners managed to hide their intellectual property for such a long time without much help from the U.S. government. They still could charge "misappropriation" if they suspected that their secrets were obtained by others through confidential relationships or illegal means.

In general, patents or copyrights have been easier to enforce than the unprotected trade secrets. However, these alternate means of protecting the intellectual property had their own pitfalls.

The Historical Origins of Patents

The term *patent* is elliptically derived from "letters patent" [q.v.], which referred to the documents which confer a privilege. In practice the term patent is exclusively used by a government to grant exclusive property rights to make, use, or sell a product. Because this results in monopolistic effects, government gave out such rights to inventors in exchange for disclosing the secrets of inventions and discoveries. This was done for a limited period. The underlying intent of this exchange was "to promote the progress of science and useful arts."

As the U.S. Constitution was being drafted in the late 18th century, the federal government was empowered to issue patents and protect the copyrights of authors. Article 1, Section 8 of the U.S. Constitution specified that congress shall have the power "to promote the progress of science and useful arts by securing for limited times to authors and inventors the exclusive right to their respective writings and discoveries."

On April 10, 1790, the first legislative act was passed to grant complete authority to a board consisting of the U.S. Secretary of State, the Secretary of War, and the Attorney General. At that time, Thomas Jefferson was the Secretary of State, and he personally examined the applications for patents for many years.

Because these high officials of the U.S. federal government were too busy with other things, only a few patents could be issued in the early years after the American Independence. A new law was passed on February 11, 1793, under which the authority for issuing patents remained under the U.S. Secretary of State, but the patents were granted for complying with formal requirements of submitting a description, drawings, a model, and the necessary fee.

This method of granting a patent without examination of novelty increased the rate of patenting to more than 600 patents a year. This practice continued until July 4, 1836 when a new act was passed. This included an "examination" in granting a patent. At that time, this practice was being followed in many other countries. Applications were examined by officials, who had the authority to accept or refuse. There were provisions for appealing the decisions.

The Act of 1836 was amended from time to time, until 1870, when the patent law was completely rewritten. Thereafter the U.S. Congress passed about 60 patent-related acts, until 1952 when the law was revised under Title 35, "Patents" of the United States Code.

The U.S. Patent Office was originally located in the U.S. Department of State. In 1849 it was moved to the department of interior just created, and on April 1, 1925 the location of the office was changed to the Department of Commerce.

The Current State of Patenting

In the 1990s, the administration for granting patents was under the United States Patent & Trademark Office, as a bureau of the Department of Commerce. The Patent Office was staffed with examiners and other staff members with specialized technical and legal background. The specifications and drawings of patents were published in a weekly *Official Gazette,* containing the claims, decisions, and other information related to a patent. The U.S. Patent Office also administered the federal trademarks, which are crucial to interstate and international trade and commerce (Note 3.11).

The Criteria and Contents of Patents

In the United States, patents were granted for the invention of "any new and useful process, machine, manufacture, or composition of matter, or any new and useful improvement thereof."

1. The word "manufacture" referred to articles which were made, and it included manufactured articles and products.

2. The word "process" was defined as meaning process or method used, and it was not limited exclusively to processes for producing articles. However, it has been ruled that processes involving mental steps are not proper subject matters for patents.

3. The meaning of "machine" needed further explanation, and the "composition of matter" included substances composed of mixtures of ingredients as well as those produced by chemical reactions. New chemical compounds, as well as processes to make them, could be patented, like various foods and medicinal products.

The application of fissionable material or use of atomic energy to create atomic weapons was, however, specifically excluded from patentable inventions. These were covered under the Atomic Energy Act of 1952.

4. The word "useful" implied that the invention must be directed to a useful purpose, and it must be capable of accomplishing that purpose. A machine that was inoperative to produce the intended result could not be called useful. The commercial (or large scale) practicality, however, was not a prerequisite.

Guiding Principles

Patents were issued based on the following guiding principles.

A. Novelty. To be patentable, the invention had to be new. Therefore, a patent could not be issued if (a) the invention was known or used in the United States (for one year prior to the date of application of inventor) or

(b) it was patented or described in a printed publication anywhere, before the invention was made by the inventor who was seeking a patent for it.

Mere prior knowledge in a foreign country did not prevent a true inventor from obtaining a patent. Similarly, prior secret knowledge and use by others did not ordinarily defeat the right to a patent.

B. Patent and Publication. Generally, patents were not issued if an invention had been in public use or on sale in the United States, or had been described in a printed publication anywhere, for more than one year before the date on which the inventor applied for a patent. The inventor's own printed publication or public use of the invention, as well as such publication or use by anyone else occurring more than one year before applying for a patent, resulted in the loss of the right to a patent. Publication or use during the one-year period did not affect the right to a patent if the inventor seeking the patent made the invention before the date of the publication or use.

C. Not Obvious. Furthermore, a patent could be denied even if the subject matter sought to be patented was new—in the sense that it was not shown exactly by the prior things known or used and included one or more differences over them. To get a new patent, the differences between the new subject matter and the old knowledge must not be obvious to a person having ordinary skill in the art. The courts have ruled that the new subject matter must involve "invention" over what was old. This requirement has caused many conflicts of opinion.

D. The Applicant for a Patent. Only the actual inventor could apply for a patent in the United States. There were some exceptions. The exceptions were when the inventor was dead, then the executor or administrator of the estate could apply; or if the inventor was insane, his/her legal representative could apply. If the invention was made jointly by two or more persons, they had to apply for a patent as joint inventors.

E. The Application for a Patent. The application for patent was expected to contain (1) a written description of the invention, (2) a drawing when the nature of the invention admitted a drawing, and (3) an oath of inventorship executed by the inventor. The patent application must be accompanied by the government filing fee.

The description of the invention must be so complete as to enable any skilled person (i.e., skilled in the art to which the invention relates) to make and use it, and it must set forth the best mode contemplated by the inventor of carrying out the invention.

The application also required that the claims particularly and distinctly pointed out the subject matter regarded by the applicant as invention. The claims were expected to define the scope of the subject matter protected by the patent. The description and claims were called the specification.

F. Need for a Model or Demonstration. Prior to 1870, the patent applications required a model with every patent application which involved a device. This was later discontinued as a general requirement. However, in the case of an invention called a "perpetual-motion machine" the patent office could still demand a working model. The inventors of these types of machines claimed that their machines could generate energy on their own and could keep moving in a perpetual motion. These inventors have caused many controversies in the past, when they inundated the Patent Office with their claims. Therefore, the patent applications for such machines were expected to be accompanied by a working model.

G. Procedures for Granting a Patent. After the patent office received an application for a patent, it was referred to the appropriate examining division of the patent office. The examiners examined the applications in the order they were received. To determine if the invention was new and patentable under the U.S. law, the patent examiner had to search through all the relevant U.S. patents issued earlier and also had to go through those patents of other countries. To establish the state of the art and also establish the novelty of the claims in a patent, the examiners searched the printed publications available to them.

H. Patent as a Property. A patent is recognized by law as a property. It has therefore all the attributes of a personal property. A patent may be assigned or sold to others. It may be mortgaged, bequeathed by will, or inherited by heirs of deceased inventors.

The patent property is assignable by law by a written instrument, either after it is granted or while it is in the application stage. If the patent application is assigned and that assignment is recorded in the patent office, then the patent will be issued directly to the assignee. Sometimes, only a part interest in a patent may be assigned. Similarly, a patentee can convey a grant for a specified territorial part of the United States. An assignment, grant, or other conveyance can be recorded in the patent office at any time. But if it is not recorded in three months, it might not serve as protection against the subsequent sale of the patent to another.

Because the patent gives the owner exclusive right to exclude others from making, using, or selling the invention, the inventor can also authorize others to do any of those things by assigning a license to them. A license is a contract between the parties, and it may include whatever valid conditions and limitations may be agreed upon, such as the amount of royalties to be paid. The patentee, however, cannot violate the antitrust laws in his/her dealings with patents. Some nations have more specific laws and formalities for the sale of patent rights.

I. Infringement of Patent Rights. The patent statute defined infringement of a patent as the unauthorized making, using, or selling of the patented invention, within the United States, during the term of the patent. Active inducement of infringement by others could also constitute infringement.

If a patent was infringed, the patentee's remedy was by a lawsuit against the infringer in the appropriate district court of the United States. The question of infringement is driven and determined by the claims of the patent, which defined the scope of the rights granted.

J. Government's Right on Patents. The U.S. government could make or use any patented invention without the consent of the patentee. But the patentee had the right to compensation which may be recovered by a lawsuit in the court of claims if not paid.

K. International Rights of Patents. The rights granted by a patent to its inventor(s) or owners were limited to the jurisdiction of the country granting the patent. Thus, an inventor who desired patent rights and protection in a number of different countries was expected to obtain a separate patent in each one of them. There was no international patent, applicable globally around the world. However, the laws of a few countries have mutual agreements that provided that under certain conditions a patent granted in another country may be registered in their own country and become effective.

Different countries of the world have their own patent laws, which differed from each other in many details. They typically varied in terms of the following.

1. The subject matter for which patents are granted
2. The conditions which defeat novelty
3. The requirements and procedure in obtaining patents
4. The length of the term and other related matters

PATENT PROTECTION: A PRIVILEGE OR A PROPERTY

Controversies Over Contentions

Even though the patent laws governing the protection of patents concerning one's intellectual property were carefully defined as described above, there was often much controversy over their contentions.

1. GM Versus Lopez. For example, in May 1993, General Motors (GM) in a highly publicized case, accused Jose Ignacio Lopez de Arriortua of stealing GM's intellectual property. Lopez was a senior executive with GM for a few years, and he earned a reputation for cost cutting. He was lured by one of GM's competitors Volkswagen. In October 1992, Borland, a major player in America's software industry, accused an ex-employee of stealing trade secrets and passing them to his next employer, Synatec. Finally, Dr. Peter I. Bonyhard left IBM for Seagate Technology and was sued by his former employer.

2. *International Trade and Piracy.* In international trade, the piracy and protection of intellectual property became a megabuck issue in computer software, entertainment, and pharmaceutical industries. According to their trade associations, in 1990 alone the annual revenues lost to international piracy amounted to about $10 to 12 billion for computer software products, about $3 to 5 billion for pharmaceutical formulations, and to about $2 to 3 billion for U.S.-made music and film tapes. These added up to a nifty $15 to $20 billion per year. Sometimes the estimates were much higher. In 1986, the U.S. International Trade Representative estimated that the U.S. businesses lost an estimated $60 billion to international intellectual piracy. This contributed to a significant part of the U.S. trade deficit.

The U.S. Trade Representative also used a $60 billion annual trade surplus with Japan in 1986 to complain to the Japanese about their Mafia-like video film piracy. As the presumed Japanese piracy persisted, the United States threatened to clamp down additional tariffs on Japanese-produced electronic goods imported to the United States. As a result, it has been often pointed out that within five years the sales of U.S.-made original video films increased to $300 million per year in an estimated $350 million annual market in Japan.

The United States has been playing a leading role in the international negotiations for protection of intellectual property in the world markets.

Somehow, these controversies over intellectual properties do not seem to end. Many observers attribute them to the way patents are granted to individual or corporate inventors. As stated earlier, such controversies resulted in a strong movement to remedy and bring about some changes in the laws governing patents.

Patent Changes in the Future

The changes, proposed to the U.S. Congress in 1997, in the patent provisions and the organization of the U.S. Patent and Trademark Office were expected to affect different parties in different ways.

1. *PTO as a Government Corporation.* Many small inventors believed that reorganizing the U.S. Patent and Trademark Office as a government corporation would reduce the government oversight of the patenting of intellectual property. They also believed that the small inventors would have less protection from the infringement of their inventions by the larger corporations.

Many other observers believed that making this change would make the patenting process more efficient. As a government corporation, it would free the Patent Office from rules for hiring, firing, and printing that govern the workings of the U.S. government. To reduce cost and speed of the patenting process, the government corporation would be able to outsource certain tasks which have been slowing them down in issuing patents faster.

2. Submarine Patents. Changing the time duration for publishing a pending patent also was a significant move. In the past, many patent applications lay hidden for years in the patent office while the applications were waiting to be processed. In principle, though not as often in practice, this caused a unique complication.

When these patents were actually issued, their owners could use their rights to sue other inventors (particularly large corporations with deep pockets) who may have made inventions infringing on their claims, while not knowing about the claims lying hidden in the submerged patent applications.

The government patent offices in Europe and Japan had already changed to publishing the pending patents after 18 months from the date of application, instead of waiting until the patents were actually issued. The large U.S. companies were forced to follow this practice when they filed their patents in these foreign countries. They wanted the same practice in the United States—to level their competitive playing field with their foreign competitors and small independent inventors locally.

The independent inventors argued that such 18-month publication would leave them with very limited protection against the infringement of large companies stealing ideas from their inventions.

3. Slow Patenting in Fast Technologies. Another major challenge was the slow patenting process in the fast changing technology sectors. In bioengineering and electronics, the product life cycles had shrunk significantly due to global competition in the 1980s. For many electronics products, the product life cycles were shrunk to a few months. A long lead time in receiving a patent defeated the purpose. The owner of the intellectual property could hardly exploit the patent, because by the time a patent was issued the technology had already become obsolete.

4. Broad Patents in Emerging Technologies. In biotechnology and other emerging areas, patents posed another dilemma. With new knowledge embodied in the patents, the patent examiners with less expertise could mistakenly issue patents to a whole field of knowledge. This could stunt further growth in technology in that area.

Overpatenting of Genetic Cotton. For example, in 1994 the U.S. Patent Office issued a patent to a biotech firm for genetically engineering all cotton plants. This implied that hundreds of researchers and laboratories working with the Department of Agriculture had to seek a license and pay royalty to the biotech firm to allow them to study alternate means to change cotton plants genetically. This could have altogether stopped further research in this area, with potentially devastating effects on the U.S. agriculture. Immediately a huge public outcry emerged with large number of complaints to the Patent Office. The Patent Office was thus persuaded to reexamine and rescind the patent.

The patent experts believed that often in the early years of an emerging technology the Patent Office could erroneously issue an overly broad patent. As technologies and industries grow and mature, the number of such instances is reduced.

5. The Software Patents. In the Information Technology Age, the computer software programs have posed new challenges to the age-old patenting process. In the 1990s, the software technology was growing so fast that the Patent Office was flooded with thousands of pending patent applications.

Beside the enormous volume of patent applications, the computer softwares were also very different from the hardware inventions patented in the past. Each software program, with many modules and thousands of lines of code, could easily infringe on hundreds of other software patents. Until a few years earlier, the computer software programs were protected by copyrights. The rules of the patenting game were changed mid-stream (Note 3.12).

In the early 1990s, there was a huge battle over the look and feel of the user interface of a desktop computer. Apple, Microsoft, and Hewlett-Packard bitterly contested with millions of dollars over a patent application in this area. Finally, Apple was issued the patent for this invention. Microsoft was allowed the right to continue to use the look and feel of its Windows program because earlier Apple had granted it a similar technical license. The litigation failed to note that Apple was inspired by the Star interface technology created by Xerox's Palo Alto Research Center (PARC).

Patenting Blues

The key challenge for the patenting process in the 21st century would be to balance the protection of an invention, with its role in the promotion of future progress and continuous nurturing of new technologies. In other words, a tightrope balance was required between the public good and the private greed behind a patent.

A BRIEF SUMMARY OF LESSONS LEARNED

Creation of proprietary know-how by various American pioneers gave birth to many new technologies and industries in the late 19th century. This propelled America to its position as the leader of the industrial world. The various alternate forms of proprietary know-how range from accidental discoveries to hard-to-swallow research alliances with the competitors. There are three main modes to protect the intellectual property. In the near future, some of the laws governing patenting would require retooling because of the unique nature of certain emerging new technologies such as genetics and software programming

SELF-REVIEW QUESTIONS

1. Compare and contrast the alternate paths to developing new technology.
2. Proposition: Competitors should collaborate to develop new technologies. Discuss for and against.
3. Should U.S. government receive more funds for fundamental research and development?
4. Discuss the new challenges for patenting procedure in a highly globalized economy.
5. Debate the implications of patents as a privilege by the State, and not as a "natural" property right of the inventor.

NOTES, REFERENCES, AND ADDITIONAL READINGS

3.1. Wise, G. 1980. Industrial research at General Electric 1900–1916, *Technology and Culture,* 21:408–415.
3.2. Mabon, Prescott C. 1975. *Mission Communications.* Murray Hill, NJ: Bell Telephone Laboratories. In particular, see Introduction by John D. DeButts, Chairman AT&T.
3.3. See a number of *Fortune* stories by Bro Uttal in 1981 and 1983. For example, Uttal, Bro. 1981. Xerox zooms towards the office of the future, *Fortune,* May 18:44–52; Uttal, Bro. 1983. The lab that ran away from Xerox. *Fortune,* September 5:97–102.
3.4. Freed, Les. 1995. *The History of Computers.* Emeryville, CA: Ziff-Davis Press.
3.5. Adapted from Gehani, R. Ray. 1997. Creating with Competitors. How a Trans-Pacific Team Developed Advanced Photo System Technology Platform? Unpublished working paper.
3.6. Adapted from Utterback, James M., and Abernathy, William J. 1975. A dynamic model of process and product development. *Omega,* 3(6):639–656.
3.7. For discussion on S curves and discontinuities see Foster, Richard N. 1986. *Innovation: The Attacker's Advantage.* New York: Summit Books. Earlier empirical research can be found in, Fisher, J. C. and Pry, R. H. 1972. A simple substitution model of technological change. *Technology Forecasting and Social Change.* 3:75–88. This was reiterated recently by Bower, Joseph and

Christensen, Clayton. 1995. Disruptive technologies: catching the wave. *Harvard Business Review,* January/February:43–53.

3.8. Patton, Robert. 1995. The digital age. *Financial Times,* May 4:12. Also see *Business Week* cover stories, Shootout, and Kodak's digital developments in July 7, 1997 issue.

3.9. Some of the discussion on patents is based on comments which appeared earlier in Gehani, R. Ray. 1993. Are patents for public or private gain? *Rochester Business Journal,* July 16:10.

3.10. Carey, John, and Port, Otis. 1997. Rumble at the patent office. *Business Week,* June 2:142–147.

3.11. A number of sources are available to cover this subject in greater detail. The U.S. Patent Office has published a number of publications for the general public. Consider: *Patent Laws* (revised periodically); *Rules of Practice of the United States Patent Office in Patent Cases* (revised periodically); *General Information Regarding Patents* (revised periodically), and so on.

3.12. A good overview of some of the emerging challenges for patenting process was provided by Shulman, Seth. 1997. Patent medicine. *Technology Review,* November/December:28–34.

CASE 3A: PIONEERS OF TECHNOLOGY

Wallace H. Carothers: The Founding Father of Polymer SCIENCE, Synthetic Fibers, and Rubber Technologies

Today we cannot visualize our lives without synthetic fabrics, rubber tires, and plastics. Yet barely 70 years ago mankind either did not know many of these useful substances or they did not understand why these substances behaved the way they did. One man's leadership and persistent research for 10 years changed all that. Managers of technology can learn a few lessons or get inspiration from other pioneers' lives. In this case we will learn from a polymer legend.

Wallace H. Carothers (1896–1937) is considered the founder of polymer science and synthetic fiber technology. He died young, taking his own life at the young age of 41 years. Yet in a short span of one decade he did ground-breaking research leading a path to the discovery of a vast number of basic materials for the 20th century. Socks and stockings, synthetic fibers, and strong tire reinforcements attribute their success to Carothers' polymerization research.

Family Origins and Education. Wallace Hume Carothers was born the oldest of four children in Burlington, Iowa, on April 27, 1896. His Scottish ancestors were farmers and artisans who settled in Pennsylvania in the pre–Revolutionary War days. Wallace's father was a teacher, and became the vice-president of the Capital City Commercial College in Iowa. His mother was from Burlington, Iowa. Wallace and his family loved music, and his sister performed professionally in a radio trio.

In his youth, Wallace spent hours experimenting with mechanical things and tools. In school he was thorough and he always finished the work he had started. He graduated from high school in 1914 and joined his father's Capital City Commercial College to study accountancy and a secretarial course. After that he moved to Tarkio College in Missouri to major in chemistry. To fund his education he worked as an assistant in the Commercial Department where later he was appointed as a chemistry instructor. This started his pioneering career in chemistry. In 1920, at the age of 24, Carothers joined the University of Illinois to work on his M.A.

A year later he graduated and returned to Tarkio College as a lecturer. He prepared his lectures carefully, though he was not considered a forceful lecturer. His mentors advised him to pursue higher studies. Carothers returned to the University of Illinois and earned his doctorate in organic chemistry in 1924. After working as an instructor for 1 year, he joined Harvard faculty in 1920 at the age of 30. By that time he had published a number of noteworthy papers.

Du Pont's Needs. By 1928 Du Pont was hit hard by the loss of orders for explosives due to the sudden end of World War I. Du Pont was ready to branch out in new nonexplosive areas. Probably significant financial resources were accumulated from the sale of their explosives first to the Panama Canal construction project and later to the World War I warring parties in Europe.

With the end of the war, Du Pont was ready to take some long-term risk. Therefore, in the footsteps of General Electric Research and Development Center and Bell Telephone Laboratories, Du Pont too decided to start a laboratory dedicated exclusively to pioneering research and development efforts. Carothers, at the age of 32 and with interest in macromolecular substances, seemed like a rising star and a good candidate to lead Du Pont's pioneering efforts.

To Carothers, the Du Pont proposal was a dream come true. He would have a university-like freedom to do his fundamental research with a corporation-like funding. Carothers' decision was Harvard's loss, but it was definitely polymer chemistry's gain.

The State of Chemistry in 1920s. Chemists in the early 20th century did not understand molecular structure of the substances. Some substances exhibited a high molecular weight and were often viscous, if soluble, in solvents. Most of the chemists at that time believed that molecules were only as long as their unit cells. All the macromolecular substances had a common molecular structure. They consisted of repetitive chemical units. Later these substances were commonly called polymers, with repeating units of monomers.

German scientist **Hermann Staudinger (1881–1965)** was one of the early chemists anywhere in the world to systematically produce the polymeric materials. In 1920, for the first time he proposed a molecular structure for polystyrene and other polymers. He also proposed that rubber had a chain-like molecular structure and that the unique elastomeric properties of some substances were due to their chain-like molecular structure. Staudinger published over 20 papers between 1922 and 1930 based on his research on rubber. Yet, he failed to convince many chemists of his era. In the 1920s, few people wanted to believe in the possibility of polymeric materials.

Focused Research. At Du Pont, Carothers focused his research on searching the theoretical basis for the way certain large-molecular substances were formed when substances with simple molecules were reacted together. He soon postulated two different types of polymerization chemical reactions. The addition polymers were produced when the molecular units of a monomer added to each other to make a long polymeric chain. Another type, which he called the condensation polymers, were produced when two dissimilar molecular units reacted with each other to form long molecular chains. Carothers developed the polymerization theory, but he and his associates needed experimental proof to support their hypothesis.

Multiple Birth of Elastomers. In less than two years since joining Du Pont, on April 16, 1930, Carothers' team was able to synthesize a condensation polymer as predicted by Carothers. The polymer material could be drawn into a useful fiber form. But Carothers needed more proof to support his theory of polymerization. He therefore shifted his attention to the addition polymerization reactions.

Starting with acetylene he produced addition polymers that gave birth to synthetic rubber. While synthesizing a trimer of acetylene on a Friday, Carothers found the material solidified over the weekend. The material bounced off the laboratory bench. A careful material analysis showed that it was chloroprene. Carothers figured out that the chlorine came from the catalyst used to produce the trimer. In 1930 polychloroprene was produced with physical properties of a vulcanized rubber, and its resistance to oxidation was superior to that of the natural rubber. Carothers and his team researched polymerization of synthetic rubber, and then he handed over the project to others in Du Pont's development laboratory. In June 1932 Du Pont marketed polychloroprene under the trade name of Du Prene.

Polymerization Theory. Between 1929 and 1931, Carothers and his team published a series of papers which laid the foundation of organic polymer science. He gave clear definitions of terms such as polycondensation and addition polymers, and he explained the polymerization of cyclic compounds by the ring opening mechanism. Based on his polymerization theory and his experiments on synthetic polymers, Carothers was able to explain the observed but yet to be understood properties of natural polymers (such as wool, cotton etc.). In 1931, at the young age of 35 years, Carothers proposed principles which are still in use more than six decades later. He urged more chemists to do research in the new field of polymer science, and he made a tremendous contribution to its growth.

Fiber Formation. To make synthetic fibers stable in useful forms, Carothers developed cold drawing and other processes. These provided the foundation for another new field of fiber science and technology. Many synthetic polymers were drawn under different operating conditions to make strong synthetic fibers. Using the X-ray diffraction studies of Herman Mark, Carothers showed that the cold drawing process oriented the long polymeric chains along the axis of the fiber. Such orientation made the drawn fibers dimensionally more stable and tougher, with tensile strengths comparable to those of silk fibers. He pioneered the discoveries of nylon and polyester polymers, two of the most popular polymers still in use today. In 1931 he received a patent for "Fiber and Method for Producing It."

Carothers was not very social in person, but professionally he was very active. In 1929 at the age of 33, he was elected the Associate Editor of the *Journal of the American Chemical Society.* A year later, at 34 he became the editor of *Organic Synthesis.* At the age of 40 he was the first industrial chemist elected to the National Academy of Sciences.

Carothers was completely dedicated to his work, and he loved reading on a wide range of subjects from politics to labor relations. He did not marry until he was 40 years old. In February 1936, Carothers married a co-worker with a chemistry background, but the marriage did not last.

Dissatisfaction of a Perfectionist. Even though the world in general, and chemists in particular, considered Carothers brilliant, he himself often felt that he had not achieved as much as he should have. He was often depressed that he might run out of good ideas to guide his research team. In January 1936 he was devastated by his sister's death. In 1937, the world was shocked when a not yet 42-year-old Carothers took his own life.

The founding father of polymer science and the synthetic fiber industry could not see the day a year later when Du Pont publicly announced the new nylon polymer product. The first truly synthetic fiber nylon was shown to the entire world at the 1939 Worlds Fair. Dr. Wallace H. Carothers must have been pleased to see that from the Heaven.

Source: Adapted from Gehani, R. 1985. Genesis of synthetic fibers: with Carothers' pioneering efforts. *Man Made Textiles,* January:12–21.

CASE 3B: TECHNOLOGY MANAGEMENT PRACTICE

Selected Technological Milestones in the First 50 Years of Bell Laboratories

Bell Telephone Laboratories, considered one of the most reputed research and development centers of the world, is a role model for the technology-driven enterprises producting proprietary know-how. Over the years the researchers at the Bell Laboratories have made many ground-breaking inventions and discoveries. Many of these discoveries not only fetched the coveted Nobel Prizes, but they also gave birth to new industries. Given below is a partial list to illustrate their technological milestones and achievements during their first fifty years of efforts.

Bell Telephone Laboratories Incorporated (1925). The Western Electric Engineering Department was incorporated as Bell Telephone Laboratories in New York City, with 3600 employees.

Hi-Fi Sound (1925). First high-fidelity (Hi-Fi) sound recording was developed under Western Electric patents. The transmission of stereophonic sound was demonstrated in 1933.

Motion Film Synchronization (1926). Audio equipment was developed to synchronize a musical score (Don Juan) with the first sound motion picture. One year later, a similar technique was developed to synchronize dialogue with motion film (*The Jazz Singer*).

Negative Feedback Amplifier (1927). H. S. Black used part of an amplifier's output as a feedback to stabilize distortions.

Matter's Wave Nature (1927). C. J. Davisson discovered that electrons diffracted from a nickel surface behaved in a wave-like manner. He shared the 1937 Nobel Prize in Physics.

Long-Distance TV Transmission (1927). Conversion of light to electric current helped develop transmission of television images over wire. Ten years later a commercial television broadcasting service was introduced.

Artificial Larynx (1929). A reed-type artificial larynx was developed to help people speak again after they had lost their voices in an accident or by a disease. A transistorized version of the larynx was introduced in 1960.

Quartz Crystal Filters (Late 1920s). The design and manufacturing process for quartz filters was developed and used in most modern communication systems.

Systems Engineering (1930s). The systems engineering approach to decision-making and evaluation of an innovation was developed. This involved technical feasibility, compatibility with existing and potential systems, generation of alternatives, and economic feasibility.

Radio Antennas (1930s). Antenna theory and various forms of antenna were developed for propagation of radio waves.

Consolidation with AT&T's R&D (1934). The Development and Research Department of AT&T was consolidated with the Bell Laboratories.

Scientific Quality Control (1935). W. A. Shewhart applied statistics and mathematics to control the quality of Bell System equipment, thus pioneering the field of quality control.

Microwave Radio Equipment Technology (1938). Waveguides, antennas, power sources, and other devices coupled with microwave propagation theory contributed to the development of radar in World War II and led to commercial microwave radio thereafter.

U.S. Navy Radar (1938). Bell Laboratories' scientists did research on fire-control and search radars for Navy ships.

FM Radio Altimeter (1939). This was used to help aircraft pilots estimate their altitude above ground by bouncing FM radio signals off the earth's surface and receiving the reflections by a broadband receiver.

Thyristor or Thermal Resistor (1940). Bell Laboratories developed a semiconductor device which changed electrical resistance with temperature, and used it to measure temperature in industrial, automobile, and communication applications.

Electronic Analog Computer (1940). The electronic analog computers were applied to control anti-aircraft guns. They helped hit 76% of the German flying buzz bombs over England.

Move to Murray Hill, New Jersey (1941). The first group of researchers moved to their future headquarters.

Magnetron (early 1940s). High-power microwaves were generated by magnetrons to develop microwave radars during the early years of World War II. They used vacuum tubes with strong magnetic fields applied to moving electrons to generate microwave frequencies with high power. Magnetrons gave birth to microwave ovens.

Transistor (1947). John Bardeen, Walter Brattain, and William Shockley invented the transistor, and they started the electronics industry. They earned the 1956 Nobel Prize in Physics.

Polyethylene Sheath Insulation (1947). The polyethylene sheath replaced lead in insulating cables and telephone wires.

Information Theory (1948). C. E. Shannon published a theory of maximum information carrying capacity for a communication system.

Field-Effect Transistor (1951). The first practically useful field-effect transistor was invented by William Shockley.

Direct Distance Dialing System (1951). Direct dialing was first introduced from Englewood, New Jersey to selected sites across the United States.

Silicon Solar Battery (1954). Bell Laboratories scientists extended their understanding of the transistor to invent the silicon solar battery, and they were able to convert sunlight into electricity.

Computer Operating Systems (mid-1950s). This software facilitated the use of early large computers, and it later evolved into a time-shared UNIX operating system for minicomputers.

Transatlantic Telephone Cable System (1956). Electronic amplifiers for use in ocean were developed to lay the first cable from America to Europe. A pilot trial was conducted to Cuba.

Laser (1958–1960). Bell scientists and consultants invented the continuously operating laser, using first helium and neon, and later carbon dioxide.

Epitaxial Transistor (1960). These transistors used material grown in thin epitaxial layers for high-speed applications.

Electronic Switching System (1960). Electronics was used to program the switching system and change service without rewiring the switching equipment.

Telstar Communication Satellites (1962–1963). Long-distance telecommunications were developed by launching Telstar I (1962) and Telstar II (1963) satellites. Earlier the Echo balloon was used to bounce coast-to-coast calls.

Touch-Tone Telephone (1964). Push buttons replaced the rotary dials to speed up dialing.

Magnetic Bubble Devices (1966). Magnetic bubble technology helped substitute solid-state materials for large-capacity memories in computers.

Charge-Coupled Devices (CCDs) (1970). W. S. Boyle and G. E. Smith invented CCDs for applications in imaging, logic, and memory applications, thus replacing complex integrated circuits.

Picture-phones (1970). Visual images were added to telecommunication, to help a caller see the receiver or to visually see other images.

Source: These achievements are only a few of the many listed in Mabon, Prescott C. 1975. *Mission Communications.* Murray Hill, NJ: Bell Telephone Laboratories.

CHAPTER 4

Product Development and Customer Trust

PREVIEW: WHAT IS IN THIS CHAPTER FOR YOU?

During the 1980s and 1990s, a number of discontinuous market trends had a profound impact on the way technology-driven enterprises competed with each other in the market. Customers around the world developed high expectations. They demanded satisfaction from enterprises interested in obtaining a loyal relationship for the goods and services that the manufacturer provided. This forced the technology driven enterprises to alter the way they developed their new product developments, and the way they marketed these products to the increasingly picky customers.

An increasing access to technology and a rising intensity of global rivalry between transnational competitors shrank the product life cycles. Products which had sustained growth for many years saw a much quicker progression to their maturity and decline stages. This took place in one industry after another. Semiconductors, computer makers, and software program sellers were the first ones to face the impact of that change. With aggressive product development by Intel and Microsoft, frequent new products were introduced. The 80286 microprocessors were quickly replaced with 80386, 80486, Pentium, and Pentium MMX microprocessors. Successive versions of software programs hit the market. In such technology-driven sectors, the competitors tried hard to get first to the customers before their competitors, by offering them their new innovative products with better bells and whistles.

New developments in information technologies expanded the customers' access to worldwide markets. Customers could easily "surf" the Internet to find out the latest offerings from producers in distant parts of the world. They could access far more information than ever before. With intense global competition and the fast-spreading Information Age technologies, the effective and fast-response development and commercialization of new product technology became the prerequisites to market success. The customer satisfaction acquired a critical significance for competitive survival and growth.

In this chapter we will benchmark, compare, and contrast some of the best practices in the areas of process of development of new products, market positioning, and cultivation of customer trust. We will study how these efforts were used by the providers of goods and services to gain customer satisfaction from their targeted customers. A brief review of how marketing evolved and gained significance over the past few decades will provide a historical context. Technology push was gradually replaced by market pull. Product portfolio and alternate product development strategies will be discussed. We will see how increasing intensity of competition put a firm's new product development competency in a strategic role in the competitive wars between global rivals. The new product development process will be reviewed. The traditional sequential product development process will be contrasted with a concurrent rugby-style process for new product development. Finally, the significance of customer satisfaction, customer trust, and customer loyalty for lifelong relationships between customers and producers will be explored.

NEW PRODUCTS AND CUSTOMERS' TRUST MATTER

In Chapter 2 we considered the role of production and operations management in technology-driven enterprises. In Chapter 3 we reviewed the different ways these enterprises generated and acquired proprietary know-how using either scientific research or a systematic experimentation. We must next explore how the new products are developed and commercialized and positioned into the market to the largest number of satisfied customers. These efforts produce the loyal customers who contribute most to the bottom line of a technology-driven enterprise. But before we do that, let us consider the role that marketing has played over the past few decades.

How Marketing Gained Significance

During the World War II years, Americans mobilized all the available resources and energies in a highly focused way. Most Americans were enraged by the Japanese attack on Pearl Harbor in December 1941. Due to the long distance to the battleground in Asia-Pacific, Africa, or Europe, coupled with a limited manpower, technology played a key role in America's arsenal. The war crisis gave birth to a number of new technologies for radar and computers, to detect and decode the enemy's whereabouts (see Case 3B on milestones in the first 50 years of Bell Laboratories). Synthetic rubber was invented and produced to fill the gap created by the Japanese occupation of rubber plantations in Southeast Asia (see Case 3A for Wallace Carothers' research on polymers at Du Pont). Also, the nuclear bomb technology and the Manhattan Project, which resulted in a better understanding about the molecular structure of an atom, were developed to end the war decisively (see Case 8A on the Manhattan Project).

Marketing Madness

After the end of World War II, the technologies developed during wartime were transformed into many desirable civilian products. These were then marketed to the eagerly waiting American customers. In particular, these products and services were aimed at the returning military personnel and the G.I.s with significant accumulated savings and a great desire to buy the new amenities for a modern civilian life. They had seen material deprivation in war-torn Europe and in the highly exploited regions of the Far East. These soldiers were eager to catch up by acquiring all they could. With a boom in birthing babies, the new and expanding families needed homes loaded with modern amenities such as radio, telephone, refrigerator, and, of course, a shining fully loaded new car for the road. With these eager buyers waiting out there, selling must have been fun. Millions of customers were willing and able to buy all that the producers could offer.

Many new products were developed based on the civilian application of the technologies developed for the wartime needs. These products used the research & development (R&D) centers that had sprung up everywhere as a result of America's war needs. Los Alamos, the Oak Ridge laboratories, and others had demonstrated that they could develop and create what was considered impossible up to then.

Many of the corporate R&D laboratories, however, were isolated from the rest of their corporate operations. Often these labs received very limited guidance or communication from their senior managers and leaders. Some of the top corporate managers with business degrees and financial orientation did not understand the intricate scientific technologies underlying their businesses. They therefore often delegated the technology-related decisions to their technical heads in the lower layers of their organizations.

Besides that delegation, in the postwar boom the market demand seemed unlimited and growing fast. The senior managers were therefore busy planning ways to increase their production capacities to generate higher economies of scale. Funding such expansions was a big time-consuming challenge. Therefore, the executives had little time for the fine nuances of tracking their customers' preferences or the market's requirements. As years passed by, it was noticed that not all technological know-how developed in-house was successfully converted into commercial products. Some of the commercialized products, launched without much market research, were rejected by the customers. Suddenly, by the end of the 1950s, many of the fast-growing companies were not growing as fast as before or as fast as they were predicted to grow.

Short-Sighted Marketing

In 1960, Professor Theodore Levitt of Harvard University wrote a ground-breaking article in the *Harvard Business Review.* He proposed that many of the growth industries had stopped growing because they neglected their customers. He called this the "marketing myopia." (Note 4.1). He postulated that a "thoroughly customer-oriented" management could keep a growth industry growing, even after other competitors declared that they had no more opportunities in that industry. For example, Du Pont and Corning were two thoroughly customer-oriented companies which continued to grow because of their focus on new market opportunities.

The R&D competency alone, when isolated from the rest of the corporation, did not sustain a high growth for an indefinite period. The growth industries ceased to grow because of their marketing myopia and their erroneous belief in the market myths listed below:

1. Infinitely expanding affluent population
2. Excessive mass production and reliance on economies of scale
3. Obsessive product-centered orientation
4. No fear for substitutes

Many of these beliefs led to the practice of selling for the seller's need. This was to exchange their products with the customers' cash. This was a transactional exchange, not a marketing effort with a goal to meet the customers' needs or service requirements. Let us illustrate these myths by looking at Henry Ford's Model T.

Model T Myopia. The most famous illustration of the marketing myopia and the "selling" orientation of business enterprises in America was Henry Ford's Model T. The Model T was priced and sold so as to keep the assembly line running. It was available in "any color" a customer wanted, as long as the color was black. The mass popularity of the car was driven by its low price and high economies of scale, and not by meeting the individual customer's satisfaction.

The Great Depression during the 1930s choked Ford's access to an infinitely expanding affluent population of customers. Ford was also not prepared when it was substituted for by its competitors' more customer-driven car products. General Motors and Alfred Sloan offered differentiated products for the different market segments of the car buying customers.

For a long time since Henry Ford's Model T, Detroit did not become customer-oriented. Customers interfaced with the auto makers for the first time either at their dealers or later at the car repair garages. But Detroit's auto makers did not consider their customers' dissatisfaction at these interfaces to be the auto makers' concerns. Something was not right from the customers' point of view.

From Technology Push to Market Pull

The technology push ceased to work. A more enlightened view of technology is as a process that produces customer satisfaction. This is distinctly different from the view of technology as the process that produces products. In the post-modern technology-driven enterprise of the 1990s, according to Peter Drucker, "manufacturing is seen as an integrated process that converts goods . . . into economic satisfactions." Using a systems approach, Drucker "embeds the physical process of making things, that is, manufacturing, within the economic process of creating business value" (Note 4.2). Therefore, the business value of a product or service, must be as perceived by the paying customers in the relevant and attractive market.

Under the marketing pull perspective, the entire organization is driven by its ability to meet customers' needs and cater to their preferences. With this perspective, the primary goal of a production process is to provide the volumes of goods needed by the customers, as and when needed. And the activities of R&D department are governed by its mission to discover and invent new ways to resolve the problems customers face from time to time. These practices are different from those of the traditional technology push organizations.

A. Evolution of Markets. The roaring 1950s and 1960s, with all the freedom to experiment and the "my way" lifestyles, posed a big challenge to the marketers at that time. Most market failures were quickly blamed on the inadequacies of the technology. What worked was credited to the achievement of marketing. The field of market research emerged to fill the gap of ignorance about customers' behavior and about what constituted the market. The market researchers collected and carefully analyzed the facts surveyed from the customers.

The two petroleum oil shocks during the 1970s demonstrated the vulnerability of many industrial companies. The prices of petroleum oil skyrocketed and quickly reduced the resources available to organizations. The companies could no longer afford to launch their new products by trial-and-error. Progressive companies, such as Du Pont, IBM, GE, Monsanto, and others, discovered that their product failures were not primarily due to their understanding of technology, but were due to the isolation of the generation of technology from the rest of the functions of these organizations. To these progressive companies, their technologies were still their key assets. They learned that a careful integration of technological excellence, coupled with customer focus, could provide a sustainable source of competitive advantage.

In the 1980s, the two oil shocks completely redefined the market conditions. With sky-high increases in the price of one barrel of petroleum oil, customers' pocketbooks started hurting. They started exerting their preferences for the fuel-efficient and compact products, whether they be cars, furnaces, refrigerators, or other appliances in their homes, offices, or factories.

B. Rising Customer Expectations. With increasing prices, customers appreciated the attention paid to them by the foreign producers. The foreign marketers were keen to penetrate the prosperous and profitable American markets. With their market penetration, meeting the customer expectations became a necessary condition to get their customers' attention. The U.S. corporations saw deeper market penetration by their foreign competitors. The German auto makers made a deep impact on American car buyers. Then the Japanese competitors made major inroads into the automobile, electronic appliances, imaging, and office equipment sectors in America. During the 1980s, the global rivalry between different transnational corporations took a warlike intensity. The fall of the Berlin Wall and the end of communism in the former Soviet Union during the 1990s accelerated the economic globalization of world markets.

C. Info-Savvy Customers. Since the early 1990s, the Industrial Age has been quickly giving way to the Information Technology Age. The use of the Internet and user-friendly browsers like America-on-Line (AOL) and Netscape has a significant impact on the way customers meet their information-related requirements. In industrialized economies, many customers frequently use simple-to-use search engines to collect information to quickly compare the competitive products offered to them. They are also able to easily find out, at their own leisure, much more information about the producers, their business philosophies, and the alternate choices.

As a result of these geopolitical, economic, and market shifts, most of the technology-driven companies were forced to proactively innovate and to frequently introduce new products and services to their customers.

New Products and New Customers

With increasingly intense competitive conditions in the global markets (including the American markets easily accessible to foreign competitors), there were very few remaining sources of sustainable competitive advantage left in the 1990s. One such source is a company's core competency in the development and commercialization of its new products to meet the needs of its most profitable loyal customers.

3M'S New Product Development Success. For example, 3M (Minnesota Mining & Manufacturing Company), is a manufacturer with over 60,000 high-performance products. 3M has derived an enormous competitive advantage because of its core ability to develop many new products (Note 4.3). The management of 3M set a goal to derive more than 25% of its sales revenue from the new products introduced by them in the previous five years. This ratio of the sales contribution of new "baby" products was used as a key strategic measure to assess the performance of different business units and their managers.

Every year 3M gives the Carlton Award, named after Richard Carlton, the company's president from 1949 to 1953. This is given to the employees who make a major contribution to 3M's product line. Thanks to its technological innovations, 3M has grown annually for many decades, even when everybody else was severely hurt by economic recession or the oil crises.

The 3M company was founded in 1902, and it survived its infancy by responding to the customers' need for a sandpaper that could be used under wet conditions. A few years later, 3M's salespeople reported to the R&D people that the auto makers were having a hard time keeping the paint from running in cars painted in two colors. This gave birth to 3M's masking tape. The company is also well known for its Scotch tape line of products. The top management at 3M, particularly CEO Lewis Lehr, encourage innovation. This initiative resulted in such products as the "Post-it" pads, invented as a bookmark for churchgoers. More than half of 3M's products are derived from their potential customers' needs. 3M always listens carefully to what the customers are saying. Most of 3M's products are derived to develop new applications and uses for their bonding, sealing, and coating technology.

The 3M employees are generally given enormous freedom to come up with new ideas. They are encouraged to spend about 15% of their time to work on their pet projects. This can be done without seeking or waiting for formal authorization. 3M employees rarely leave the organization for jobs elsewhere, and the headhunters always have a hard time in persuading them to do so.

New Product Mortality

The new products create the wealth and growth for a company. But, unfortunately, a large fraction of the new products also die young. A low rate of survival of new products is typical of many industries. According to a 1982 study by the consulting firm of Booz, Allen and Hamilton, the U.S. organizations derived more than one third of their product revenues from their new products (Note 4.4).

To illustrate, let us consider the food products. We are all target consumers for the various new products introduced every year by the processed foods and beverage industry. Though these food products seem simple and easy to consume, this industry uses considerable technological sophistication in food processing, refrigeration, preservation, and packaging. The life and death of a company in this industry is determined by the steady stream of new products that the company must introduce from time to time. However, of the 2000-5000 new products introduced by this industry in a year (the actual number depending on the state of economy), only 5% (i.e., 100–150 products) survive beyond their first year of existence. That is, 95% of the new products introduced in the market die in their infancy, before they are even one year old.

Schumpetarian Constructive-Destruction

In 1937, economist Joseph Schumpeter proposed that free market capitalism was driven by the innovative entrepreneurs, who from time to time introduced new products and radical new technologies which constructively destroyed the conventional market boundaries (Note 4.5). The "constructive destruction" changed the intensity as well as the rules of competition. More recently, Professor Michael Porter of Harvard University reiterated that the technology-driven firms must identify and commercially exploit the unfulfilled customer needs by developing and introducing their differentiated products (Note 4.6).

Example: Constructive Destruction in Communication Product and Service Technology. Human history is full of a variety of new products introduced from time to time to facilitate more accurate and faster communication.

Before the invention of writing, writing products (such as paper), or printing, we communicated with each other by the spoken word. Almost no products or technologies were used as an intermediate medium. Kings and governments made their announcements by gathering everyone in the center of the town, and then a crier yelled out the announcements. Naturally, this had certain limitations. We could not yell across long distances. Some smarter tribesmen developed a new product—the drums—to communicate sound-coded messages over longer distances.

Then came the development of another new product—the written word. With the written word we did not have to be in close proximity to communicate ideas to others. First we wrote on all kinds of natural products, such as the papyrus in Egypt, wooden and porcelain seals in China, and the bark of trees in the Indus Valley. Next, came the development of printing press with moveable type. This launched other new products, such as books and biblical writings, to propagate the written word over longer distances to a larger number of people. Writing products started dominating our communication.

As civilization spread farther to distant parts of the world, an elaborate mail system was developed in these different parts of the world. In the United States, the Pony Express mail service took letters from the original colonies on the East Coast, across the wilderness of the Midwest prairies, all the way across to the West Coast settlements in California and Oregon.

The primary responsibility for developing communication service and products, however, continued to be with the national governments. This changed with the invention and transmission of a telegraph message from Washington to Baltimore in 1844 by Samuel Finley Breese Morse. Private entrepreneurs saw the potential for making money in providing competitive telecommunication products and service. In 1876, Alexander Graham Bell filed his patent for the first telephone, barely 2 hours before Elisha Gray did too. From here onwards, the competitive race for new products became intense. The market forces decided the choice and evolution of communication products and services.

The commercialization of communication technology gave birth to Western Union, Bell Telephone system, and other telecommunication service providers. The telegraph was originally targeted at the fast-spreading railroad markets. Railroad operators needed telegraphic communication to coordinate and schedule train timings. To avoid trains crashing into each other, they had to be timed accurately. This was a challenge across a nation of continental size with multiple time zones.

Telegraph technology was great for certain essential and emergency services. But this one-way communication was too slow and monotonous for day-to-day mass use. The telephone was originally introduced as an appliance to aid hearing or to listen to musical concerts. It soon substituted the telegraph and became the dominant mode of day-to-day communication.

In the 1990s, telephony faced new challenges and threats for its constructive destruction. The Internet, an easy-to-use search network, provided a more economical and faster way of communicating from one to many or from many to many. We will use this overview of communication products to illustrate the alternate product development strategies available to a technology-driven enterprise.

NEW PRODUCT DEVELOPMENT MATRIX

A firm's new product development strategy depends on its existing product portfolio and its future product objectives.

Exhibit 4.1 shows a nine-cell 3 × 3 grid, indicating the alternate new product development strategies that a firm could pursue. These nine potential strategies are based on three product choices a firm could pursue. A firm could (1) use its existing products and expand them (2) improve them incrementally to related products, or (3) extend product families by developing entirely new product platforms. On a newness scale of 1 to 7, these three choices represent scores of 1, 4, and 7, respectively.

Similarly, a firm had three choices for the markets it could target. On the degree of market newness scale of 1-to-7, their corresponding scores are 1, 4, and 7 for the existing markets, related markets, and entirely new markets, respectively.

What Is a New Product?

One question that often arises in the development of such a taxonomy of new product choices is the definition of the newness of products and markets. Are the new products new to the specific firm under consideration? Or, are the new products entirely new to all the firms and therefore, to the society? For the sake of our discussion, we will primarily consider new products as new to the firm under consideration. These new products may or may not be new to the society as a whole.

EXHIBIT 4.1. New Product Development Strategies

New Markets	1 × 7 New market extension		7 × 7 New platform development
Related Markets		4 × 4 Related business improvement	
Existing Markets	1 × 1 Product expansion		7 × 1 New Product development
	Existing Products	**Related Products**	**New Products**

Next, we will discuss (a) the four corner strategies—namely, the 1 × 1, 7 × 1, 1 × 7, and 7 × 7 strategies—in the nine-cell grid and (b) the 4 × 4 strategy in the middle of this grid. This should provide a good perspective about the range of choices managers have for new product development strategies. Each of these five strategies represents a different degree of risk. However, the risk often corresponds with the potential return. Like "no pain, no gain," new product development strategies follow the "no-risk, no returns" dictum.

1. (1 × 1) Product Expansion Strategy. With this 1 × 1 new product strategy, an enterprise decides to stay conservative and take minimum risks. It chooses to expand its existing products by expanding it with minor incremental improvements. The enterprise primarily sticks with its current product successes. These products continue to be targeted at its current customers in its existing markets. The enterprise implements this new product strategy by making minor modifications or incremental improvements in its packaging. Capacity may be added for the same or similar product. Economies of scale may provide the company competitive edge over the competitors.

If all the competitors in an industry pursue 1 × 1 strategy with product expansions, and the size of the total market "pie" remains the same, then an increase in the market share of one firm is generally at the expense of a loss of market share of the other firms in the same market. Or, an increase in expanded total supply could reduce the price of the product. In other words, it is a zero-sum game.

Even though this seems like a no-risk and "safe" product development strategy, it carries with it certain hidden risks. These risks come from the threat of the substitute products and services. The substitute products are products which can meet the same requirements. The threat of substitutes depends on (a) their relative price-to-performance ratio and (b) the cost of switching for customers.

For illustration purposes, let us consider the communication products and services we introduced earlier. Most people use easily accessible telephone technology. For the past many decades, telephones have gradually substituted the telegraph and have even replaced surface mail as the primary mode of day-to-day communication.

Since the breakup of Ma Bell AT&T into the regional Baby Bell companies, the customers in the United States have been subscribing to long-distance telephone services offered by AT&T, MCI, Sprint, or other long-distance carriers. Let us assume that these long-distance telephone service providers made only incremental improvements and that the total market demand for telephony products and services remained constant. Then, a gain of an additional percentage point in market share of a competitor, say MCI, would come at the expense of a loss of one percentage point market share of AT&T or Sprint.

In the early 1990s, the major threat to the growth of AT&T, MCI, or Sprint came not from other long-distance telecom competitors, but from the Internet communication providers. Increasingly larger number of telephone subscribers started shifting some of their long-distance communication from the telephone lines to the Internet connections available between a network of computers.

2. (7 × 1) New Product Development Strategy. With this strategy, the managers are willing to explore the market potential of the existing established markets by targeting them with extended new products; this is the "7 × 1" New Product Development Strategy. These new products may be commercialized to complement or replace the current products of the firm. A firm introduces the new products before their old products are substituted by superior products from the rival competitors.

Intel corporation, a world leader in the semiconductor microprocessor technology, has often introduced new and higher-performing microprocessor computer chips, even when their earlier chips were still selling well. Intel believed in cannibalizing its own products with its own new products, rather than have them cannibalized by Intel's competitors.

3. (1 × 7) New Market Extension Strategy. This is a strategy to identify new markets for the existing products of a technology-driven enterprise.

A classical example of the (1 × 7) New Market Extension strategy was the classic introduction of a very popular brand of Arm and Hammer baking soda (produced by Church & Dwight) in new nonfood markets. With the success of baking powder in the conventional food baking sector, the company leveraged the brand recognition and repositioned the existing products in the refrigerator deodorizing market, and then it did the same in the toothpaste market.

4. *(4 × 4) Related Business Improvement Strategy.* This is a hybrid form of new product development strategy. The enterprises choosing this strategy are willing to take limited risk. This could be due to either resource-related or other constraints. A firm may not have a lot of financial or human resources to invest in exploring untried and risky new products targeted on untried new markets. The enterprise instead could try to achieve growth by developing related products and then target them at markets related to the firm's existing markets. This could result in some improvement in the business performance.

The (4 × 4) Related Business Improvement Strategy can be better understood by illustrating new products in two subgroups. These are based on (A) related diversification of products or (B) vertical integration.

4A. Use-Related Diversification. The first subgroup of Related Business Improvement strategy involves related diversification products. They are related new products because they used some common competencies in the overall technology value-adding system. For example, a telephone service provider like AT&T could introduce a cellular telephone service using its switching network, or it could introduce a cable service by phone lines using its telecommunication lines. Similarly, an automobile maker such as Ford Motor Company could use its passenger car production assembly lines to produce related products such as mini-vans or pickup trucks targeted to a different but related market segment.

4B. Use-Related Vertical Integration. The second subgroup of related business improvement (4 × 4) strategy involves developing new products that are related to its existing products by either a forward or backward vertical integration relationship. For forward integration, a new product closer to the ultimate consumer of a product or service is marketed. For example, a semiconductor manufacturer, such as Motorola, could develop new related products for cellular telephony. Similarly, TV and computer monitor manufacturer Hitachi could introduce forward vertical-related products, such as a desktop computer. This could be targeted at either (a) customers who are related to Hitachi's TV and electronics consumers, or (b) customers belonging to market segments along a value-adding chain.

5. *(7 × 7) New Business (Platform) Development Strategy.* This is the most challenging and risky choice out of all the new product development strategies. In this case the firm takes maximum new product risk by developing entirely unrelated, radical new products. The firm also compounds its risk further by targeting these risky new products at risky new markets and customers. Yet, the successful enterprises are able to make such transformations and then move to the next platform of growth and profits. The platform transformation is like the birth of a butterfly from a caterpillar. To illustrate this radical departure, let us review how a small paper supplies company became a global giant in the office equipment industry by using a new business platform.

Example: From Haloid to Xerox. In 1906 the Haloid Company was incorporated in Rochester, New York, the home of photographic giant Eastman Kodak, to manufacture and market photographic paper. Haloid got interested in photocopiers with its 1935 acquisition of the Rectigraph Company.

In the late 1940s, the Rochester-based Haloid Company was still primarily a small paper supplies company in the Upstate New York region. Very few customers ever heard about this company. The company supplied paper products to other companies. This was a cutthroat commodity business. Yet, its chairman, Joe Wilson, was a visionary leader. He was willing to take some risks on radical new products. He had heard that Chester Carlson, a patent attorney in New York City, had received a patent for a photocopying image transfer process. This yet to be market-tested invention for dry writing (or xerography in Greek) by Carlson had been backed by the Batelle Memorial Institute. It was an untried technology for an unknown market. Carlson had been working on the photocopying process since 1937. Carlson had approached the giant companies like GE, IBM, RCA, and others, but they were not interested in commercializing the electrophotography technology.

Joe Wilson and Chester Carlson got together, and Wilson decided to risk some resources for the development of a new business (platform) for Haloid Company. The two men together passed through many hurdles. Leading market research companies had surveyed their potential market and estimated a total maximum requirement of less than 5000 copiers worldwide. Photocopier copying faced the challenge of competing against much cheaper carbon copies. Even though the proposed dry photocopying machine was a marked improvement over the wet ammonia-based photographic units used at that time, many customers were not interested in making large new investments on an office equipment.

Eventually, with persistent efforts, the Carlson–Wilson pioneer team prevailed in popularizing photocopiers. Haloid commercialized the dry photocopying process and launched the Model A photocopier in 1949, followed by the Xerox Copyflo in 1955. By 1956, the xerography-related new products contributed to about two-fifths of Haloid's sales revenues. To reflect this product shift, in 1958 Joe Wilson changed the name of the Haloid Company to Haloid Xerox.

In 1959 the first user-friendly photocopier was introduced in the market. It was called model "Xerox 914" because of the 9-inch × 14-inch paper it handled. The new copier product was a great success, and it sold in unprecedented quantities. The Xerox 914 photocopier was adopted by its customers, defeating other competing technologies such as the mimeograph technology provided by A. B. Dick, the thermal paper based copying technology sold by 3M, and Kodak's damp copy technology.

The popularity of the Xerox 914 copier showed an enormous untapped market gap for photocopying required to be done at the destination—that is, the copying of documents done after a receiver received a document. The market researchers had estimated only the market demand for copying at the source, where the document was generated. Since then, it has been clearly demonstrated that the industrial world makes 10 times more copies at the destination of a document (once a document is ready) than at the source (when the document is still being generated). To the customers of carbon paper, this photocopying market at destination was incomprehensible at first.

To reflect its commitment to the new business platform, the Haloid Company changed its name to Xerox Corporation in 1961. In 1960, the Xerox revenue was barely $37 million. Within five years, this revenue skyrocketed to $268 million. In the 1960s, the Xerox Corporation diversified into publishing and computer businesses. These businesses were, however, later discontinued or sold.

From the above discussion, we see that there are five different new product development strategies with different amounts of risks involved in each. A firm could choose its new product development strategy depending on the resources and the core competencies it has, and can afford to risk. Joe Wilson of the small Haloid Company was willing to take a big risk on a new business platform of photocopying technology, whereas many giant companies of his time were unwilling to do so. The high returns reaped by the Haloid Company, later renamed the Xerox Corporation, corresponded to the risk and the persistence by the company's senior management. Next, we will look at the process of new product development, from the generation of a new concept to its successful commercialization.

NEW PRODUCT DEVELOPMENT PROCESS

Efficient and Effective Process

Once a business firm has decided its new product development strategy, the firm must develop an efficient as well as effective new product development process. The product development process should be efficient because considering the high "infant" mortality of new products, more new product introductions imply a better probability for market success. On the other hand, an effective new product development process would identify the "right" new products meeting specific untapped market needs. Fitting the product needs will lead to market success. The trade-off is determined by the turbulence in the markets.

An Hourglass Model of New Product Development Process

Typically, the scope of a systematic new product development process corresponds to the shape of an hourglass. The new product development process begins by considering a broad range of operations—as broad as possible. Through successive stages of the new product development process, the choices are reduced from many possibilities to a limited few potential candidates—like the narrow middle portion of an hourglass. Once the limited few potential product concepts are decided, then they are fanned out with more detailed execution—in engineering, production, and market testing. This implementation part corresponds to the bottom half of the hourglass (see Exhibit 4.2).

Let us review the Hourglass Model of the development process of a new product more carefully. The complete new product development process can be divided into the following seven stages.

Stage 1. Generation of New Product Concepts
Stage 2. Screening of Generated Ideas

EXHIBIT 4.2. The Hourglass Model for the New Product Development Process

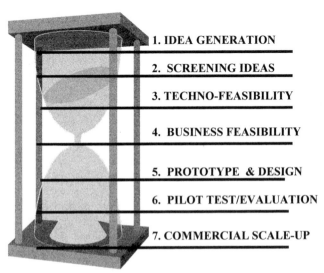

1. IDEA GENERATION
2. SCREENING IDEAS
3. TECHNO-FEASIBILITY
4. BUSINESS FEASIBILITY
5. PROTOTYPE & DESIGN
6. PILOT TEST/EVALUATION
7. COMMERCIAL SCALE-UP

Stage 3. Technological Feasibility Analysis

Stage 4. Business Feasibility Analysis

Stage 5. New Product Prototyping

Stage 6. Pilot Testing and Evaluation

Stage 7. Scaling-up and Commercialization

Stage 1. Generation of New Product Concepts. The new product development process begins with the collection and compilation of innovative ideas for new product concepts. These innovative ideas, to improve products or develop entirely new products, can come from a variety of sources. Some customers complain on the customer hot lines about the ways they are dissatisfied with the products offered. Some customers in their anger even offer suggestions to improve the products. Sales representatives and field representatives sometimes receive innovative suggestions from their customers regarding product improvement. Some employees may also have suggestions to improve or develop new products.

Using an Employee Suggestion System. Many Japanese organizations are famous for their suggestion systems. On average, a Japanese employee offers hundreds of suggestions per year. Unlike in the American corporations, Japanese offer these suggestions to make their jobs easier and smoother to do, rather than to reduce the cost or get a cash award from the company. This then results in enormous savings for the company and generates many ideas for improved and new products.

TOYOTA IDEA SUGGESTION SYSTEM. For example, the Toyota Motor Company is considered by many observers as the company that revolutionized the automobile production technology worldwide. Toyota's legendary production system is generally defined by its Just-In-Time inventory and production system and its flexible automation with low setup time (see Note 4.7). However, a critical component of Toyota's production system is the Toyota Creative Idea Suggestion System. This system has generated more than 20 million innovative ideas over a period of 40 years or more. Interestingly, the suggestion system is an idea which was observed by Toyota's senior managers at the Ford River Rouge Plant in 1949. The world-famous Japanese management practice of Kaizen, or continuous improvement, is a derivation from the original suggestion system.

KODAK'S FIRST SUGGESTION. According to some Japanese managers, the first documented suggestion in a business organization was received in 1898 at another American company, namely, Eastman Kodak. In the late 19th century, Eastman Kodak was known for creating its Brownie camera, which used flexible photographic film and for which the catch phrase "Kodak Moments" was coined. A highly paternalistic company based in Rochester,

New York under the leadership of pioneer-founder George Eastman, Kodak received its first suggestion from an employee to wash the windows to make the workplace look brighter. Though simple, this was significant considering that the winter weather in the Upstate New York is gray and gloomy for many months in a year. Furthermore, Kodak used an integrated chemical process, with in-house production facilities and exhausts from refining of silver ingots to the finishing of the photographic films and prints. Until 1975, Kodak recorded 800,260 suggestions during 78 years, and awarded US$1,620,000 for those suggestions (see Note 4.7).

Toyota's Creative Idea suggestion system is based on suggestions that help reduce waste or improve a product/process. Each suggestion is evaluated, implemented, and rewarded based on a number of criteria. For easier understanding, these criteria are classified under three different categories of (I) Impact, (II) Attributes, and (III) Sources. Given below is the list of criteria used by Toyota to evaluate the suggestions received.

I. Impact of a Suggestion. The impact was measured by the following five criteria.

1. Degree of contribution/benefits in meeting organization's goals.
2. Feasibility and ease of implementation.
3. Scope and applicability in other areas.
4. Positive or adverse effects on others.
5. Likely duration of the benefits of the suggestion.

II. Attributes of a Suggestion. The suggestions were defined by the following criteria.

1. Completeness of suggestions, and its readiness to be implemented.
2. Creativity and originality of a suggestion, or improvement of known ideas.

III. Sources and Resources for a Suggestion. These were based on two more criteria.

1. Efforts invested or trials tested to develop the suggestion.
2. Relationship between the suggestion and the suggestor.

The Brainstorming Technique or a Structured Suggestion System. The important point to remember in the concept generation step is that it should be carried out as extensively and comprehensively as possible. One tool to do this systematically is the use of the brainstorming technique. This is a quite simple but powerful way to generate a large number of new ideas in a fairly small period of time. However, to make this technique work effectively, some structure and order are required to implement a technique that is based on spontaneity and creativity.

Brainstorming technique usually should take place in two separate sessions. These two sessions should be scheduled preferably on different days, but not too far apart. The first session should last for 1 to 2 hours and is used to generate and capture as many number of ideas as possible. The brainstorming team should be selected in such a way that there are representatives from different constituencies. This should include customers, suppliers, technical operators, and experts. The nonexperts with good common knowledge about the task on hand should also be used to gain fresh perspective.

The first suggestion-giving session should begin with goal definition. The entire group should collectively develop a goal statement for the brainstorming exercise.

In this first session, each participant is asked to make a quick suggestion. This goes on in a round-robin manner, like the bidding in a bridge card game. As in the card game, each participant makes a suggestion that is captured verbatim at a visible place for all to see. Each participant also has a choice to "pass" his or her turn. The suggestion-giving stops when all the participants say "pass."

The second brainstorming session could last a little longer, for about 4 hours. This session begins with a review of the goal. The recorded suggested ideas are also reviewed. The purpose of this follow-up session is to capture the ideas which were not captured in the first session. During the intervening period between the first and the second brainstorming session, some of the participants may have come up with new ideas. These should also be recorded in the second session. Any "outright" ridiculous ideas are purged. Suggested ideas which seem similar or duplicate may be consolidated together.

Stage 2. Screening the Generated Ideas. The suggested ideas for new product concepts, generated either by the suggestion system or by the brainstorming sessions, need screening. This is done by first classifying them according to whether their scope is (a) controllable, and permissible within the team's domain of influence, or (b) not controllable, or not permissible and beyond the domain of control of the team.

Each feasible idea is then classified and rated for its alignment with the commonly agreed goals. For each suggested idea, each brainstormer assigns a value on a 1–5 scale (with 5 for the maximum fit with the overall goal). These assignment scores are tabulated for all the participants, for all to see. Their mean values are calculated.

Each person then gets a second chance to change his or her rating score. Some participators may like to elaborate on their criteria behind their rating score in order to convince others and persuade them to change their scores. Finally, when all participants are satisfied with their rating scores, the suggested ideas are sorted in a descending order of scores. Their standard deviations may also be recorded next to the means. The brainstorming team can then decide which ideas for new product concepts should be short-listed and recommended to be considered further.

Stage 3. Technological Feasibility Analysis. In this step, the recommended suggestions for new product development are evaluated for their technological feasibility. At this stage, technical factors are brought in—for example, availability of alternate raw materials required for a new product, product uniqueness, compatibility of the proposed new product with the existing product portfolio, and so on. One new product idea may require handling of hazardous materials. Another idea may require a rare raw material. At the end of this stage, the potential candidates for new product development have been further reduced.

At Eastman Kodak the ease of large-scale manufacturability of a new product concept is considered a critical requirement in the screening stage.

Stage 4. Business Feasibility Analysis. The surviving suggestions for the new products are then analyzed for their business case viability. One new product may have a very high break-even point requiring high volume production. Another new product development idea may require too much capital. Or the plant and equipment needed to produce a product may be available only from a limited few foreign vendors.

In stages 1, 2, and 3, the number of alternatives were narrowed down. In stage 4 onwards, the scope of investigation for the few potential surviving candidates for new products is expanded for implementation in the successive stages. The surviving product concepts are subjected to questions regarding the details. What will be the cost of manufacturing for each potential new product? What is the potential demand for its total market, and what is likely to be the firm's market share in that total market? What is the growth potential for the total market demand, and what is the intensity of rivalry among the current competitors? What is the likelihood of threat from new competitors entering the same market? The likely sales growth, and profitability under different market scenarios are also estimated. These questions are explored and addressed for the analysis of business case viability.

Stage 5. Prototype Development and Design of New Products. Technology-driven enterprises are increasingly striving hard to design and develop new products by integrating their technological strengths with their customers' expectations. A traditional technology push approach can cause havoc on a firm's competitive position in intensely competitive markets. Technology trajectories have been moving fast, and sometimes discontinuously. So have the customers' expectations. Therefore a technology push strategy insensitive to these expectations can drive profitable customers to rival competitors. A sustainable competitive advantage can be gained by a cross-functional integration of customer expectations and preferences with the firm's technological capabilities.

Quality Functional Deployment. Professor John Hauser and Professor Don Clausing of the Massachusetts Institute of Technology popularized the management process used for cross-functional integration, called Quality Functional Deployment (QFD). This management tool originated in 1972 at the Kobe shipyard of Japan's Mitsubishi Heavy Industry. It was developed further by Toyota (Note 4.8). The cross-functional integration was portrayed graphically by the "house of quality." The Japanese manufacturers and their counterparts in the United States have used the house of quality in a variety of industrial and service sectors (see Exhibit 4.3).

Quality functional deployment involves a variety of algorithms used to facilitate cross-functional integration in the designing, manufacturing, and marketing of products and services. In Chapter 5 we will discuss the quality promise a technology-driven enterprise makes to its targeted customers with the goal to gain their satisfaction, trust, and loyalty.

In Chapter 3 we learned that prior to the functional specialization ushered in by the Industrial Revolution, the classical craftsman single-handedly integrated the marketing, manufacturing, and design functions in his head. The medieval knight in the Middle Ages contacted a local craftsman when he needed the armor for the next Crusade. The craftsman customized the production of the armor according to the knight's body measurements and other requirements.

Such integration and customization was compromised as the industrial organizations grew in size to gain economies of scale. This disconnected the different functional departments. But if these different departments or their representatives were to sit together, how could they systematically integrate their concerns? The house of quality is a tool that can facilitate such cross-functional integration of customer expectations into the design of features and characteristics in a product or service.

Quality functional deployment or the house of quality requires the following steps. These steps are shown in Exhibit 4.3.

EXHIBIT 4.3. Quality Functional Deployment Summary

Step 1. What: identify the customers' wants.
Step 2. Rank the customer wants by importance.
Step 3. How: Identify product's engineering/design attributes.
Step 4. Relate customer wants with engineering attributes of a product.
Step 5. Who: Compare the competitors' advantages for each of the customer wants.
Step 6. Competitive objective engineering measures.
Step 7. Interrelationship between competing design technical characteristics.

Step 1. What: Identify the Customers' Wants. A marketing team starts QFD by asking target customers about the attributes that they would like to incorporate to define a product's characteristics. They could do so by surveying the customers directly or by observing their buyer behavior. If Eastman Kodak was to develop a new and advanced photographic system, they may survey customers about their wants in a camera. The amateur photographers would mention attributes such as easy loading, compact, lightweight, multiple formats, good focus or auto focus, reliable exposure, user-friendly camera equipment, and ease of reordering. The customers' voices are captured in their own words. These customers' wants produce the memorable "Kodak Moments" for the photographers.

The customer wants can be identified in a multitiered manner. The photographers may broadly list their wants as user-friendly camera and high-quality pictures. These are their first-level wants. The customers need to be coaxed further to find out what they mean by each first-level want. Thus the first-level wants can be elaborated further by the second-level wants. The first-level want for a user-friendly camera involves easy or automatic film-loading, auto-focus, and lightweight. The first-level want of a high picture quality may be broken down into second-level wants such as high-resolution film, no double exposures, and a smart film that records the conditions at the time of picture-taking. If time permits, and depending on the scope of the quality function deployment project, the second-level wants can be further elaborated by the third-level wants.

In a house of quality the customers' wants are listed in the front yard of the house.

Step 2. Rank the Customer Wants by Importance. The customers are also asked to rank their desirable wants by the importance they attach to each of their wants. Some of the customers' wants are mutually conflicting or even inconsistent. An elaborate auto-focusing want in a camera may result in a heavy camera which the customers do not want. A small light lens may not provide good resolution. The ease of a point-and-shoot auto-focus camera may compromise the focus of objects in different fields of vision. The customers are therefore asked to give their preferences and are then asked to rank their wants in the order of their relative importance. Once in a while a creative solution can meet different conflicting wants.

In the house of quality the relative rankings are noted along a chimney between the front yard and the wall.

Step 3. How: Identify Product's Engineering/Design Attributes. The product design team clusters the attributes of a product or service in terms of its component design or engineering attributes. In the case of a photographic technology, this could include: photographic film, lens and exposure system, film loading mechanism, battery life, camera weight, photo-finishing equipment, and so on. A clear resolution requires a heavy glass lens. Auto-focus requires distance and light-sensing components. Automatic film-loading cameras require drive and rewind motors. The film formats can be either 16 mm or 35 mm. The Advanced Photo System is in a new 25-mm format. These are some of the camera's engineering/design attributes.

In the house of quality, the product attributes are noted on the roof wall of the house.

Step 4. Relate Customer Wants with Engineering Attributes of a Product. A cross-functional product development team then relates (a) the customer wants identified by the marketing team with (b) the product's technical design and engineering attributes identified by the design team. For example, a lightweight camera would require smaller and high-resolution film and a compact camera. Out-of-focus pictures can be avoided by an auto-focus exposure mechanism. The picture quality can be enhanced if the conditions at the time of picture-taking can be recorded on the film itself, and then communicated with the photo-finishing equipment at the time of developing the prints from the film negatives.

The relationships between the customers' wants and the product's technical design and engineering attributes are noted on the main house wall, or the relationship matrix of the house of quality. Each relationship between a customer want and the product's technical design attribute is qualified by the intensity of strength and the sign of the relationship. These relationships may be strong, medium, or weak in intensity and may be positive or negative in sign.

These relationships may be attained by different means. They may be empirically established or simply intuitively assigned.

Step 5. Who: Compare the Competitors' Advantages for Each of the Customer Wants. The cross-functional team or the marketing team may also want to assess how the company and its competitors rate on each of the customer wants. For each customer want, the customers are also asked about their perceptions about the rankings of different competitors. Thus, Kodak cameras may be compared with cameras offered by Fuji, Canon, Minolta, Nikon, and others. Each of these cameras may be ranked relatively by customers on individual customer wants, such as the user-friendliness, weight, price, and so on.

These perceptions by customers are listed on the backyard wall.

Step 6. Competitive Objective Engineering Measures. Each of the product's technical design characteristics is measured by objective measures and compared with the measures for the leading rival products. The resolution of a film may be measured by pixels per square inch. Weight is measured by grams or ounces.

Step 7. Interrelationship Between Competing Design Technical Characteristics. In the house of quality the interrelationships between the different technical characteristics in the product's design are noted in the arch of the roof in the house of quality. For example, an elaborate auto-loading mechanism can add to the weight of the camera. A narrower film can help save the weight of the camera. But a narrower film can adversely affect the photo resolution as well as clarity. These relationships between different design technical characteristics are also noted for the intensities as well as the signs of their relationships. If a particular design characteristic adversely affects many other design characteristics, then it may be prudent to leave that design characteristic alone.

APPLICATION OF QFD. The house of quality, or the quality function deployment, is a useful way to help marketers communicate with the product designers and technical engineers. The customer wants are related to the design characteristics.

As a comprehensive graphic tool, QFD helps decide priorities between the different alternatives. The trade-offs are clearly and explicitly spelled out. The sources of competitive advantages are identified from the customers' perspective.

The greatest advantage of using QFD is that the voice of customers drives the new product development and gets the attention it deserves. This process can help set the targets for sustainable competitive advantages.

A series of houses of quality can be linked to make a series of cross-functional decisions.

ROBUST DESIGN FOR X. Increasingly the technology-driven enterprises are recognizing that a proper design plays a critical role in many downstream activities of the product development process. The design for engineering, design for manufacturing, and so on, generally referred as "design for X," elaborate the need to address upstream design concerns for downstream activities such as engineering, assembling, and marketing. A careful design can eliminate the need for expensive engineering changes, reduce cycle times, and avoid other quality problems.

Robust design is a design which consistently produces high-quality finished products even under normal variations in raw materials and processing conditions. For examples, the manufacturers of electronic home appliances have learned that by reducing the number of components and subassemblies in a product, they can reduce the opportunities for the product to fail. This can save significant amounts of waste and cost.

RAPID PROTOTYPING. With the shrinking of product cycle times in many technologies there is growing need for faster prototyping. Advanced photo sensitive polymeric materials can directly take output from computer-aided design (CAD) softwares to develop prototypes for detailed testing, evaluation, and test-marketing. Thus, three-dimensional prototypes are produced rapidly from theoretical concepts for test-marketing and other uses.

Stage 6. Pilot Test-Marketing New Product. Test-marketing is a stage that many new products go through in their journey towards large-scale commercialization. Test-marketing is expensive and time-consuming. The information obtained through test-marketing has a significant impact on the subsequent stages involving larger commitments of resources. Therefore, this stage must not be handled in a casual manner.

Goals of Test-Marketing. The test-marketing stage has primarily two sets of goals.

1. It acts as a simulated laboratory where many different alternative characteristics of the new product, along with its marketing mix (product, price, promotion, and placement), can be tested. The interdependence between the four Ps of marketing can be tested during the test-marketing. Test-marketing highlights any problems that the market or the customers may have regarding the new product.
2. As a sample, the test-market results help the managers project and forecast total national or international sales based on an extrapolation of the sales results from the test-market areas.

Risks of Test-Marketing. The benefits of test-marketing accompany some risks. There are two types of risks: (1) the actual and measurable cost of test-marketing and (2) some intangible losses. The measurable costs of test-marketing include the cost of a plant and equipment required to produce the products for test-marketing. Additionally, there is the cost of the marketing force needed to guide and carry out the test-marketing and advertising of the new product. Besides these there is the additional opportunity cost incurred by the attention and shelf space that the current products lose while the new product is getting the focused attention for test-marketing.

Given these benefits and the costs associated with test marketing, how do we decide whether or not to test-market a new product? There are a number of factors which can help managers answer that question (Note 4.9).

1. **Low Risk of Failure.** If the potential risks and costs of a failure of product launch are low, then it may not be necessary to do the test-marketing.
2. **Need for Full-Scale Technology.** There are products which require process technology and investment for producing new products for test-marketing that are almost the same as the technology and investment needed for a full product launch. For such cases, the test-marketing provides a limited advantage. On the other hand, if the test-marketing can be

done by a process technology or investment that is significantly smaller than the technology and investment required for full launch, then it may be prudent to do the test-marketing first. This would help collect the information which could help reduce the larger risk.

3. **Competitors' Time to Catch Up.** Test-marketing has the unavoidable adverse effect of leaking out the future plans of a company to its rivals, and sometimes the stronger competitors. However, if the competitors are likely to take a considerably long time to develop rival products that would compete against the test market, then the damage caused by the test marketing of new product is limited. The test-marketing companies must also be aware that the competitors also monitor the test-marketing and its likely outcomes. Conducting test-marketing would be suicidal if the company's competitors are so agile that they could successfully launch a rival product, and gain a time advantage, without going through a test-marketing stage.

4. **Impact on Company's Image.** Finally, before conducting a test market, the company must estimate the adverse effects of failures of test-marketing on the image of the company. If the reputation of a company is in a critical stage, then the test-marketing may be avoided. A failure in test-marketing would adversely affect the competitive survival of such a firm.

Selection of the Test Markets. Once a company has decided to proceed with test marketing, the next challenge is to pick the site or sites for test-marketing very carefully (Note 4.10). Some of the selection criteria for test markets related to product, promotion, placement, market regions, and so on, are given below.

1. Metropolitan versus nonmetropolitan market.
2. Large versus small representative population.
3. Demographic diversity versus homogeneity by age, religion, family size, and so on.
4. High versus low per capita income.
5. Isolated versus mainstream region.
6. Test-marketing few versus many product categories.
7. Variety of distribution channels.
8. Use of a few versus many communication channels.
9. Complete versus limited control of company on test-marketing.

Results of Test-Marketing. Once the test-marketing is conducted, there are a number of ways the results could be used. The data and information collected must be carefully scrutinized. The results from the test-marketing experiment cannot be directly used to predict the outcomes of the future launch of the test-marketed new product. The data as well as the design of the test-marketing experiment must be carefully analyzed. For example, the potential profit margins of a product are severely distorted during the test-marketing of new products. Unit costs of test marketed products are high because of the diseconomies of scale in test-marketing. Another limitation of test-marketing is that it cannot evaluate a fully developed distribution system.

The test marketers must be aware of the fact that the product life cycles have been shrinking rapidly since the 1980s. Most competitors in the industrial economies have almost equivalent access to technology. The competitors can quickly imitate the test-marketed products by reverse engineering their own rival products. The results from test-marketing must be always validated by checking for any abnormal competitive activity. Some competitors can sabotage a test market by excessively increasing their own promotion efforts or deep discounting of rival products during their competitors' test-marketing. This could pollute the information collected from test-marketing.

Stage 7. Scaling-Up for Commercial Launch. Finally, the new product is ready to be scaled-up for commercial launch. To do the scale-up, first the potential sale for the product must be forecasted. A number of forecasting techniques may be used for forecasting (a) the demand for the new product and (b) the needed production capacity for meeting the same.

Forecasting Methods. A number of quantitative and qualitative methods are available to forecast future events based on the knowledge about the past or current events. The surprise of the future events can be reduced only by studying more about the past trends. These future events could be demands or requirements of a new product. Some of the forecasting methods are qualitative, while others are quantitative. Given below are the brief descriptions of four qualitative forecasting methods followed by short introductions for two quantitative forecasting methods.

1. *Leading Indicators Forecasting.* One way to forecast the demand for a new product is to look at the changes in certain variables which lead and are good at predicting the future changes in the demand of the new product. For example, changes in business and industrial inventories lead the demand for business and industrial products. As the inventory levels decrease, a demand surge is likely to follow that change, and vice versa. The increase in average work week of production workers generally leads the money available to workers to buy consumer goods. College graduates will lead in the purchase of new homes a few years down the road. One must remember that these are projections, and actual demands for new products may vary somewhat from their predictions.

Unlike the leading indicators, there are some other indicators which either lag or change simultaneously with the demand for the new product. For example, the unemployment rate has a simultaneous effect on retail sales. These variables do not help project or anticipate the future demand of a new product.

2. *Expert Jury Panel.* Experts, who may be the senior executives of the firm, are asked to give opinions about the potential future demand for a new product. Responses of a panel of such experts may be averaged. The internal executives may be drawn from different functional areas, such as marketing, research and development (R&D), engineering, production, and so on. This method is relatively inexpensive and quick to do. The disadvantage may be that the internal executives associated with a new product are likely to be biased in favor of the new product.

3. *Buyer Focus Group.* The past, present, or potential future customers of a new product are invited in a focus group. They are asked about their likely purchase of a new product. These potential customers may be surveyed by telephone, mail, or other means. The potential customers may be asked to compare competing products and rank their desirability. The disadvantage of this method is that the potential customers tend to exaggerate their future buying projections. They claim that they will buy more than what they will actually do.

4. *Delphi Method.* This method seeks the opinion of a panel of experts in an iterative way. The Delphi method is used to forecast the sales of a new product that is based on a radical new technology. Only a few experts can project and foresee the demand or potential uses of such products. These experts are asked a series of elaborate questions on the new product's future sales potential. Their responses are summarized. They are then shared with the experts to allow them to modify their individual projections. These iterations are repeated until a stable pattern of respondents' values appear. Sometimes the experts' responses are used to modify the questions for the subsequent iterations.

5. *Time-Series Forecasting.* This is a quantitative method of forecasting. The time-series-based models assume that the future demand for an event can be projected based on an extrapolation of the present and past demand trends over evenly spaced events or data points.

5a. *Seasonal Cycles.* Many products exhibit repetitive patterns in their demands. For example, the sales of photocopiers and cars exhibit cyclical patterns. They show peaks in demands at the end of each financial quarter for the photocopiers. Similar peaks in demand near the end of calendar months are also observed for the car dealers. These cyclical changes may be tied to the way companies allocate their budgets, or the way individuals get their paychecks. There are seasonal variations which vary with the seasons in a year. The demand for air conditioners is expected to increase in the summer months and decrease in the winters. In a similar manner, the demand for furnaces or swimming pools also show seasonal variations. While projecting the future demands for these products, such seasonal variations are taken into account. (This is called the decomposition of a time series.)

5b. *Moving Average Smoothing.* Sometimes the erratic variations in a trendline can be smoothed out by averaging the projected demand over multiple periods. A simple moving average, used to project the demand for the next month, can be calculated by summing the monthly demands over the past N months. The value of N can be 3, 4, or 5. This is also called simple moving average. Here, all the past N months are equally weighted.

A weighted moving average puts different weights on different periods. These different weights are developed by experience. For example, a furniture store may forecast its future demand in the Nth month by using a 3-month weighted moving average. By experience, the store manager may have learned that the demand in the most recent $(N - 1)$th month must be weighted most, say by 3. The demand for $(N - 2)$th month is weighted by 2, and the $(N - 3)$th month is weighted by a weight of 1. Then, the forecast of Nth month by the 3-month weighted moving average will be

$$\text{Demand } (N) = [\text{Demand } (N - 1) * 3 + \text{Demand } (N - 2) * 2 + \text{Demand } (N - 3) * 1] / 6$$

where the denominator 6 is the sum of weights $(3 + 2 + 1)$ and $*$ means multiplication.

5c. *Exponential Smoothing.* To forecast the demand for the next period, this forecasting method uses the difference between the projected demand of the last period and the actual demand for the last period. This method therefore requires very limited historical records. The difference between the actual and the forecast demand is weighted by an exponential smoothing constant E, with a value somewhere between 0 and 1.

If the demand for period N to be forecast is $F(N)$

and

the demand forecast for period $(N - 1)$ was $F(N - 1)$,
the actual demand for period $(N - 1)$ was $A(N - 1)$, and
the exponential smoothing constant was E,

then,

$$F(n) = F(N - 1) + E * [A(N - 1) - F (N - 1)]$$

This equation can be used to develop the demand for the successive period.

For example, let us say that the number of actual hotel guest-days in March in the Global Hotels was 8648, and the predicted occupancy of the hotel for the same month was 8464. Then using the smoothing constant of 0.4, the predicted occupancy for the month of April can be obtained by using the following equation.

$$F(\text{April}) = F(\text{March}) + E * [A(\text{March}) - F(\text{March})]$$
$$= 8464 + 0.4 * [8648 - 8464]$$
$$= 8464 + 0.4 * 176$$
$$= 8464 + 70.4$$
$$= 8534 \text{ (rounded)}$$

This procedure can be repeated to forecast the expected hotel-guest days at Global Hotels in the successive months.

6. *Regression and Correlation Analysis.* When an extensive amount of data is available from the test-marketing of a new product, the data are statistically analyzed by using regression and correlation analysis. These methods generate relationships between the projected demand of new products and some other variables.

If the projected demand of new product is represented as the dependent variable Y, while the other variable may be classified as the independent variable X, then the relationship may be expressed by the regression equation, $Y = A + B * X$. B is the correlation coefficient, while A is the intercept of projected demand when the value of X is equal to zero. Such regression equations are fitted to paired data using a least-squares statistical method. These methods are discussed in greater detail in Chapter 5 under quality tools.

The Adoption and Diffusion of Products. Adoption is the mental process by which a user becomes aware of a new product innovation and adopts the innovation. Diffusion is the process from conception of an innovation to its ultimate use by buyers. Diffusion is the market aggregation of individual customers' adoption processes.

During the 1940s there was a high interest in making farmers adopt new farm technology (Note 4.11). America was fast industrializing, and each farmer was feeding more Americans than ever before. There was therefore keen interest to improve agricultural productivity. The new farm technologies developed in laboratories had to be diffused in the field to many farmers. Similar diffusion studies were conducted in a variety of fields including the marketing of consumer and industrial products.

THE ADOPTION PROCESS. The process of adoption for a new product or innovation can be broken down into its different components (see Note 4.10).

These are as follows:

A1. Awareness. The potential customer is aware of an innovation, but does not have full information. This may be because customers are not fully motivated yet to adopt an innovation. The target customers are bombarded with promotional information to lure them.

A2. Interest. The potential customer becomes interested enough that he or she wishes to find out more. At this point he or she is getting more curious and has need for more information. Infomercials play an important role in meeting this need.

A3. Evaluation. The interested customer mentally starts evaluating the innovation. The innovation is evaluated for the potential benefits and the likely cost. Many marketers of new products share comparative evaluation of competing products, from their own perspective.

A4. Trial. The individual customers are motivated enough to go and try the new product or innovation. A demonstration would help them too. The software sellers send their demo discs to meet the potential customers' need to explore. These demo discs are truncated versions of the whole software program, with enough examples to demonstrate the different features of the full program.

A5. Adoption. The customers liked what they got, and they are ready to fully adopt the new product or the innovation. At this point they see that the potential benefits of acquiring the new product will exceed the pain of parting with their money to pay for it. They believe that acquisition of the product will provide them a net value-added gain.

THE DIFFUSION PROCESS. The diffusion process includes adoption of the new product by a large number of customers. The process has four elements. These are given below.

D1. The Innovation or New Product. The product is new to the buyer. Earlier we developed the classification of products by their newness. The radical new products generate a bigger anxiety in the potential customers. The highly priced products also cause more anxiety than the lower priced new products.

D2. Communication. This involves communication from one adopter to the next adopter. Satisfied adopters tell other potential adopters about their experience with a new product/innovation. In the same manner, the bad news moves even farther from the dissatisfied customer to other potential adopters. The cumulative effect of such communication is significant.

D3. Community or Relevant Customer Segment. These are all those individuals who collectively decide to adopt the innovative new product. They could be the potential customers in a test market. The information about the diffusion process gained from test-marketing can be used to design and develop the new product for the full-scale market launch.

D4. Time and Evolution of Adoption Process. Diffusion—that is, adoption by different adopters—does not take place instantaneously. It takes time for the innovative product to be adopted by all potential customers. First a few innovators try the new product innovation. Then a few early adopters adopt the innovation. They are followed by early majority and late majority. Finally, the laggards join in.

The product managers can use their understanding of the adoption and diffusion process by satisfying customers' needs in these individual stages. In one stage, the customers may be interested in finding out more information about the potential benefits and side effects from the trials of a new product. In another stage, they may be interested in access to the innovative product.

In 1996, Advanced Photo System (APS), a new photographic system, was developed by Eastman Kodak and four Japanese co-developers. APS had many significant advantages over the conventional 35-mm-film cameras. The APS technology used a compact, auto-loaded, high-resolution 25-mm film, producing pictures in three alternate photo formats. However, some distributors' complaint was that they could not get enough APS cameras and films to meet the curious customers' demands. The adoption and diffusion processes influence the life cycle of a product.

PRODUCT LIFE CYCLE

Products have lives like other species. They are born or introduced into the market. Then they grow, mature, and finally decline. The life of a product is divided into these four stages (see Exhibit 4.4). At any time, a company

EXHIBIT 4.4. Product Life Cycle

Stage	Sales	Profits
1. Introduction	Low (Growing slow)	(− high) Reducing
2. Growth	Medium (Growing fast)	(− to + low) Increasing
3. Maturity	High (Slowing)	(+ high) Increasing
4. Decline	Low (Declining)	(+ low) Decreasing

may have a portfolio of different products at different stages. We will first discuss what happens during each stage of a product's life cycle. Then we consider the use of portfolio analysis (Note 4.12).

1. Market Introduction Stage. This is the phase when a new product is first introduced into the target market. There are many uncertainties surrounding the new product. At this point in time the new product's demand has not been proven yet. Nor does the firm know much about why customers prefer this product over other competing products. Other aspects of the buyer behavior are also not clear. Customers have to be motivated to switch from buying the existing products to buying the newly introduced products. Sales in the introduction stage are low, and they grow at a slow pace. Profits from the new product do not exist, because of the enormous fixed cost of developing a new product. In this stage, the efforts and resources required to make customers accept and adopt the new product outweigh the revenue generated by the product's sales.

Professor Theodore Levitt suggested that a well-designed product that clearly meets customers' needs would take much shorter time in the introduction stage. He was amazed at some companies pursuing the "used apple policy." Instead of pioneering a new market by being the first to introduce their new product, they preferred to wait and watch. They let other companies and competitors take the lead in pursuing attractive market opportunities. They let them take risk and have the first bite into the market apple. When the first movers' market introduction works, then the followers come in for the less risky but big second bite into a proven market apple. They do not take the risk of biting into a poisonous apple. For the "used apple" strategy to work, the follower firm must be fast in responding to the discovery of new apples by others (see Note 4.12).

2. Market Growth Stage. By this point in time in a new product's life cycle, the demand for the new product has picked up speed and acceleration, while the overall size of the market for the product starts growing fast. Some marketers also call it the take-off stage. The product is therefore generating more revenue. But, this success is also motivating the firm to invest more resources in building its production capacity to meet the new product's future demand. The firm in this stage may be still laying the business infrastructure needed for the growth of the new product. This stage involves activities such as the development of distribution channels and expansion of the plant and production capacity. Therefore, even though the sales revenue from the product is increasing rapidly, the net profit, after deducting the fixed cost incurred in developing the new product, starts increasing. The net profit may be still slightly negative or somewhat positive.

3. Market Maturity Stage. At this point in the life of a product, some market saturation starts taking place, and the demand starts leveling off. The sales are not increasing at the rate at which they did in the growth stage. For example, in the 1980s, most of the U.S. households had washing machines, dishwashers, microwave ovens, and so on. Therefore it became harder for home appliance manufacturers like Whirlpool to generate additional sales revenue in their domestic U.S. market. They therefore diverted their attention to the opportunities in the international markets.

4. Market Decline Stage. By this time, the product has started losing its appeal to customers. The sales of this product may start declining. This could happen because of a number of factors. There is a new, improved or a radically different product which meets the customers' needs more effectively. For example, when Henry Ford's Model T car was introduced, the demand for horse-driven coach buggies started declining rapidly. The demand for manual typewriters declined rapidly as the electric typewriter became more popular. And the demand for electric typewriters declined rapidly as the new personal computers and word processors entered the market. In the decline stage, the company reads the writing on the wall and invests very few new dollars to the future of this product. Some minimum amounts of resources may be invested in the maintenance of the current plants, but no new plants are likely to be commissioned.

Lives of Products

Life of a product from introduction to decline may take decades, as in the case of white bread and bedsheets. Or a product's life cycle may span a few years, as in the case of pocket radios, video recorders, and other home entertainment appliances. In the case of some products, the life cycle may span a few months, as in the case of the seasonal fashions and software programs, or even a few hours, such as for the newspapers. Managers can benefit by being aware of the different life-cycle stages of their products.

Significance of the Product Life Cycle Pattern

The managers who are aware of the typical pattern of a product life cycle can use that information to plan their marketing efforts for each stage ahead of time. This can provide significant lead times in anticipation of a rapid growth or a decline in a product's demand.

For example, how long will the product development stage or the introduction stage for a new product be? The length and slope of this stage depends on the newness of the new product. A radical new product requires enormous customer education effort before a sizable adoption of the new product takes place. The innovator of a radical new product must anticipate and plan for such time delays in its widespread adoption. Recall that during the introduction stage of a new product, the firm would not see significant increases in its profits. This could stretch out the financial resources needed to tide over its introduction stage and get to the growth stage. The length of the introduction or market development stage also depends on the number of persons involved in making the purchase decision for a new product. For example, the purchasing of office equipments involves many people, and therefore one can expect that the introduction stage of a new office equipment will be long. On the other hand, the fashion goods are bought impulsively, and therefore the fashion goods and other fads have very short introduction stages. Professor Theodore Levitt also postulated that "newer the product, the more important it becomes for the customers to have a favorable first experience with it" (see Note 4.12). This requires that the new products be launched very carefully. In the case of the introduction of washing machines, the retailers who reassured the customers and ensured the product's correct utilization gained the biggest share of the total market.

The first few flights of Federal Express overnight delivery service traveled long distances with only a few packages on board. The total number of packages during the early introduction stage were so few that these could be sorted by hand, without the help of the elaborate sorting facility constructed for that purpose in Memphis, Tennessee. Yet Federal Express had to demonstrate the reliability of the new overnight delivery service to the potential customers.

Extending the Product Life Cycle of Nylon

Professor Theodore Levitt also demonstrated how a product life cycle can be extended and stretched out to make a new product live longer.

Nylon fiber, one of the early new synthetic fibers, was developed by Du Pont for its higher strength compared to that of the natural fibers like cotton, wool, and so on. With its low moisture absorption characteristic, nylon was an ideal material for applications in parachute ropes during World War II. As World War II ended, the demand for parachutes disappeared. But nylon lived a second full life with its new application in the women's stocking and other hosiery products. With the devastation of Japan's silk stocking industry during the World War II, the demand for nylon stockings grew rapidly. Du Pont nurtured the demand for nylon stockings by promoting and advertising the need to wear stockings all the time. The company also developed stockings in different colors, as "fashion statements." As the demand for nylon stockings started showing signs of leveling, Du Pont developed new products based on nylon, such as nylon carpets, nylon tire cord for reinforcing truck tires, and nylon engineering plastic resins for bearings. Thus the engineers and product developers at Du Pont were able to extend the life of nylon again and again.

Like Du Pont, General Foods Corporation has been able to extend the product life cycle of "Jell-O," and 3M was able to do the same for its "Scotch" tapes. They extended the product life cycles by (1) promoting

EXHIBIT 4.5. Alternate Ways to Extend Product Life Cycles

Approaches	Nylon	Scotch Tape	Jell-O
Manufacturer	Du Pont	3M	General Foods
Original use	Women's hosiery stocking	Tape	Dessert
	Extension Ways		
1. Frequent usage	All-the-time use stockings	Handy tape dispensers	More flavors, multipack
2. Varied use	Body-colored, patterned, tinted for fashion statement	Colored, patterned, water-proof, invisible and write-on tapes	Salad use
3. New users	Early teenagers and pre-teenagers	Varied widths for painters and industrial masking	Weight loss appeal
4. New uses	Stretch socks, nylon carpet, tirecord plastic resin	Double adhesive tape, reflective tape	Bone-builder, finger-nail strengthening

frequent uses of their products, (2) developing varied usage of their product, (3) targeting new users of the product, and (4) by developing new uses of their products (see Exhibit 4.5).

PRODUCT PORTFOLIO

Over many years, a firm can accumulate a portfolio of many products. At any time, these products may be in the different stages of their life cycles. Each product in this portfolio, depending on its stage in its product life cycle, generates revenues and consumes resources in different amounts. The senior managers in a firm are interested in monitoring and finding out whether they have a balanced or an unbalanced portfolio. One management technique to do so was the market share–growth portfolio matrix developed by the Boston Consulting Group (BCG).

The Market Share–Growth Portfolio Matrix

During the late 1960s and early 1970s, the U.S. companies were rapidly developing and acquiring new products and new businesses. The Boston Consulting Group (BCG) developed a very popular portfolio matrix to help

senior managers assess whether these companies had a balanced or an unbalanced portfolio.

BCG's portfolio matrix divided the firm into a number of strategic business units (SBUs). The SBU was a pseudoindependent business entity,
which could be acquired or sold without affecting the entire enterprise.
Each SBU had its own product(s) and a target market. The BCG portfolio
matrix was developed to look at the balance between the SBUs, but it can
be adapted here to analyze the portfolio of products (see Exhibit 4.6).

The BCG portfolio matrix is based on two dimensions. Along the horizontal axis is plotted a product's relative market share, from high on the left
to low on the right. This is the market share of a product (or product family)
divided by the market share of the largest competing product (or product
family). The BCG matrix uses the relative market share as a measure of the
firm's relative competitive advantage in the market.

For example, if Eastman Kodak's new camera product has a market share
of 20% while its rival Canon camera has a market share of 40%, then the
relative market share of Kodak's camera product is 20/40, or 0.5. On the
other hand, Kodak's photographic film product is a market leader. It domi-
nates the market with an 80% market share, and its largest rival Fuji has a
market share of 16% only, thus the relative market share of the film product
is 80/16, or 5.0. So, when a product is a market leader, its relative market
share is greater than or equal to 1.0.

Along the vertical axis, the BCG portfolio matrix plots the growth potential of the relevant industry. The growth potential of a product can be measured by the annual rate of growth of the product's industry. This is calculated by taking the industry's total sales in a year 199X, and comparing it
with the same industry's total sales in the previous year (199X − 1). The

EXHIBIT 4.6. Product Portfolio

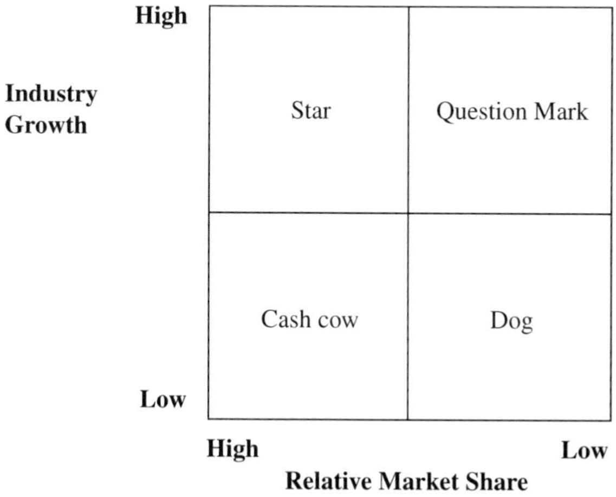

growth rate of an industry may be compared relative to the growth of the entire economy.

The BCG portfolio matrix is developed by plotting the relative market shares of different products along the horizontal axis and plotting their industry growth rates along the vertical axis. As mentioned before, the horizontal axis is plotted from a high growth rate on the left to a low growth rate on the right. The vertical axis is plotted from low to high, going up. The matrix is divided into four cells, each cell representing a different type of product. Each product in the firm's portfolio is represented by a bubble, whose size or diameter represents the size of that product measured by its sales revenue.

1. Question Mark Products. Let us begin at the upper right-hand cell. This cell corresponds to products which have low relative market shares and which are in industries with high potential growths. This cell can be correlated with the products in their introductory stages. These products are called question marks, because their future is somewhat uncertain. If the question marks are nurtured carefully, they improve their relative market shares and move from right to left to become star products. Because of their low relative market shares, the question market products generate limited revenues. On the other hand, the high potential industry growth for these products requires the senior managers to invest more resources to the growth of these products. Therefore, the question mark products generate negative net cash flows.

2. Star Products. As question mark products turn into their star products, their relative market shares have improved significantly from low to high. These products with high relative competitive advantage have started generating large revenues.

The industry of a star product is still in a growth stage. This motivates the firm to continue to invest aggressively in a star product's future. Therefore, the star products continue to consume large amounts of resources. The net cash flows of star products can be either somewhat positive or negative.

3. Cash Cow Products. As the star products continue to evolve, their industries may star saturating. The growth of star products attracts more competitors. This may start saturating a product's market and reduce its industry growth from high to low. However, if a product is able to maintain its high competitive strength in the market, the product is classified as a cash cow. These products, like cows, consume a smaller amount of resources (eat less cow-feed) and generate more resources (cows give milk). Therefore, the cash cow products generate a net high positive cash flow. This is why the senior managers love to have all their products in this category.

4. Dog Products. Unfortunately, cows grow old. Then they produce less milk while they continue to consume the same amount, or a somewhat smaller amount, of resources. In the same manner, the cash cow products evolve with time, and their relative competitive strengths, as measured by their relative market shares, start shrinking from highs to lows. With still low industry growth rates, such products are classified as the dog products.

Balancing the BCG Portfolio. Senior managers in a firm can graphically represent all their products in a portfolio matrix. This gives them a comprehensive snapshot of their portfolio of products in the organization. Each cash cow product generates a surplus cash, whereas each question mark product consumes net additional resources. The star products can also require additional resources if the firm is bullish on the future potential of these products. On the other hand, the dog products do not hold much future, and can be a drain on the cash position of a company. Typically, the net cash generated by the cash cow products pays for the growth of the firm's question mark and star products. The star products are carefully nurtured because they have the potential to become the cash cow products in the future. Too many cash cow products can produce an abundance of net cash flow. However, with time this could cause all the cash cows eventually turning into dogs sometime in the future. This could cause the firm to lose its financial solvency.

Link Between Product Life Cycle and Portfolio Matrix. One can see a few similarities between the product life cycle model discussed earlier and the product portfolio matrix. The question mark products are similar to the products in their introductory stage. The star products correspond to the growth stage products in the product life cycle model. The product in the mature stage in the product life cycle corresponds to the cash cow products. And finally the decline stage products and the dog products share some common characteristics.

MARKETING MIX

One of the foundation pillars of marketing management is the marketing mix, popularly known as the four Ps of marketing. The four Ps are product, price, placement, and promotion. This concept was introduced by E. Jerome McCarthy in 1960 (Note 4.13). McCarthy's contribution was to offer these four parameters as a concise paradigm for defining the functions of a marketing manager for marketing consumer goods in the late 1950s.

P1: Product. Product-related marketing decisions involve a definition of the core product, as well as the optional "bells and whistles" of an extended product. A desktop Dell computer is defined not only by its core elements such as its microprocessor, random access memory, hard disk capacity, and connectivity, but also by its other elements such as expandability, compatibility, and multimedia capability for audio and video input/output. A Toyota Camry car is defined not only by the size of its engine and passenger capacity, but also by options such as the bucket leather seats, anti-lock brakes, passenger-side air bags, and other optional features. Many times, a product's options attract a bigger customer appeal than the core product does. Sony desktop computers and Toshiba laptops are sold for their color and sleek appearance as well as for their core functions such as capacity and speed.

P2: Price. The price parameter in the market mix is the pricing of the product. This includes the price of the product as well as the terms of sale. Therefore a sofa sold at $795 on cash basis has a different price than the sofa sold at $795 with 2.5% interest financing payable in 1 year. The discounts and other payment terms change the total price of the product.

Managers in technology-driven enterprises face many decisions. Should a new product, such as a car, be priced so that even a college student can afford it, or should it be priced so that only the millionaires can think of buying it? Should the new Ray-Ban sunglasses be priced near the average price of other similar products, or should it be priced slightly below or slightly above the average price? These are important pricing decisions which can significantly influence the demand and the market success of a new product.

P3: Placement. The placement parameter in the marketing mix paradigm includes the decisions about the variety of distribution channels used to place the product within its customers' access. Should IBM sell its desktop computers in the company owned stores, or should these smart machines be "placed" with mass merchandising retail outlets such as, Sears, Kaufman's, K-Mart, or J. C. Penney? Or should these computers be sold at the specialty stores such as Computer City, CompUSA, Radio Shack, and so on. Can these computers be sold through 1-800 telephone lines, or by mail-in catalogs? Why are so many new household appliances offered by TV infomercials and not sold in stores?

P4: Promotion. Finally, the fourth P is the way a firm informs and educates the target customers about its product offerings. For Harley–Davidson motorcycles should a television promotion be preferred over its print media advertisements? If print media, then should the advertisements be placed in *Time, Newsweek, Fortune,* or a trade publication for motorcyclists? Where should Harley–Davidson promote to create the image that Harleys are not just for bandits, but for professionals and women riders too?

Marketing Approaches

The marketing mix can be used in one of the following three alternate ways.

1. Undifferentiated Marketing Mix. The first approach is to use an undifferentiated marketing mix for the entire market. Henry Ford offered the

Model T car to all the segments of American society. One car product in one color with standard product features was offered to all strata of customers, through one dealer distribution channel, and promoted uniformly to all. Coca-Cola also is a world product which is marketed in a fairly uniform way all around the world (except that in the developing countries, Coca-Cola enters into joint ventures with local partners who help in interfacing with the native customers and the local governments).

Differentiated Marketing Mix. In this approach, first the total market is segmented into distinct but commercially viable segments. Each segment consists of customers who are similar in certain ways. For example, the geographical segments combine customers in one particular region of the United States. The U.S. markets may be segmented as the East Coast, the West Coast, the Midwest, and the South segments. The markets may also be segmented by demographic characteristics. The total market can be divided into the preschool children segment, the preteen segment, the young adult segment, the 18- to 54-year-old segment, the senior citizen segment, and so on. Or male–female gender segmentation can be made, depending on the product under consideration.

Once the market is appropriately segmented, a differentiated marketing mix is developed for each market segment. The taste of Kentucky Fried Chicken (or product) may be different in certain segments of a city. The price of a KFC meal as well as its promotion may be varied. The luxury cars are targeted at high-income customer segment, and the product includes many customized and luxurious accessories. An easy-to-load and point-and-shoot camera may be targeted at young families expecting their first children. Eastman Kodak offers wedding packages bundling a number of not-so-expensive single-use cameras, which are placed on each table at a wedding reception. Kodak urges members of the wedding party to use a camera to shoot pictures and leave their cameras behind for the host to develop.

3. Niche Marketing Mix. Some firms, with limited resources, may decide not to spread their resources thin across all the segments of a market. Instead, they first segment the total market and then decide to focus on only a single niche segment of the overall market. A clothing retailer may decide to target only the teenage girls segment or the plus-sized women segment. The different magazine publishers may focus only on the business segment or on the health-conscious senior citizen segment. Once a firm has identified a gap in the total market served, it then exclusively targets this segment with a suitable marketing mix.

Evolution of the Marketing Mix

Professor Julian Yudelson of the Rochester Institute of Technology examined the different ways by which the four Ps of marketing mix have evolved since the 1960s. He proposed that over the past four decades, the focus of marketing function has shifted from the exchange transaction for a producer's sales department, to the way a product meets the customers' needs, wants, and other requirements. In the case of the former approach, the customer was assumed to be a blank slate over which the marketer imprinted a brand's image and other messages. In the case of the latter, the customer is put at the center stage. Furthermore, the marketing function has gradually evolved from consumer products to industrial and business-to-business marketing, service marketing, and nonprofit marketing. For example, the nonprofit marketing undermined the money-based selling transactions and involved value-based marketing to persuade others for a safer environment, smoke-free workplace, or abortion-free society.

Yudelson explained that with the advent of Total Quality Management in the 1980s and the 1990s, the four Ps of marketing mix were redefined as the means to gain customer satisfaction and long-term loyalty. With this perspective, a product can be redefined as the pleasure, or all the real or perceived benefits which the customer gains. This shifts the emphasis from "what is produced" by a producer, to "what is desired" by a customer. The price can be redefined as the pain or parting that a customer has to experience to obtain the earlier mentioned pleasure. This extended definition of price would include the time and effort wasted in exchanging or replacing defective goods. The promotion can be redefined as the perception, along with the communication needed to create the desirable perception. Finally, the place in the marketing mix can be redefined as the process of interaction between the sellers and the buyers in a marketing relationship. This process of interaction can take place over multiple channels of distributions, as well as at different places and times. This modified framework of "new and improved four Ps" reflects the current state of marketing products and services more accurately (see Note 4.14).

ENTERPRISE DESIGN FOR NEW PRODUCT DEVELOPMENT STRATEGY

In the preceding section we saw how new product development and marketing strategy are formulated for an enterprise's survival and growth.

An efficient and effective implementation of a new product development process involves coordination and cooperation across many different functional areas. Newly emerging principles embodied in proprietary know-how owned by other enterprises may be legally acquired to develop the next generation of products. The cooperation of production operators and supervisors is required if the product must be cost effectively produced on a large manufacturing scale.

Such cross-functional interdependence requires a careful design of the enterprise. Different enterprises do it differently, as explained below.

1. Sequential-Silo Functional Organization. Many enterprises work with well-defined rigid silos for each of their different functional areas. Each department does its own thing, with little interaction with other departments. The silo members tightly guide their turfs. Therefore, in such organizations, at any time only one department is actively involved in the new product development process.

This narrow functional specialization is based on the "scientific management" approach proposed by Frederick Taylor around the turn of the 20th century. To improve productivity, such approach favors functional specialization of industrial work. Taylorism proposed a "one best way" to do each task. This gave birth to the functional departments with specialization in different industrial practices. The marketers knew the "one best way" to do marketing of products for the enterprise. On the other hand, the statistical quality inspectors were trained with the expertise on how to achieve superior quality.

With Taylorism, the relationship across different functional departments is commonly governed by the principles of ideal bureaucracy proposed by Max Weber. With their own rules and procedures, the different departments operate almost independently of each other. The product development process must therefore move across different departments.

The National Aeronautical and Space Agency (NASA) developed a sequential phase program for new product development process. It operates like a relay race. It runs with well-defined accountability for responsibilities. Only one department at a time has the "baton" of the task of developing potential new products.

Let us say that a market representative identifies an unfulfilled customer need. He shares that with his supervisor, who talks about this to her manager. The idea slowly percolates up the marketing silo. Then the vice-president of marketing passes the marketing concept for a potential new product "over the wall" to the head of research and development (R&D) department. The head of R&D shares that with subordinates. The market concept for a potential new product slowly comes down the R&D silo. More technical details are worked out and the potential new product "fans" out. Then the product concept goes up, back to the head of R&D silo.

Once again the head of R&D throws the technical concept for a potential new product over the departmental wall facing the engineering department. The engineering department operates the same way as other departments, and they pass the engineering designs over the wall to the production department.

A unique feature of the sequential-silo approach to new product development process is that a lot of interchange of ideas takes place between professionals within a departmental silo. But a perfect "code of silence" is maintained with respect to sharing such information with the professionals from other departmental silos. As a result, the professionals in the downstream silo are often surprised when they first receive the idea from their upstream department.

2. The Volley-Ball Seesaw Product Development. As a result of the above-mentioned surprises and the typical mistrust across different departments, the sequential new product development process does not always move in the forward direction. This is particularly so when there is no fire-like emergency facing the enterprise and its people. The "relay race" turns into a "volley-ball" or "ping-pong" match.

The designers receiving the technical concept for a new product from the R&D silo find it impractical to conceive a design product concept based on the technical product concept they received from the researchers. They return the product proposal back to the R&D researchers, accusing them of not being in touch with the practical reality. On the other hand, R&D staffers might accuse the marketers of asking what is physically impossible by the laws of nature.

As a result of such interdepartmental counter accusations, the new product process turns into a "volley-ball" match, going back and forth between the different functional silos. Often such exchanges are counter productive, adding little value to the improvement of the new product. Each time such exchange takes place, some precious time is lost. This adds delays to the total cycle time for the total new product development process from concept to commercialization.

In earlier years, such delays in cycle times for new product development did not matter. The customers, with limited alternate choices, patiently waited for years. They kept buying the Model T Ford in black because they did not have other choices. This was no longer the case in the 1990s because of globalization of world markets. The customers became impatient, and if a particular product brand was not available, they were willing to switch to other brands. The competitors too were eagerly looking forward to attract such dissatisfied customers on the lookout for better deals. In the 1990s, an extended cycle time in new product development often meant lost customers.

3. Simultaneous Rugby-like Product Development. During the 1980s, the Japanese manufacturers trying to penetrate into global markets knew that they could not afford the luxury of losing their customers because of their long cycle times. Furthermore, to capitalize on the market gaps appearing in the American or European markets and to respond quickly to changing market preferences, the Japanese producers developed their products differently. They used a more flexible and "parallel" product development process, where the different departments coordinated their work for a new product development process in a rugby-like manner.

They shared information across different departments, well before passing the "baton" from one department to the next one. The departments and the work groups worked concurrently. Their responsibilities were sometimes overlapped, for smoother and faster hand-overs.

Cycle Time and Market Performance

This simple work-related innovation, from sequential relay-race product development to simultaneous rugby-like product development, made a big difference in the market competitiveness of technology-driven enterprises. For example, the photocopier pioneer Xerox had long cycle times for product development, and they took 4 to 5 years to develop a new product from its concept. Its more nimble competitors introduced their new products into the market in less than 2 years. Xerox had over 80% market share of copiers in 1978, but only 45% market share in 1982. Under the guidance of CEO David Kearns, Xerox focused its efforts on its "Leadership Through Quality" program and reduced their cycle times significantly. The company went on to receive the Malcolm Baldrige National Quality Award for superior customer service and became a benchmark for many other companies (Note 4.15).

CHANGING ROLE OF CUSTOMERS IN MARKETS

In the past a "better mousetrap" of technology attracted enough customers that the product embodying such "rare" technology sold by itself. This was called the technology push. Market competition changed all that. As new competitors offered competing products and services, customers became the kings. A paying customer became the main reason why a business organization emerged, grew, or survived in its competitive markets. The willingness of customers to pay for the products and services helped the stockholders, directors, executives, managers, workers, and others to draw their benefits from that business organization. However, the customers' voice was very often so feeble that the stockholders, the managers, and others in a high position in the hierarchy of a large business organization failed to hear it.

In the 1980s and the 1990s, due to an increasingly intense competition, customers were offered new computers, cars, and communication services from a large number of competitors around the world. Many customers interested in goods and services of technology-driven enterprises were put off by those organizations which overlooked their demands and preferences. In such cases, customers "talked" by taking their business elsewhere, and they "walked" away from those companies which did not pay "reasonable" attention to their requirements.

In one industry after another, customer satisfaction and retention became the necessary conditions for survival and growth of a firm.

Relationship Marketing

As new information and communication technologies, such as the Internet and World Wide Web, became popular with millions of new subscribers and potential customers, the relationship between the customers, producers, and providers changed further. The interaction and relationship between a potential customer exploring his or her choices on the Net and the vendors demonstrating their offerings on the Net became more personalized. The new information technologies provided and demanded a change from the traditional marketing of goods and services at a statistical demographic "market," to a more "relationship"-oriented marketing with "individualized" customers.

In the rest of this chapter we will investigate these mega-shifts to customer focus and relationship marketing, and we will review some of the "best practices" by some progressive companies. We will study how they identify new customers, listen to their voices, and retain them for life in different technology-driven manufacturing and service industries. How do these companies go beyond customers' satisfaction and delight them with innovative offerings beyond their own expectations? How do these enterprises not only target the current customer, but keep future customers in mind too?

Market Development by Customer Trust

Intensely Competitive Fast Markets. In the last few years of the 20th century and the first few years of the 21st century, the global markets for technology-intense products such as consumer electronics, computers, and cars are most likely defined by the high intensity of rivalry among global competitors. In many technology-driven industries such as aircraft manufacturing, telecommunications, semiconductors, office equipment and imaging, and even automobiles, the competition is primarily between a few equally large-scale competitors. Due to the global mergers and consolidations in the 1990s, many smaller companies were forced to close or were acquired by the bigger firms.

Unlike in the Industrial Age economy, in the post-industrial Information Age the large size of a corporation alone will not provide a competitive advantage with the customers. The customers, with easy access to user-friendly information technology, could hardly be swayed by corporate promotions roughly targeted at a "demographic" segment. Instead the "fast" organization that is able to get its products and services to a large number of individualized customers fast will have a competitive advantage.

Information Age Customers. With the emergence of an increasingly user-friendly information technology, a number of significant strategic mega-shifts have been taking place in the way customers prefer to buy and take care of their requirements. Because customers are able to frequently and economically transfer large amounts of information across distant parts of a globalized world economy, they are becoming extremely aware of their options and choices. Therefore, they routinely are likely to optimize their purchases for the highest value they can get from anywhere in the world. For example, high quality and reliability of goods and services have already become the minimum requirements for doing business in these industries anywhere in the world.

These mega-shifts put some new demands on the producers for the development, acquisition, and retention of their customers. With rapid industrialization in different parts of the world, the technology for producing products and services is no longer limited to a few large corporations in the Western world. Customers are now offered increasing number of choices of more technology-intensive products. For example, in the computer and electronics industry a cutthroat competition emerged between the top-10 global brands from IBM, Compaq, Apple, Dell, Gateway, Toshiba, Sony, Hitachi, Samsung, and others from America, Europe, and Asia. In the auto industry there are the Big Three U.S. auto makers competing side by side with the Big Six Japanese auto makers and an equal number of Big European auto makers. The competition for tires was intense between rivals Goodyear, Bridgestone, and Michelin.

In the 1990s, satisfying customers has become one of the major challenges of business managers. Even if a business organization had a well-recognized brand (such as Apple or IBM, Goodyear, Kodak, Philips NV, or Federal Express), they were continuously assaulted by rivals and new entrants on one hand, and substitutes on the other hand. The desktop computer makers IBM and Apple faced Compaq, Dell, and Gateway compatibles on one hand, and Hitachi laptops and Sharp palm tops on the other hand. Professor Richard D'Aveni has called this *hypercompetition* when the intensity as well as the rules of market competition change rapidly (Note 4.16). In hypercompetitive markets, customers have a large choice of products and services to choose from. For the producers there is a constant disequilibrium and large-amplitude change due to frequent redefinition of the market rules.

In the Information Technology Age, the customers have more access and information about the product and service offerings than ever before. The Amazon.com bookstore in the cyberspace has millions of books to offer compared to a Borders bookstore in the physical space. The current and potential customers are still targeted by promotions via telemarketing, direct mail, TV, and radio advertisements. But millions of customers can "surf" their home pages and almost instantaneously get extensive information about a company's historical heritage, its social responsibility record, and the different families of products it offers. The customers could get these goods and services in mega-malls, strip-malls, or in the basement of their homes via a television monitor or a personal computer.

Customers are changing fast, and therefore marketing to them too has to change in radical new ways. Many competitors have already innovated new ways to adapt to these intensely competitive markets and the new information-intense customers.

Information Age Marketers. Like the customers' access and use of user-friendly information technology to make their selections, the producers can also leverage the sophisticated new developments in information technology. With a rapid increase in microprocessor speeds, more networking capabilities, and cheaper storage memories, producers and providers can collect and use information about their current or potential customers in far more ways than what they could do in the past. They can interact and integrate with customers more intimately by keeping track of their individual needs, wants, and preferences.

Let us illustrate this by reviewing the practices of four innovative companies responding to this new challenge. We will briefly look at two examples from the manufacturing sector, as well as two examples from the service sector. Levi's and Dell offer customized jeans and computers. Arthur Andersen and Federal Express offer information-rich customized consulting and overnight delivery services.

1. Levi's Custom-Fit Jeans. Levi's jeans are perhaps the most popularly recognized pants in the world. For more than one century, Levi Strauss sold the original denim jeans, described by its lot number 501, without any major changes. Customers, male or female, bought similar 501 Levi's jeans. The company was started by Levi Strauss, a German merchant's son when he went to San Francisco to cash in on the needs of the miners participating in the Gold Rush of 1849. The miners had a hard time keeping their trousers and pockets from tearing in the rugged mining environment. In 1873 Levi Strauss and a tailor partner stitched together trousers from heavy-duty canvas. They used rivets on pockets to keep them from falling apart. They called it jeans because their cloth was originally made in the city of Genoa. Since then, only minor changes were made in these jeans based on customers' feedback. For example, the rivets on the back pockets were removed in 1937 because they scratched saddles of cowboys and school seats. Some rivets on the front side were also replaced because they got hot when a person stood for a long time in front of a fire or a furnace. Other than these minor changes, for more than 100 years Levi's jeans were marketed to customers worldwide, men and women, almost unchanged.

In the 1990s, for the first time Levi Strauss used information technology and flexible manufacturing system to introduce custom fitted jeans specially designed for women customers. In the past, women customers were forced to accept a few discrete sizes of mass-produced jeans designed for rugged men. Levi Strauss discovered that their women customers needed jeans of their own. So for an extra $10, a woman could try one of the sample jeans at a store. The sales clerk would note the changes required and enter that information into the company's database. A computer-integrated fabric cutting system took that information and scheduled custom production of jeans with a production lot size of only one. Within a few days, the custom-stitched jeans for the lady customer would arrive at her store, and she could pick up her custom-fitted jeans (see note 4.17).

2. Dell's Customized Computers. Dell computer did with computers what Levi's did with women's jeans. Specific requirements of a computer buyer were taken over Dell's famous 1-800 lines, or the new Internet connection. Orders for monitors were then placed directly for delivery from a preapproved supplier. Instructions were issued to the production shop floor to custom assemble a computer. Delivery vendors such as Federal Express or UPS would be instructed about the delivery schedules for the computer and the monitor. The two shipments starting from different sources would arrive at the doorsteps of a customer at the same time (see Note 4.18).

3. Arthur Andersen's "Best Business Practices" Consulting. During the 1980s and the 1990s, new developments in enterprise-wide information systems posed new challenges to the public accounting practice of large public accounting and consulting firms (for details see Case 4A). Development of automatic accounting systems made the hefty fees of public accountants less justifiable to the resource-challenged small and medium-sized corporations. On the other hand, the turbulent 1990s increased the demand for their consulting expertise. However, to justify their high fees, the consulting firms had to bring in added value knowledge to their customers.

Arthur Andersen, a pioneer in breaking new grounds among the proverbially conservative public accountants, started a proprietary "Best Business Practices" knowledge base for the benefit of their clients. Once their consultants were called in by a client for consulting service, the consultant brought along tons of additional information and knowledge to benchmark with. The client gained the benefits of Arthur Andersen's cumulative knowledge base of "Best Business Practices." This consisted of extensive cumulative information on 13 major modules of business processes which were considered generic to companies in many different industries. In a multitier hierarchical organization of information, the Best Business Practices knowledge base provided relevant information to a client at the desired level of detail.

The Arthur Andersen consultants also held an international Enterprise Awards competition for sharing the Best Business Practices[SM] between its current and potential future clients. The winners of the Enterprise Awards for Best Business Practices were recognized at a 2-day knowledge-sharing symposium at the Arthur Andersen Worldwide Center for Professional Education in St. Charles, Illinois. As a subsequent educational program, the applicants and other interested executives were encouraged to attend the Knowledge Network Series held in different cities for sharing their knowledge. The goal was to help current and potential clients achieve superior performances by improving their business practices based on self-examination, discussions, and dialogue.

4. Federal Express: Technology and People-Intensive Customer Service. The story of Federal Express is the story of an American Dream come true (see Case 4B). In 1997 FedEx was the world's largest express transportation company: It handled 2.5 million packages daily, using 37,000 vehicles, 530 aircrafts, 1400 service centers, and 124,000 employees worldwide. The 24-year-old FedEx increased its revenue from $6.2 million in 1973 to $10.3 billion in fiscal year 1996. In a nutshell, Frederick Smith conceptualized Federal Express's information-intensive business based on a promise to customers for a reliable guaranteed overnight service. He operationalized this customer promise based on a novel business concept of a hub-and-spoke network and an extensive transportation fleet. The key ingredients to the implementation of a reliable service were FedEx's commitment of resources in the state-of-the art information technology and a highly motivated world-class human capital resource (see Note 4.19).

Leveraging Technology. Each package collected or received from any one of the 750,000 FedEx daily customers contained a carefully designed airway bill form that was filled in by the customer and attached to the package. The airway bill uses a unique bar-coded number which is automatically scanned into a computer using a hand held scanner. Other bits of critical information, such as the source and destination zip codes, type of package and the lot size, the type of service (morning, evening, or weekend delivery, etc.) are also entered into the Customer, Operations, Management, and Services (COSMOS) computer system. This system is integrated with a Digitally Assisted Dispatch System (DADS) unit installed in more than 20,000 vans and business centers supported by the service representatives. The DADS units are hooked to COSMOS via radio contact.

From the moment a package is obtained from a customer, and through each successive hand-over, the updated information is available on the integrated computer system. The service representative could tap into the system and provide the latest status of a package to its customer. In a day, on average more than 20,000 customers make follow-up inquiries to find out the status of their valuable packages.

For reliability during the flying part of a package's journey to and from Memphis SuperHub, FedEx installed very sophisticated electronic control systems for landing, auto-piloting, and takeoff under all types of weather conditions. A central control station keeps track of more than 200 planes in the air. FedEx also commissioned its own weather analysis for early warnings in case of bad weather.

At the SuperHub, bar-coded information directs the flow of packages through the automated package handling and sorting system. The automated sorting process, used during a narrow time window of 12 midnight to 3 a.m., is operated by a large number of part-time student employees.

Since 1984 many large-volume customers were provided free PowerShip terminals to allow them direct access to a large part of COSMOS information system. They could therefore monitor and verify not only the receipt and delivery of packages, but also the accuracy of their billings. A few years later, FedEx gave away dial-up "FedEx Ship" software to help customers convert desktop computers with modems to hook into the COSMOS network. By the end of 1996, more than 440,000 customers used FedEx Ship software, and over 100,000 PowerShip systems were installed. More than 60% of FedEx transactions were placed electronically without the paperwork (see Note 4.20).

ZapMail Mishap. Even though Federal Express has been great in leveraging information technology for servicing its customers, in one instance an emerging technology gave FedEx a hard time. In the early 1980s the telecommunications technology saw a booming growth and posed a threat to substitute FedEx service. Chairman Fred Smith thought FedEx could instead tap into the telecommunication market and convert it into an opportunity. Thus large amounts of resources were allocated to the development of ZapMail, an electronic mail facsimile transmission system. ZapMail service was designed to meet the need for a superfast 2-hour delivery of documents. This was faster than FedEx's "absolutely, positively overnight" delivery.

ZapMail leveraged the use of the COSMOS–DADS network. Instead of just sharing information about the package, the network was also used to transfer customers' documents from one point to another point. In the first phase, FedEx installed ZapMail facsimile terminals at its over 300 Business Service Centers. Customers who wanted a document transmitted to a distant destination in guaranteed 2 hours would call FedEx, and a courier would pick the document. The package would be brought to a nearby Business Service Center where it would be transmitted to a Business Service Center near the final destination of the document. There a courier would deliver the transmitted document—all within guaranteed 2 hours. The price was set at $35 for 10 pages or less, and a dollar extra for each additional page.

Subsequently, FedEx rented a ZapMail terminal to a heavy user for $525 per month, and it charged $0.25 per page for transmission and $10 for delivery. In 1984 more than 500,000 ZapMail terminals were installed, and another 100,000 were put in use by March 1985. Japan's electronics manufacturer NEC was contracted to manufacture the ZapMail machines.

The ZapMail service business was based on the assumption that the ownership and use of personal computers would not spread at a fast pace. Strategically, it made sense at the time ZapMail was conceived. At the end of 1984, Federal Express was the dominant leader in the domestic express package service. FedEx's $1.7 billion revenue was more than two times ahead of the $675 million revenue of United Parcel Service (UPS). Another competitor, Purolator Courier, collected $590 million in revenue per year. Airborne and Emery had much lower revenues of $305 million and $175 million, respectively. FedEx was the leader and had to innovate to maintain that leadership position. FedEx decided to target ZapMail service at the lawyers, architects, and other professionals who frequently needed to send documents urgently to their clients.

Unfortunately, ZapMail business did not succeed as planned. For one quarter after another it lost a lot of money because of the new developments in fax machine technology. Prices of fax machines came down dramatically to about $300 for a simple machine. The ZapMail business lost a total of more than $325 million, and the operations were written off at a matching amount (see Note 4.21).

From Transactional to Relationship Marketing

In the fast emerging Information Technology Age the overall objectives of companies like Dell, Arthur Andersen, Federal Express, and Levi Strauss have been to create, cultivate, and sustain ongoing relationships with their customers. To do so, these companies, like their customers, leverage the new developments in information technologies. They maintain records of their individual customers and their past purchases (either with them or elsewhere) to understand the pattern of their preferences. These companies would then offer their customers extremely focused offerings, with a high probability of acceptance. They are no longer interested in "broadcasting" their marketing offerings to a statistically aggregated target customer. Information technology enabled these producers to produce mass customized goods, specific to an individual's needs, and market to the millions of individual customers on one-to-one basis.

Integration with Customers: A Few Best Practices in Building Customer Trust. Relationship marketing means integrating closely with customers. The goal is to gain the customers' trust. The technology-driven enterprise is thus driven by customers' wants, expectations, and satisfaction. The successful companies have learned a few important lessons in this new area.

1. The Power of Human Capital in Customer Service. Unlike many organizations, FedEx did not install sophisticated information technology at the expense of its manpower and human capital. On the contrary, people are considered the primary resource to ensure the company's promise of a highly reliable service. Founder Fred Smith and many of his direct reports actively practice "People–Service–Profit" philosophy, in that order of priority. The very best people are hired, and then they are provided or encouraged to retrain and relearn on an ongoing basis. Each employee can seek the attention of his or her supervisor by invoking Guaranteed Fair Treatment (GFT) policy. If necessary, grievances of employees are heard by a jury of their own peers. If not satisfied at the supervisor level, the employees are encouraged to go to their manager, executive, or even chairman.

FedEx regularly seeks suggestions and participation of employees for making major decisions. Self-governing high-performance teams are rewarded and recognized for their valuable contributions. In the early years, FedEx offered highly profitable profit-sharing and stock-option benefits.

With these progressive practices, Federal Express has been frequently rated as one of the top 10 companies to work for in America.

2. The Full Logistics Service. For the Information Age customers, FedEx as well as its rival United Parcel Service (UPS), are racing to become a full-service logistics arm for their customers. Both offer to monitor and manage the clients' flow of packages, parts, and information between the clients' customers, suppliers, retailers, and producers. Managers at FedEx describe themselves not as a courier company, but as a double network system. One network handles physical goods using vans, planes, and other transportation means. The other network moves bits of information using DADS terminals and a COSMOS system. FedEx is good at orchestrating both these networks with equal ease and reliability (see Note 4.20). The two networks are intimately linked, the way the double helical structure of a human DNA is. FedEx Logistics, a new subsidiary, was launched to do similar orchestration of information and goods for the external clients.

Based on the success of a highly reliable delivery service, clients were willing to trust FedEx to manage their logistics. For example, fashion retailer Laura Ashley signed a $250 million 10-year deal with FedEx Logistics to handle its logistics. FedEx was asked to take care of Laura Ashley's logistics ranging from taking mail orders to filling orders and distribution.

FedEx is also looking forward to using the World Wide Web and the Internet to satisfy its Information Age customers on a one-on-one basis. FedEx's Web site www.fedex.com was set up in 1994 to help customers directly track their packages. Using the COSMOS network, a customer could also use the Web to order FedEx shipment.

FedEx is ready for the Information Age customers. According to Fred Smith, "The Internet is going to make it very difficult for anybody in a middleman position to stay in business. I cannot imagine how you can sustain any kind of a price premium when you'll be able to send an intelligent agent out to go price shopping from Bombay to Boston." FedEx's version of the future is an electronic market of cyberstores run by managers focusing on "marketing, customer service and counting the cash" while FedEx runs their double network systems of physical goods and information (see Note 4.20).

3. Kodak's Digital New Product Launch. Let us consider another example of an innovative company reaching out to its potential customers in cyberspace for launching the company's new products and technologies. On August 1, 1997 Eastman Kodak kicked off its first major launch into cyberspace marketing by selling DC20 digital camera photography kit from its home page on the Internet (Note 4.22). The photographic giant considered this a natural fit for the "cybernaut" customers. Kodak's goal was "to drive more interest in doing photography online," according to Kathleen Sheehan, director of strategy and business development for Kodak's Internet Marketing Group. The DC20 Web Camera promotion kit was priced at $219 plus tax and shipping, and it included a $199 DC20 camera bundled with a $49 flash, special software, a camera bag, and a copy of *Fun With Digital Photography* by Rochester-based consultant John Larish. Kodak's cyber-customers could buy the package either electronically on-line or by fax and toll-free numbers if they were not comfortable using credit cards with electronic commerce over the Internet.

Kodak believed that selling the DC20 on the Internet was a natural fit considering that it was an easy-to-use digital camera with a 200,000 pixel sharpness. This was particularly appropriate for those Internet customers who wanted to include their pictures on their e-mails and Web pages. Its lower resolution (compared to the $999 DC120 zoom camera with one million pixels) made its pictures faster to download from the Internet. By marketing on the Internet, Eastman Kodak could directly target its best customers.

Since the beginning of the Web boom, Kodak has been trying to pioneer the use of Internet as a marketing tool. Kodak has used its home page to test new digital-imaging technologies, to invite customers to chat with senior executives such as CEO George Fisher, and as an educational tool to help more customers learn about digital imaging. The Kodak Web site was also used to promote and sell racing-related merchandise for the Kodak Gold Chevrolet Monte Carlo car sponsored for NASCAR.

The next step for Kodak was to advertise banner messages on other sites including the search engines. The company used software tools from AT&T Corporation to customize features for future marketing programs. According to Sheehan, they would start collecting data on personal preferences of visiting customers who were willing to fill questionnaires, but Kodak would respect the privacy of those who did not do so.

Customer Trust: Satisfaction and Loyalty Connection. Market research is repeatedly showing that satisfied customers do not often make repeat purchases—not until they trust the provider. For loyalty the customers must be delighted. A Xerox study in early 1990s indicated that even satisfied customers often defect to competitors. They do so unless they are absolutely delighted with outstanding products or services. Customer loyalty increases nonlinearly with customer satisfaction (see Note 4.23).

If customer satisfaction was plotted along a horizontal axis on a 1–5 scale, from completely dissatisfied to completely satisfied, and customer loyalty or likelihood of repeat purchase was plotted along the vertical axis on a 1–5 scale of completely disloyal to completely loyal, then the loyalty does not increase proportionally with customer satisfaction. As customer satisfaction increases from 1 to 3, loyalty increases only slightly, say from 1 to 2. The loyalty takes off only after customer satisfaction exceeds 4 or "satisfied level." The completely loyal customers are extremely satisfied customers (see Exhibit 4.7).

EXHIBIT 4.7. Customer Satisfaction—Loyalty Effect

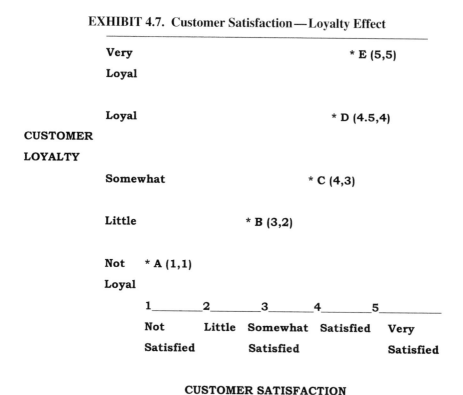

CUSTOMER SATISFACTION

Sources of Customer Loyalty. How do companies get and keep their loyal customers? There are a few important ways by which a company can succeed in keeping its customers loyal in an intensely competitive market.

1. Competitors Limited by Government Regulation. In the past, airlines' routes were regulated, and the airlines benefited from their "regional monopolies." But because of airline regulation, Federal Express had to lobby hard to get the Federal Aviation Administration to change its rules for the maximum weight of cargo that its Falcon business air-jet could carry to fly unrestricted routes. In many regions the utility suppliers, like the electricity and gas providers, are somewhat regulated. These regulations increase the barriers to new entrants. The customers have no choice but to remain loyal to such "local monopolies."

2. Restricted Competition due to Proprietary Patents and Trade Secrets. Levi Strauss and his co-founder tailor gained a patent on their riveted jeans stitched from canvas cloth. Otherwise all other tailors could have copied their unique design. Patents, as discussed in detail in Chapter 3, allow an inventor a monopoly for a limited time in exchange for disclosing the secrets about an invention with the public at large. Some companies, on the other hand, prefer not to strike that bargain. They hold on to their secrets without sharing them with anyone else. They are willing to face the risk of someone else coming up with a matching idea and patenting it first. But, companies like Coca-Cola have done quite well by holding on to their own trade secrets.

3. Increase Switching Cost. If customers have to shell out lots of money, or be very inconvenienced to switch to a competitor, they are less likely to make the switch. In the world of personal computers, for many years before Microsoft Windows, the IBM PC and Apple/Macintosh computers had different and mutually incompatible operating systems and softwares. A customer owning an IBM PC, when attracted to the graphic interface and user-friendliness of an Apple II, Lisa, or Mac, had to consider spending $2000–3000 to switch to the Apple and Mac world. In the same manner, a Mac user interested in running more softwares available with IBM PC had to budget a matching amount to switch to the IBM PC environment.

4. First-Mover and Superior-Location Advantage. Sometimes early entrants get an advantage over competitors by locking into a better and more profitable location. In the retail market a better location attracts more customers. An early entry may be considered a "better" location in the minds of the customers. Ford Model T had a first-mover advantage over General Motors. But Lotus lost its first-mover advantage in spreadsheets to Microsoft. Microsoft MSDOS gained an advantage over other operating system by getting installed in an IBM PC computer as its original equipment feature. All MS-DOS customers continued to use Microsoft software thereafter.

5. Convert Customers to Become Relatives. The "hard way" to a sustained profitable relationship with customers is to intimately know them as well as their needs, wants, and desires. As mentioned earlier, only deep customer satisfaction generates their loyalty.

Customer loyalty is often the result of customers consistently satisfying interactions with a producer or service provider. Frederick Reichheld, author of *The Loyalty Effect,* found that Lexus car buyers are loyal not only because of (a) their pleasant experience with a Lexus dealer at the time of buying their car, (b) acquisition of a superior and reliable product, (c) receipt of a high value for the price they paid, but also the repeatedly satisfactory service they receive every time they have to deal with somebody related to their proud possession.

Many companies work extremely hard and invest enormous resources at building a superior brand. Experts suggest that there is an opportunity to build a "superior brand image" every time a customer interacts with somebody associated with your organization. Every encounter with the customer can be turned into a "brand" event. For example, traveling customers interact with an airline at a number of points. They interact when they make reservations, get their tickets, check their flight schedule, arrive at the airline counter, board the plane, sit in their seats, and so on. Each of these interactions is an opportunity for the airline to create a long-lasting relationship.

Branding each customer encounter is not easy. According to Thomas Stewart of *Fortune* magazine, Canadian Pacific (CP) Hotels, which owns 27 properties including prestigious places such as Toronto's Royal York and Sky Dome hotels, wanted to target individual business travelers. These business customers are the most demanding and discerning customers in the hotel business. Canadian Pacific Hotels asked these customers what they would like to have by way of personalized service from CP Hotels to get their repeat business.

To implement this strategy of personalized customer service, the CP Hotels started collecting information on every guest encounter from checking in to checking out, and everything in between. To operationalize on the promised personalized service the CP Hotels had to innovate a new operating system. They trained employees to be flexible in servicing a Japanese sumo wrestler needing an extra-length bed, then catering to the needs of a group of 50 Eastern European salesmen huddling together in a training session for 3 days with a continuous requirement of cigarette supplies. It was not easy. The new organization structure had to be put in place with more cross-functional responsibilities. But the company's payoff came in a 16% annual growth when the market as a whole increased only by 3%. More than 25% of the customers decided to use CP Hotels whenever they could, and they stopped trying out other hotels (see Note 4.23).

These best practices indicate a sense of commitment needed by technology-driven enterprises to convert their one-time customers into long-lasting relatives and loyal customers. In a globalized world the economic advantages of scale are being quickly replaced by the economies of speed. The agile and customer-driven providers of goods and services are likely to be the new winners in the postindustrial Information Age of globalized markets.

A BRIEF REVIEW OF BENEFITS FROM THIS CHAPTER

The review of best practices in the areas of new product development and customer trust, covered in this chapter, should offer many advantages to the managers of a technology-driven enterprise.

First, the concepts covered in this chapter should provide a good overall grasp of the role of new products and customers in the newly emerging and intensely competitive markets of the postindustrial Information Age economies.

Second, the concepts covered in this chapter should help companies reduce their struggle to retain the satisfied customers for a long time and should instead help to develop a mutually profitable long-term relationship with them.

Third, the concepts in this chapter could help a technology-driven organization save on the total cost of developing or commercializing new products as well as process technologies.

Fourth, the new product development and selection process, segmentation of markets, differentiation of products, and targeting of customers for long-term relationship are essentially all chaotic and risky processes. This chapter could help current and potential managers reduce their anxiety for such chaotic yet crucial and often creative efforts.

Fifth, choosing a new product or technology demands managerial open-mindedness. But such open-mindedness can sometimes lead to excessive indecisiveness. This managerial indecisiveness can lead to cost overflows, delayed schedules, and complacency. This chapter should help managers of technology to be aware of such trade-offs and should enable them to make better and faster decisions in this strategic area. The techniques and models discussed in this chapter could help in significantly cutting down the cycle times from the generation of concepts to the commercialization of successful products and positioning in the market.

Finally, our review should help develop a long-term plan and road-map for the portfolio of products that a technology-driven enterprise would like to strive for to get loyal delighted customers.

SELF-REVIEW DISCUSSION QUESTIONS

1. Compare and contrast the technology push and the market pull perspectives for technology-driven enterprises.

2. How has marketing evolved in America over the past few decades after the end of World War II? Discuss Levitt's marketing myopia.

3. Explain the role new products play in the Schumpetarian constructive destruction of markets. Illustrate with examples.

4. How will you analyze new product development strategies?

5. Elaborate the different stages in the hourglass model of new product development process. Give examples of the tools you can use in different stages.

6. What are the different stages in a product life cycle? How will you extend a product's life?

7. Describe a product portfolio matrix, and discuss how you will use it for a technology-driven enterprise.

8. What do you understand about the sequential (relay-race) and simultaneous (rugby-like) organization designs for new product development? Give examples.

9. How have the recent advancements in new information technologies influenced the marketing of new products?

10. Discuss the link between customer satisfaction and customer loyalty in relationship marketing.

NOTES, REFERENCES, AND ADDITIONAL READINGS

4.1. Levitt, Theodore. 1960. Marketing myopia. *Harvard Business Review,* July/ August; also see author's *The Marketing Imagination.* New York: The Free Press, 1983, 1986.

4.2. Drucker, Peter. 1990. The emerging theory of manufacturing. *Harvard Business Review,* May/June:94–103.

4.3. 3M. For details see 3M's Web page on the Internet at www.3M.com.

4.4. An earlier version of some of the discussion presented here on new product development process was published in Gehani, R. Ray. 1992. Concurrent product development for fast-track corporations. *Long Range Planning,* 25(6):40–47.

4.5. Schumpeter, Joseph. 1937. See Note 4.4.

4.6. Porter, Michael E. 1980. *Competitive Strategy: Techniques for Analyzing Industries and Competitors.* New York: Free Press. Also see Porter's application of his models to technology-driven organizations, in Porter, Michael E. 1988. The technological dimension of corporate strategy, in R. A. Burgelman and M. A. Madique, ed. *Strategic Management of Technology and Innovation,* Homewood, IL: Irwin.

4.7. Yasuda, Yuzo. 1991. *40 Years, 20 Million Ideas.* Cambridge, MA: Productivity Press.

4.8. Hauser, John R., and Clausing, Don. 1988. House of quality. *Harvard Business Review,* May/June:63–73.

4.9. Cadbury, N. D. 1975. How, when and where to test market? *Harvard Business Review,* May/June:96–105.

4.10. Hisrich, Robert, and Peters, Michael. 1978. *Marketing a New Product: Its Planning, Development, and Control.* Menlo Park, CA: Benjamin/Cummings Publishing.

4.11. Rogers, Everett M. 1962. *Diffusion of Innovations.* New York: Free Press.

4.12. There is an enormous amount of literature on the product life cycle. Some of the pioneering concepts were introduced by Levitt, Theodore. 1965. Exploit the product life cycle. *Harvard Business Review,* November/December: 81–94; also compiled by the author in *The Marketing Imagination.* New York: The Free Press, 1983, 1986.

4.13. McCarthy, E. Jerome. 1960. *Basic Marketing: A Managerial Approach.* Homewood, IL: Richard D. Irwin. Also see 4th edition published in 1971.

4.14. Yudelson, Julian. 1997. The Four P's, Cute Pnemonic or Gnostic Insight? A working paper submitted to the *Journal of Marketing Education.*

4.15. Reiner, Gary. 1990. Cutting your competitor to the quick, in David Asman (ed.), *Wall Street on Managing: Adding Values Through Synergy.* New York: Doubleday Currency.

4.16. De Aveni, Richard. 1995. Coping with hyper-competition: utilizing the new 7S's framework, *Academy of Management Executive,* 9(3):45–60.

4.17. See Levi Strauss home page. Also see Mitchell, Russell, and Oneal, Michael. 1994. Managing by values: Is Levi Strauss's approach visionary—or flaky? *Business Week,* August 1:46–52.

4.18. See Dell's home page on the Internet.

4.19. For additional information contact: Federal Express Corporation, 2005 Corporate Ave., Memphis, TN 38132. Telephone: (901) 369-3600. Internet: http://www.fedex.com/.

4.20. Lappin, Todd. 1996. The airline of the Internet, *Wired Magazine,* December.

4.21. Trimble, Vance. 1993. *Overnight Success,* New York: Crown Publishers.

4.22. Patalon, William, III. 1997. Kodak hits the Web for sales, *Democrat & Chronicle,* August 1, Friday: 12D, 12C.

4.23. Stewart, Thomas. 1997. A satisfied customer isn't enough. *Fortune,* 21 July:112–113. Also see Jones, Thomas and Sasser, W. Earl, Jr. 1995. Why satisfied customers defect. *Harvard Business Review,* November/December: 81–94.

CASE 4A: TECHNOLOGY-DRIVEN ENTERPRISE

Arthur Andersen's Consulting: Value-Adding Knowledge for Customers' Satisfaction

Technological innovations and developments in enterprise-wide information systems in the 1990s posed new challenges to the large public accounting firms such as Arthur Andersen. Their customers' needs shifted so much over the years to management consulting that some observers thought that they were no longer "accounting firms." In the 1990s, Arthur Andersen's business strategy to succeed was to apply knowledge and information to help clients perform better. This case illustrates how changes in customers' requirements could significantly impact the organization structure of a firm.

In the mid-1980s, stockholders and creditors of some failed companies held their public auditors accountable for failing to report the financial weaknesses of their dying clients. To remain in business and grow, the public accounting firms had to provide increasingly value-added consulting service to their clients. By the end of the 1980s, the public accounting firm of Arthur Andersen, for example, earned almost 40% of its revenue from consulting assignments—often setting up computer and information technology systems. Their consulting projects resulted in average fees of the order of $25 million, whereas the auditing assignments averaged $4 million or so.

As one of the first companies to start management consulting practice and publish its financial results for the customers, Arthur Andersen took pride in breaking new ground for the public accounting firms. Arthur Andersen, for example, also pioneered setting up its Center for Professional Education, an excellent 800-room training and conference facility in St. Charles, Illinois. The state-of-the art facility was used to train the customers as well as the employees of the firm.

From the very beginning, Arthur Andersen was sharply focused on the external environmental changes. According to the founder Arthur Andersen, who started the firm in Chicago in 1913, "Had there been a stronger accounting (audits), there wouldn't have been . . . (the Great) depression."

In the 1950s, this public accounting firm pioneered the new field of third-party management consulting. Their first assignment was to help General Electric set up computer operations. As computers became increasingly favorites for managements keen to reduce their reliance on manpower by installing computer integrated operations, Arthur Andersen was able to tap into the enormous new opportunities. Gradually, the consulting side of the Arthur Andersen practice developed into eight consulting niches. These included, besides computer systems, other related areas such as cost accounting, operations research, and production control. However, most of these niches were disconnected and heavily dependent on the expertise of a few partners here and there.

In the early 1960s, Arthur Andersen made a break from its tradition. This caused some grief at that time, but eventually helped this public accounting firm gain its worldwide leadership in later years. Until then the company considered itself primarily a public accounting firm. All newly hired consultants were required to serve for minimum of 2 years on the accounting audit teams. But the customers were requiring fewer audit services and seeking more consulting services from Arthur Andersen. As a result of this market shift, many young employees, trained for 2 or more years as the audit staff, jumped ship to the consulting side of the Arthur Andersen business. Many senior auditor partners complained about this to the higher management. The company, to reduce turnover and save resources, decided to eliminate the 2-year audit requirement for the new entrants. Thereafter the newly hired consultants started working directly in the Management Advisory Services (MAS) section of the firm.

During the 1970s the market shift to consulting continued and even accelerated. More customers started questioning the increasingly high fees charged for their public audit services, and they demanded more consulting advice to help them compete effectively in the fast changing market environment. On one hand the increasing intensity of competition among public auditing firms reduced the audit fees, and on the other hand a few audit failures resulted in massive lawsuits against public accounting firms. With the first oil shock in October 1973 a number of new stock issues suffered, and the demand for such services declined significantly. The public accounting industry found itself shifting from a growth stage to a mature stage.

SEC Accounting Release 264. In 1979 the U.S. Securities and Exchange Commission (SEC) raised the question of the "independence" of "third party" public audits while a firm also provided the management advisory services to the same clients. The SEC's accounting release 264 urged the public accounting firms to limit the advisory part of their practice that was not related to their "independent" audit function. Arthur Andersen was most affected by the accounting release 264. Already more than 20% of its total revenue came from MAS, whereas its competitors earned only between 5% and 10% of their revenues from nonauditing practices.

Arthur Andersen's brilliant chairman, Harvey Kapnick, presented a "split-the-firm" strategy to Andersen's partners' council in September 1979. The auditor partners feared that the lack of support from their consulting arm would put them at a disadvantage over their competitors with broader audit practice. Kapnick had a great vision looking far into the distant future. But this vision did not get communicated to his partners, who had a right to vote on his proposal. The partners rejected the chairman's proposal, and Kapnick decided to resign in October 1979. The turmoil made partners pick vice chairman Duane Kullberg, a cautious consensus builder, as the next CEO starting in 1980. The auditor partners as well as the consultant principals looked forward to more autonomy under his decentralized way of decision-making.

In the meantime the outside customers kept coming with a growing demand for consulting services which continued to contribute a higher profit margin for Arthur Andersen. Even though consultants contributed more to the firm's bottom line, the consultants were structured as "principals" with limited rights to vote on decisions affecting the auditor partners. The consultant principals were allowed to cast ballots for "the information" of the auditor partners.

The Executive Exodus. In December 1986 a senior consultant and well-respected managing partner Victor Millar left Arthur Andersen to join as a CEO for the London-based Saatchi & Saatchi Consulting. As a reaction to his departure, CEO Kullberg appointed a committee to define the mission of the firm. The committee, dominated by the auditor partners, passed the verdict that Arthur Andersen would continue to act and decide like an audit firm. Three alternate proposals were made: (1) to spin-off the MAS practice, (2) to spin-off the audit practice, and (3) to spin-off only the large-scale systems part of management consulting practice.

As a response to these recommendations, some consultants decided to write their own mission statement for the consulting practice of Arthur Andersen. The consultants' actions, such as discussion of alternate solutions for restructuring Arthur Andersen's consulting practice, were considered "insubordination" by the auditor partners in charge (PIC) of the local offices. After much tension, in September 1987 chairman Kullberg approved the formation of the consulting side of Arthur Andersen's practice as a national business unit. Consultants were now free to service a customer by consolidating resources from more than one office, without first going to a PIC for approval. The consultants were still not completely autonomous under an independent financial profit center though a new post of the managing partner of U.S. Consulting was created.

The perceptions of internal conflicts between the consulting principals and the auditor partners started an attack by outside headhunters to lure the high-performing consultants in Arthur Andersen. In October 1987, Andersen's Board of Partners reacted to these external threats by recommending a "noncompete rule" to restrict an outgoing consultant's ability to get employment outside the firm. The consultants hated the noncompete clause, and some consultants joined hands to make a representation to CEO Kullberg. Some high-performing consulting principals considered leaving Arthur Andersen to start their own practices. Those who left were sued for breach of partnership agreements. In May 1988 a board meeting considered the consultants' petition and withdrew the new amendment. Furthermore, in the winter of 1988 the fast-growing consulting practice was spun-off as the new Andersen Consulting division and it was approved by partners in January 1989. CEO Duane R. Kullberg tried hard to reconcile the two sides, and he finally decided to step down in March 1989.

Lawrence A. Weinbach, at the young age of 49 years and with an impressive track record at Arthur Andersen, took charge of the company as the new CEO. After earning an accounting degree from Wharton Business School and developing an expertise in the high-ticket practice based on mergers and acquisitions, he had become a partner in 9 short years.

Customer Responsiveness as Competitive Advantage. When the customer needs changed, Arthur Andersen was one of the first public accounting firms to meet customers' expectations. This often had a significant impact on the organization structure of the firm. Similar tremors were also felt in other big public accounting firms.

In the late 1980s and early 1990s, many potential customers started globalizing their operations as well as their customer bases. They needed consulting help for doing that effectively. In many competitive bids, Arthur Andersen often won a business contract based on its truly unified worldwide practice. This often was a big issue for multinational corporate customers who needed consulting advice at different sites around the world. Unlike other competitors, which sometimes grew their foreign practices by rapid franchising, Arthur Andersen opened foreign offices with Andersen people, and he trained them at Arthur Anderson's St. Charles facility.

Arthur Andersen also submitted new bids based on a customer's ability to afford and pay for the required services. Over years of association with Arthur Andersen, a client often felt a sense of kinship. They became friends and almost colleagues. The clients often retained Arthur Andersen's services because Andersen was willing to renegotiate its fee structure on a competitive basis with market prices prevailing at a time. Some other consulting firms took for granted that they would continue to get annual fee increases from their audit clients.

Customer Satisfaction Added-Value Study. Arthur Andersen was also on the look out to add more knowledge value to their services.

Here is one example of how Arthur Andersen gained and retained its customers by providing added-value knowledge.

In August 1995 Joe O'Leary, a partner in Arthur Andersen's Business Consulting Practice, concluded a 12-month study with a focus on the best practices in customer satisfaction. This study was part of Arthur Andersen's ongoing research under its Global Best Practices[SM] initiative. Its focus was on the best practices in customer satisfaction, and it was conducted at seven global companies with a reputation for superior customer satisfaction and financial performance. The study indicated that "a single-minded focus on customer satisfaction does not guarantee profitability." For customer satisfaction to be a profitable management tool, it must be integrated with all levels of a company's interaction with the customers.

The successful companies did this in the following ways:

1. They gained an intimate understanding of customers by motivating their employees to understand their needs and wants.
2. They encouraged employees to effectively communicate and interact with customers.
3. They sought customers' input and advice as a partner in improving their existing products and developing new products.
4. They motivated their employees to "own" customers' problems, and they encouraged them to form teams with customers to discuss issues of common concerns.
5. They discovered improved ways to align their own processes to meet the customer needs.
6. They set "stretch" goals for customer satisfaction in the overall objectives of their corporation as well as for their senior management. Such commitment stimulated employees to creatively reinvent their key processes.

Arthur Andersen built its Customer Satisfaction Service as a part of its Business Consulting Practice. The objective was to help medium and large companies to improve their business processes and technologies. Arthur Andersen consultants offered their extensive industry experience and coupled it with the Global Best Practices[SM] Knowledge base. This knowledge base had the cumulative experience of Arthur Andersen's previous interactions with their clients worldwide. As of August 31, 1996, Arthur Andersen had about 80,000 customers serviced by more than 380 offices in about 79 countries and about 100,000 associates. Arthur Andersen Worldwide, the world's largest professional services organization, has had two business divisions since 1989: Arthur Andersen and Andersen Consulting, which together generated

EXHIBIT 4A.1. Arthur Andersen Growth

Year:	1991	1992	1993	1994	1995
Total personnel:	33,085	34,235	36,349	38,936	43,500
Total revenues ($ billion):	2.7	3.0	3.2	3.5	4.1

a total revenue of over $9.5 billion in 1996. In 5 years, the Arthur Andersen consulting division increased its revenue from $2.7 billion in fiscal year 1991 to $4.6 billion in fiscal year 1996 (see Exhibit 4A.1).

Source: For more information, contact Arthur Andersen & Co., 69 West Washington St., Chicago, IL. 60602. For Enterprise Award and Knowledge Network Series call the hotline at 1-800-222-5257; also see Stevens, Mark. 1991. *The Big Six.* New York: Simon & Schuster.

CASE 4B: MANAGING TECHNOLOGY-DRIVEN ENTERPRISE

Federal Express: Definitely, Positively Customer Satisfaction

Frederick W. Smith, the founder of Memphis-based Federal Express, built a multibillion-dollar company around a market gap in customer satisfaction and the need for a reliable service for overnight package delivery. He also promoted the philosophy that the satisfactory service was delivered by the people, and that this was a prerequisite to corporate profits.

Before starting Federal Express, as a young businessman in his twenties Fred Smith ran Arkansas Aviation Inc., a hangar service, aircraft parts, and maintenance company. There, as a customer he was often disappointed and frustrated by the unreliable delivery of parts he ordered. Even when he ordered parts by air express and on an urgent basis, they would arrive anywhere between 2 and 7 days later. Often a million-dollar business aircraft would sit idle in the hangar because a $10 part was not delivered in time. This brought back to Fred Smith the urge to implement a hub-and-spoke idea he had conceived 4 years ago as a college junior for an Economics paper at Yale University.

Freight Forwarders. Before Federal Express started in 1973, the poor reliability of the express delivery service of small packages was rooted in the way this service sector was organized.

Around 1960 the commercial airlines, like American, Delta, Eastern, and United, started using large aircrafts like Boeing 747s, Lockheed L-1011s, and the DC-10s. They therefore decided to carry cargo of small packages booked by freight forwarders on "availability of space" basis. The priority of the airlines was to load their planes to the fullest extent, and not on the service provided to the cargo customers. Customers preferred to deliver their packages late in the evening, by 5 or 6 p.m. They also wanted the packages delivered in the morning, preferably before lunch break, even if the destination was thousands of miles across from the East Coast to the West Coast. Freight forwarders provided only airport-to-airport service, and they preferred to deliver their packages to the airlines late in the evening after 9 p.m. or so. The commercial airlines, on the other hand, flew most of their flights during the daytime. There was mounting civilian pressure on them to reduce off-peak hour flights taking off or landing after 9 or 10 p.m.

The airlines started their cargo businesses as separate profit centers, and they did not generate much revenue from them. With increasing competitive pressure, the commercial airlines were forced to cancel their minor routes and consolidate different flights. One after another commercial airlines reported losses on their cargo operations and started scaling them down. The customers received a raw deal when these operations were cut down. But they had no choice. They were at the mercy of the freight forwarders and the airline executives.

The Hub-and-Spoke Network Concept. In 1965, as a college junior at Yale University, Fred Smith was assigned to write a 12–15 page term paper for his Economics course. He decided to write about a "hub-and-spoke" distribution network for door-to-door delivery of small packages across America. The network service was targeted at customers in high value adding technology-driven industries, who wanted a reliable service for delivering their small packages and documents overnight across America.

Fred Smith explained in his paper that the hub of the delivery network would be located in Middle America, say in Little Rock, Arkansas or Memphis, Tennessee. All the sorting would take place at night at this hub. The "spokes" from this hub would reach out to the major markets around the country, say in Boston, New York City, Washington DC, Miami, Dallas, Los Angeles, Seattle, Chicago, Cleveland, and so on.

Process Mapping. Pickup vans would pick up small packages from the doorsteps of different customers at the end of the "spokes" (say in Cleveland) until 6 p.m. or so. The van operators would load them into planes waiting at their local airport. The planes would take off and land at the "hub" before midnight.

Between midnight and about 3 a.m., the packages collected from around the country (including the one from Cleveland) would be sorted for their respective destinations. These packages would be reloaded into the waiting planes which would take off and land at their destination before dawn. For example, the Cleveland plane would bring back packages bound for Cleveland from customers around the country. If the Cleveland package was bound for Los Angeles, the Los Angeles plane would take the Cleveland package, along with other packages bound for Los Angeles. The delivery vans waiting at the Los Angeles airport would take the sorted packages and deliver at the doors of customers. Since the "hub-and-spoke" arrangement was dedicated to the overnight delivery of packages only, there would be no delays because of passengers, as in the case of cargo service provided by the commercial airlines.

Fred Smith did not get a great grade for his innovative idea. But that could have been because of the form and style of the term paper rather than its innovative content. He could have improved his grade by linking the "hub-and-spoke" concept with "the traveling salesman" or linear programming theory.

Market Feasibility. Seven years later, in 1972 on way to seeking investors for his overnight delivery business, Fred Smith hired two New York–based market research companies. Aero Advanced Planning

Group (AAPG) and A. T. Kearney & Company were to evaluate the feasibility of the "hub-and-spoke" concept, as well as confirm customer acceptance for the overnight delivery of packages. Both of the consulting groups charged $75,000 and confirmed the market gap Fred Smith had postulated. They estimated a potential $1 billion demand gap for on-time express delivery of small packages. Their research indicated that customers needed a door-to-door delivery service and were willing to pay a premium price for a guaranteed delivery service.

Launching a New Service Industry. On March 12, 1973 Federal Express inaugurated its overnight delivery service. They could spare only six planes to service six market sectors in Little Rock, St. Louis, Kansas City, Cincinnati, Dallas, and Atlanta. They targeted freight forwarders and large volume shippers through local trade publications. This was the norm of this service industry in 1973. During the test-launch they collected a total of only eight packages. It was clear that the customers would not use their overnight delivery service unless there were more cities in the network.

The FedEx managers redesigned their service network and expanded it to include 22 cities. New cities like Columbus, Miami, New Orleans, Pittsburgh, and Rochester, New York were included. The service was relaunched on April 19, 1973. This time a more reasonable total of 183 packages was collected, sorted, and delivered.

Gaining Customer Trust. As customers developed trust in their service, the number of packages increased steadily. By the end of the first year, Federal Express was servicing at the rate of about one million packages per year. This was still below their break-even point. In fact until 1975 Federal Express lost money on their innovative delivery network. In 1975 their net loss was $11.5 million or about $1 million a month, somewhat smaller than a loss of $13.4 million in 1974.

In 1975 the marketing managers at Federal Express carefully reviewed their performance from their customers' point of view. They realized that FedEx was too concerned with "selling" freight-forwarders its process of collecting and delivering of packages. FedEx was not focusing on the needs of their customers. They launched a fresh campaign with a focus on the ultimate consumer of their service. Federal Express realized that anybody in a business organization could be their potential consumer for the overnight delivery service. They decided to broaden their target market from just the "freight-forwarders" and shipping clerks to everyone in the organization.

A new campaign was developed around positioning the FedEx service as the "absolutely, positively overnight" service for anyone in an organization looking for reliable guaranteed service. This became one of the most successful advertising campaign of its time.

Customer Satisfaction and Market Success. The new campaign was a hit with a broad base of direct consumers. The overnight delivery shipments for FedEx started growing at an unprecedented rate of doubling every 2 years. The annual revenue of Federal Express increased from $17.3 million in fiscal year (FY) 1974 to $43.5 million in FY1975 and $75.1 million in FY1976. And in FY1976, for the first time Federal Express turned in a net income of $3.6 million.

The customer focus by the company brought in significant new customers and revenues to the young company. The annual revenue of Federal Express exceeded $75 million in FY1976, $100 million in FY1977, $160 million in FY1978, $250 million in FY1979, and more than $415 million in FY1980.

By the end of the 1970s, even after the second oil shock when the price of one barrel of petroleum oil went sky high, Federal Express was an undisputed business success. This brought in new entrants to competition offering matching overnight delivery services. For example, the U.S. Postal Service leveraged its existing elaborate network of post offices and post boxes, and it offered express service for the customers who dropped off their packages at a post office. After a rate hike, this express service was priced at $10.75.

Federal Express decided to compete head-on with their new big competitor. In 1 year Federal Express opened 60 FedEx collection centers in 28 major market areas. These centers were located in areas convenient to a large number of their customers. Customers were offered a $3 discount coupon for dropping off their packages at a FedEx pickup center. An average package price was $11, barely a quarter more than that of the U.S. Postal Service. Initially, the dropoff service offered by FedEx only cannibalized its pickup customers.

Educating Customers and Employees. In September 1985 FedEx launched a new campaign called "Batting the Bird." This campaign was targeted at the customers as well as the FedEx employees. The employees were considered a primary asset to gaining customer satisfaction for FedEx. They were very actively involved in the launch and implementation of the new campaign. Employees signed and sent 'thank you' cards to customers using the information provided by the customers on their discount redemption coupons. Employees were given detailed information packets with competitive comparisons between FedEx and other competitors. At each collection center, a TV monitor and a VCR was installed explaining the "Battle the Bird" program.

Employees quickly realized how important they were in the mind of the FedEx management. Their morale was high, and within 6 months the new campaign added 4.3 million packages and an additional $68 million in revenue.

People Produce Profits. The founding CEO Fred Smith, from the very beginning of starting Federal Express, believed in hiring good people who were willing to work together and carry their weight without creating elaborate bureaucracy. There was a high level of "U.S. Marines-like esprit de corps." As a Marine, Fred Smith had learned about leading his troops with personal courage. Even though he was much younger than many CEOs, he demonstrated the unique ability to understand and deal with each person individually.

At the Memphis hub the sorting of incoming packages was done mostly by part-time college students. To keep schedules in control, first the managers tried all kinds of control mechanisms. Yet every now and then the sorting would run late, because that meant workers would earn more money. Then FedEX announced that if the workers finished their work early before a certain time, they could go home early and still get full wages. Within days, the sorting was completed ahead of schedule.

"People–Service–Profit" Priority. The Federal Express philosophy was always "people–service–profit," in that order. Fred Smith believed that at Federal Express, decision-makers always balanced and considered the needs of employees, customers, and stockholders. In making decisions, the effects on people were taken into account first. This was done with the understanding that if the company took care of its employees, they would deliver a superior service which the customers will repeatedly and frequently use. This would then add to shareholder wealth and profits. Only by making profits would Federal Express continue to exist, tap new opportunities, and employ people.

Many employees who had joined Federal Express in its early years had a pioneering spirit. For many years Federal Express had a "no layoff policy," and most of the promotions were made from within. There were no unions, yet employees were paid competitive wages, with some profit-sharing benefits. The company's corporate thrust was in three principal areas:

1. Training and motivation of employees to improve productivity.
2. Intensive marketing to create broad awareness and high volume demand for air express service.
3. A continuous investigation of all new technological developments which could help improve the expanding Federal Express operations.

To keep the employees satisfied, corporate managers were asked to visit different field centers. Employees were polled for their confidence in the fairness of management, and whether their jobs were leading to the kind of future jobs they wanted. Fred Smith started a U.S. Marine-like "Bravo Zulu" campaign for "the job well done" by an employee. An in-house newspaper candidly discussed the company's progress so far, the problems it was facing currently, and the future prospects.

National Quality Award. In January 1989, Fred Smith was so confident about the satisfaction of Federal Express customers that he decided to apply for the Malcolm Baldrige National Quality Award in the service category. A 1000-man team was organized to internally audit the actual level of service and customer satisfaction. New high targets were set. A new 12-item Service Quality Indicator (SQI) was used to help monitor the on-time service performance on a day-to-day basis. These items were divided into three groups, with each group assigned a different weight, depending on the customer's view of FedEx service.

The first group with a maximum 10-point weight measured (1) damaged packages, (2) lost packages, and (3) missed packages. The second group, with a 5-point weight, included measurement of (1) complaints reopened, (2) lost and found overgoods, and (3) wrong-day late deliveries. The last category, with a weight of 1 point each monitored (1) abandoned calls, (2) international issues, (3) invoice adjustments requested, (4) missing proofs of delivery, (5) right-day late deliveries, and (6) traces.

In 1990 FedEx entered the U.S. Department of Commerce quality competition. They were selected for site visit and were carefully audited by experienced examiners. In October 1990 FedEx was announced as the winner of the National Quality Award in its service category.

The founding CEO Fred Smith was particularly proud of one particular citation by the examiners. The examiners commented that in the previous 5 years, 91% of FedEx employees said that they were proud to work for Federal Express.

Source: Adapted from George, Stephen and Weimerskirsch, Arnold. 1993. *Total Quality Management.* New York: John Wiley; and from Trimble, Vance. 1993. *Overnight Success: Federal Express and Frederick Smith, Its Renegade Creator.* New York: Crown Publishers.

CHAPTER 5

Promise of Quality: Products, Processes, and Management

PREVIEW: WHAT IS IN THIS CHAPTER FOR YOU?

High quality of products and reliable processes have become the minimum expectations of customers during the past few years. Technology-driven enterprises must provide a competitive quality promise or they could lose their market shares to competitors who proactively do the same.

In this chapter we will develop a definition of quality using four P parameters. The changes in the perspectives of quality from producer-defined quality to the competition and customer-defined quality will be reviewed. The enigma of total quality management will be demystified using the nine lives of quality management movement. The concept of the quality value chain will be proposed to metasynthesize the different champions of quality who pioneered new frontiers. This includes Frederick Taylor's inspected quality, Deming's statistical process control, Ishikawa's prevented quality, Taguchi's design quality, Xerox CEO David Kearn's competitive quality, and Mazda Miata's creative quality. This will be followed by a description of continuous quality improvement using different quality tools. Finally, a number of quality assessment tools will be described, such as the Malcolm Baldrige National Quality Award, ISO 9000 series of standards, ISO 14000 environmental standards, and auto makers' QS standards.

QUALITY FOR COMPETITIVENESS

Do quality-related efforts add cost to the production operations? Or does pursuit of quality reduce wastage, improve product appeal and customer satisfaction, reduce product related liability costs, and thereby improve competitiveness? As in the case of many other management practices, the key to success lies in the way the quality-related efforts are implemented. Disjointed and "program of the month" type of quality-related efforts add expenses but do not contribute additional profits to the bottom line. On the other hand, sincere efforts with active involvement of all layers of an enterprise, including the uppermost echelons, can revitalize the entire organization and provide significant competitive advantage.

Motorola, the world leader in mobile communication technology, gained and sustained its world-class leadership because of the quality-related commitment provided by Chairman Robert Galvin. To focus the attention of all employees on its quality, Motorola's managers used the Six Sigma process reliability program, wherein the defects (or values outside the specified limits) are kept below the level of 3.4 per million opportunities.

Xerox, the pioneer of photocopying technology, also used a quality-based "Leadership Through Quality" program to regain its competitive position under assault from the Japanese photocopier makers. CEO David Kearns used a "cascade method" of training to educate thousands of Xerox employees in the basic quality tools and problem-solving techniques.

These companies, and many more, used their sincere quality efforts to gain market share, growth, reputation, and competitiveness in their highly competitive markets. But before we discuss various quality management practices, let us define what we mean by quality.

QUALITY MOSAIC

During the 1990s, it became very clear to manufacturers and service providers that the customers had come to expect very high levels of quality from the goods and services they purchased and received. This was expected no matter what the price was. With increasing intensity of competition, by the mid-1990s, quality became a minimum requirement, rather than something which customers appreciated when provided and did not mind when not available. But before we explore this subject deeper, we must have a clear understanding of what quality stands for.

Defining Quality

According to *Webster's New Dictionary*, quality is "that which makes a thing what it is; nature; kind or degree of goodness or worth; attribute; degree of excellence . . . ," and so on. Such a definition, though comprehensive, does not help very much in resolving a heated dispute over quality of a product that may take place between a customer and an auto producer. Nor does it provide help in a conflict over contractual obligations of a warranty between an auto maker and its suppliers. However, as we will soon see, the actual operationalization of such a universal definition of quality in an organization has varied from time to time. This has resulted in a complex and enigmatic mosaic of quality-related management paradigms.

A common definition of quality used in professional circles is "fitness to use." A quality product is fit to be used in its intended use. Thus, a quality car is fit and ready to run for its common use of normal passenger transportation. A computer should be fit to be used for word processing or Internet surfing when it is turned on.

This definition raises another question. Who decides the intended use: the customer or the producer? The customers would like to see their cars run forever through cross-country, as well as on icy roads under rain, storm, snow, or sleet. The manufacturer, on the other hand, would prefer to have the customers buy their new car model every year and also buy another 4 x 4 four-wheel automobile for cross-country driving. There is therefore a modified definition of quality: "conformance to requirements." The producer is willing to offer a product with a quality promise that meets the mutually agreed requirements. If the customer requires cross-country driving, the customer must pay for the sports utility vehicle.

Perspectives of Quality

The definition of quality has evolved over time. In the past, the promised quality was exclusively defined by the perspective of producers of goods or providers of services. Henry Ford offered a very reasonably priced Model T car in "any color" as long as it was black. Certain services were denied to some segments of the society because the owners decided to do so.

With the entry of new competitors, the quality focus shifted from producer-defined quality to the second: product quality perspective. The products' features and ability to satisfy a need defined their quality. This had another flaw. By the time the finished products were produced and their quality was inspected, they had already consumed considerable amounts of resources. The machines were used, and manpower was consumed in producing even the defective goods. Many times it is too late to discover the bad quality in the finished products. The better way was to produce quality goods the first time. Therefore, over the years, the focus of quality shifted from the product quality perspective to the robust quality of processes used to produce quality products. The underlying belief was that a robust process perspective produced a quality product consistently, even when there were variations in the quality of inputs.

Eventually, many senior executives started demanding that quality efforts must lead to gaining a competitive advantage over competitors. Thus, the quality was defined by its contribution to competitive performance in the market. This fourth competitiveness perspective of quality determined the allocation of resources and other decisions. This evolution will be discussed later. First, we must return to the different parameters that define quality.

Parameters of Quality

Let us consider an example of a simple product of daily use: the paper plate. Most of us use paper plates when we go on a picnic or when we host a party for a large gathering of people. Paper plates are handy because they do not require cleaning after they are used. They can be thrown away after the use. What are the attributes of a quality paper plate? The following attributes come to mind.

1. Light but strong
2. Liquid resistant
3. Tear-resistant
4. Low price but not cheap-looking
5. Compact
6. Easily available
7. Pretty
8. Easy to dispose of
9. Environmentally friendly
10. Available in different sizes
11. Available in different shapes and patterns
12. Should not bend easily
13. Should not cause cuts

We can perhaps go on and add many more desirable attributes of a quality paper plate. We notice that some of the attributes are in conflict with each other. We want the paper plate to cost less, but we do not want it to look cheap. We want the paper plate to be strong, but it should not be too heavy to transport or carry while eating.

The American Society of Quality Control (ASQC), an association of quality professionals, has defined quality as "The totality of features and characteristics of a product or service that bear on its ability to satisfy given needs." This definition would expect us (as providers of quality promise) to develop a list of attributes, such as the one we developed earlier, and hope that the totality of these attributes would satisfy our target customers. The challenge is to identify the totality of different attributes that define quality completely.

The Four Ps of Quality Parameters

Professor David Garvin of Harvard University, in his 1987 article in the *Harvard Business Review,* offered us further insight into the definition of quality. Garvin defined quality products or services by eight dimensions. These dimensions can be classified into four Ps of quality parameters.

P1. Quality of Product/Service Performance. Each product or service is expected to "perform" its function by meeting certain needs. The performance of a product or service can be divided into the following two subgroups.

P1.1. Core Function. In the case of the paper plate example, the product is expected to help us carry our food. Paper plates do that because of the thickness of the paper they are made from. A car is expected to take us from one place to another. Cars have four wheels, an engine, and a transmission connecting them. These mechanical parts of a car are covered with comfortable seating. A computer is expected to help us word process our documents for research papers, run our spreadsheets, and surf the Internet to retrieve useful information. It should do so with a fast microprocessor, enough hard-disk capacity, and suitable random access memory (RAM). A microwave oven helps heat foodstuff and make popcorn in a few minutes. These are some examples of the core functions these products are expected to perform.

P1.2. Supplementary Features. From most products and services, we expect certain extra bells and whistles than just their core functions. In the case of paper plates, we want them to be light. Cars should have air bags, antilock brakes, extra coffee cup holders, and so on. Sometimes these bells and whistles satisfy a customer much more than the core functional features of a product or service. We prefer a tower instead of the flat computers which take up a lot of desk space.

P2. Quality of Process and Design. Even though users are not directly concerned with how a product (or service) is assembled or produced, the production process and design define a significant part of the quality of products and services. At least three quality attributes arise out of process and design of quality products and services.

P2.1. Reliability. This attribute provides a measure for the likelihood of repair. Paper plates are disposable items, and we do not expect to repair them. If a paper plate is defective—that is, it leaks or rips apart—the host apologizes to the guest and quickly replaces it with a new one. But cars and computers are different. We expect them to run without requiring frequent repair. Cars should not have parts falling apart as they are driven out of the dealers' showroom. In fact, Japanese auto makers take pride in providing cars with minimal repair required during the first 5 years. Auto makers offer warranties for a certain number of years and/or a certain number of miles. During this period of warranty, the auto manufacturers would promise to do all the necessary repair. Most manufacturers of electronic appliances would provide at least 1 year of warranty for any repair required during that period. Most products come with some warranties.

P2.2. Serviceability. This attribute measures the ease of repair should the purchased product or service fail to keep its promise of reliability and require repair. With miniaturization, as well as reduction in the number of parts in electronic appliances, it is generally becoming harder to repair products. Canon photocopiers therefore combined all the moving parts of a photocopier for small home offices into a copying engine. In case of the need for repair, the small business owners are recommended to replace the entire engine subassembly. The replaceable copier cartridge also improves the ease of servicing.

P2.3. Durability. Life of a product or service is an important quality attribute. We expect quality products to last long. The light bulb designers have struggled with this question a lot. What should be the life of a light bulb? If a light bulb lasts for a short time, the customer is not likely to purchase the same brand again. On the other hand, if a light bulb lasts forever, it means lost revenue for the bulb manufacturers, particularly because the customers are not likely to pay a large price for the initial purchase of a light bulb. After much market research, bulb manufacturers have arrived at the optimum minimum life of 100 hours. If a bulb fails in less than 100 hours, the customers complain about it and do not buy the same brand again. On the other hand, if a bulb lasts longer than 100 hours, then the customers do not particularly recall when they installed the bulb in the first place. The next challenging question for the producers is, should the bulb last long, or save energy, to delight the customer by exceeding his/her expectations? Some light bulb manufacturers have developed energy-conserving light bulbs that last about 15% longer. These are therefore recommended for outdoor use in hard-to-reach places.

P3. Quality of Promise Conformance. This attribute of quality measures how well a product or service is able to meet its promise. In the case of industrial products, the quality promise is clearly spelled out in the product specifications. The manufactured product is expected to match perfectly with the previously agreed specifications. The same is the case with custom-building of a home. The owner and the contractor agree on the plan and the fittings of the house. The contractor must meet that specification fully and meet the quality promise. The contractor cannot add more things on his own and expect the home owner to pay for the additions. In the case of consumer goods and services, where products are not produced to customized specifications, the quality promise is measured by how well the product is fit for its intended uses. In other words, does the product's ability match the needs for which it is expected to be used?

P4. Perceived Quality. This subgroup of attributes of quality assesses the personality of the products or services. This includes touchy-feely aspects of a product or service. These are generally hard to measure, but they make a significant impact on the purchasing pattern of a customer.

P4.1. Confidence Generating. Certain products and services, along with their providers, provide a sense of confidence. An IBM product or service gives a perception of certain kind of dependability. Customers believe that should anything go wrong, say in Thinkpad laptop computers, Big Blue would stand by its products and services and fix the problem promptly. In a similar manner, Sony of Japan gives a sense of confidence that its Walkman, as well as its other types of audio equipment will include the state-of-the-art sound technology. Federal Express, a relatively young company providing overnight delivery service, based its business concept on the "definitely, reliably" overnight delivery of customers' precious packages.

P4.2. Aesthetics and Looks. In the case of some products, aesthetics and the way a product looks or feels defines its quality. European cars and fashion are expected to have that unique aesthetic look worth their premium prices. Swiss watch-makers compete on the feel and look when competing with high-tech digital watches from Japan, Hong Kong, and so on. Prior to World War II, the Japanese products had a "cheap" and "shoddy" image. During the 1970s and 1980s, the Japanese manufacturers of electronic appliances, cars and motorcycles, interested in penetrating the global markets, had to overcome that perception of poor quality.

Together these "four P" parameters and the eight attributes provide a comprehensive definition of quality for products and services. A useful exercise will be to brainstorm the four P parameters for defining the quality of laptop computers. We should try defining these four Ps of quality for services such as restaurant service, hotel hospitality, health care facility and so on. However, in order to fully grasp the current state of understanding of quality promise, we must first consider how our understanding of quality management has evolved over the past few decades.

FRONTIERS OF QUALITY MANAGEMENT MOVEMENT

Quality Value Chain

During the 1980s, as the quest for quality spread through the different U.S. industries, many managers were exposed to a variety of terms used in reference to quality promise. This variety often confused some busy managers. We will therefore review the historical evolution of quality movement as it spread across the world, and we will identify the champions who discovered new frontiers of quality from time to time. Each champion of quality formulated a new paradigm based on his own unique background and perspective.

This brief review will show how the U.S. managers in quest for quality have often gone offshore, particularly to Japan, to seek additional perspectives. It will also show that foreign quality champions such as Ishikawa, Taguchi, and others have contributed as significantly to quality management movement as did their counterpart U.S. gurus such as Deming, Juran, Crosby, and others. However, even though the U.S. companies seemed slow to respond to new quality initiatives, the Japanese gurus have often humbly admitted that they were inspired by the practices developed in the U.S. industries.

From an implementation point of view, each frontier of quality will be integrated in a quality value-chain framework (Note 5.1). From this new framework it can be seen that successive frontiers of quality sought bigger challenges by attempting to integrate domains farther removed from the core production or operations of an organization.

Resolving the Enigma of Quality

The interest in quality has spread fast in Corporate America. Since the introduction of the Malcolm Baldrige National Quality Award in the United States in 1988, the requests for its applications quadrupled yearly from 12,000 in 1988 to 51,000 in 1989, and then it tripled to 180,000 in 1990. During the 1990s, a number of criticisms started appearing about the lack of effectiveness of such quality-related programs (Note 5.2). Such conflicting evidence implied that the general understanding about quality had become increasingly confusing and enigmatic. For some new players to a rather pretty old game of quality (starting in World War II), the utility of pursuing quality and what it stands for was becoming very confusing.

Different people's perceptions about quality tend to vary with their respective contextual backgrounds and expectations. To the customers, quality may stand for dependability and desirability of products or the services that they have paid for. To the manufacturers, quality is likely to be a reduction in variability of goods and deliveries made by their suppliers. To the suppliers, quality is reliability of designs and specifications that they get with their orders from makers, as well as how promptly they receive their payments from the makers.

Besides grappling with these different and often competing expectations of quality, an organization and its managers may also wonder when and how a pressing need for quality first emerges, and whether this need changes over time. They must assess, in a dynamic manner, their customers' current expectations associated with quality of their goods or services. There are other quality-related dilemmas involved. Does one have to catch up and match up with the quality offered by competitors? Or, is it beneficial for an organization seeking a sustainable competitive advantage to go beyond their competitors' quality offerings and their customers' expectations?

One must answer some of these questions to resolve the enigma about quality. In this chapter we will try to untangle this enigma of quality by systematically tracking the historical evolution of quality movement. We will look at the roots of quality movement by identifying different champions who spearheaded and discovered new frontiers of achieving and improving quality. We then offer to metasynthesize their diverse contributions by proposing a "quality value chain" framework, based on a common core dimension of type of integration proposed by each champion in his quality paradigm. Exhibit 5.1 gives a metaperspective of overall quality movement, with nine frontiers of quality pioneered over the past few decades. These

EXHIBIT 5.1. Frontiers of Quality Management Movement

Frontier Champion	Quality Value-Chain Frontier
1. Taylor	Inspected quality
2. Deming	Statistical process control quality
3. Juran	Company-wide quality
4. Feigenbaum	Total quality control
5. Ishikawa	Prevention quality
6. Taguchi	Design integrated quality
7. Crosby	Cost of quality
8. Kearns/Xerox	Customer integrated quality
9. Mazda	Market creation quality

Source: An earlier version was published in Gehani, R. Ray. 1993. Quality value-chain: a meta-synthesis of the frontiers of quality movement. *Academy of Management Executive,* 7(2):29–42.

quality frontiers span from the early developments by Frederick Taylor in achieving quality by inspection, to the newest frontier based on Mazda stressing quality based on innovation and creativity.

We hypothesize that the solution to grasping the full meaning of quality lies in understanding this multiple-layered mosaic of quality which has developed as a result of superimposition of different frontiers of quality. Some of these frontiers of quality were developed and pioneered in other parts of the world, under conditions different from those prevalent in the United States. An obvious caveat for such a Herculean attempt to metasynthesize a managerial revolution in one chapter is to acknowledge that it is impossible to recount contributions of every single researcher ever involved in the field of quality and its management. We must also recognize that many quality champions have amended and extended their original observations over years. Some of their contributions have merged over time. We therefore list only the typical contributions made by each champion and cite these quality champions for the purpose of illustration rather than for their comprehensive coverage (Note 5.3).

For a global organization competing in the East or in the West, a wider and consistent buy-in by the different segments of an organization will depend on a comprehension and convergence of the different meanings of quality. Professor Michael Porter of the Harvard Business School, in his efforts to articulate competitive advantage and competitive strategy, has used the concept of "value chain" (Note 5.4). According to the value-chain framework, different activities of an organization can be classified as primary and secondary value-adding activities. The primary value-adding activities are those activities which directly contribute to the core transformation process, marketing or delivery of outputs of an organization. And the secondary activities are all the supporting activities supplementing and providing infrastructure for the primary value-adding activities. The quality of a product and its associated value, either as assigned by its producer or as perceived and paid for by its customer, are closely related. Thus, this value-chain framework has been used here to bring about a metasynthesis of contributions of various champions of quality paradigms. The "quality value-chain" framework proposed here provides a metasynthesis and an integration between the product quality and other operating variables of an industrial enterprise. The value-chain framework for quality can be systematically developed by introducing and interpreting the contributions made by different quality champions.

The Quality Management Movement

Over time, the quality management movement has changed its form and scope. The "natural" birth of the original quality movement took place in the United States as a result of the emerging needs of rapid industrialization in the early part of the 20th century. Modern quality, as it is commonly understood today in global markets, was nurtured most effectively in the 1950s and 1960s by the Japanese manufacturers obsessed with a desire to recover from their postwar destruction, and a prewar image of "producers of shoddy goods." After World War II, the Japanese producers and suppliers united as a nation and used the engine of quality to catch up with the Western levels of industrialization and quality of goods.

In the United States, the birthplace of quality, or in Japan, the place of its most widespread growth, the quality movement steadily evolved from one form to another. In the United States, quality evolved from strictly a shop-floor issue that was considered by the sectional foreman in the 1940s, to become in the 1990s a key component of strategic planning and mission defined by the upper echelons of an organization. Similarly, in Japan, quality emerged as an economic weapon to help Japan sustain its global leadership in one high-tech industry after another. In Japan of the 1990s, the traditional competition-based view of quality was considered obsolete and obvious or *atarimae hinshitsu*. Since the mid-1980s, the Japanese manufacturers started concentrating their efforts on (a) *mirhyoku-teki hinshitsu* or admirable quality and (b) market creating quality of products and services.

Classical Craftsmanship Quality

Before we consider the role of quality in modern industrial production and manufacturing practices, it is interesting to point out that prior to the industrial revolution, the classical craftsman produced goods using a highly integrated approach. The craftsman interacted directly with the patron customer, who was either a mercantile buyer or a consumer, to assess needs and expectations of goods desired to be produced. The quality of a craftsman's output also embodied integration of creativity, managerial decision-making, cost control, and all those other aspects of quality management which large organizations today strive very hard to integrate. Such process-wide integration is the primary objective of the most modern advanced manufacturing technologies (AMTs) using flexible manufacturing systems (FMS). It is therefore important to consider how quality-related managerial practices changed over time and traversed a full circle with respect to the degree of integration in a production process.

There was a downside of the craftsman's all-round integration. Such an integrated effort produced only a limited number of goods per craftsman. The productivity was low. With market success and growth in demand, the craftsman was forced to disintegrate the production process and specialize in operations. He was forced to delegate more and more functions to others. One approach to meet such growth was by the functionwise simplification of tasks and departmentalization of organization. The Industrial Revolution ushered in a major paradigm shift with mass production centered in "manu-factory," where a worker-craftsman worked for a manager-owner rather than for himself or herself in a household.

Nine Lives of Global Quality Management Movement

1. The Tayloristic Frontier: Inspected Quality. The birth and blame for America's age-old practice of attaining quality by inspection has been increasingly attributed to Frederick W. Taylor, the father of "scientific management." In 1911, Taylor, an industrial engineer by training, innovated the field of time-and-motion study. He proposed that by using a systematic analysis, any operation could be divided into simpler tasks. Each of these tasks could be performed by a predetermined "one best way." In Taylor's scheme of industrial management, workers were to be assigned these simplified and standardized tasks, while a supervisor and manager's job was to plan and inspect their performance (Note 5.5). Thus, the supervisors also integrated postproduction inspection with the quality of "scientifically" produced goods.

While some management theorists are still debating the validity of Taylor's data and his findings (Note 5.6), most industrial planners concur that the woes of declining U.S. industrial competitiveness in the 1980s originated in Taylorism and its low concern for integration. According to Myron Tribus, former director of the Advanced Engineering Center at Massachusetts Institute of Technology and later the head of the American Quality and Productivity Institute (AQPI), the blame for America's quality downfall lies squarely on Taylor's theories of scientific fragmentation of work practices, particularly production of goods and their postproduction inspection.

At the turn of the 20th century, Taylor's scientific task differentiation produced revolutionary increases in productivity in different industries. These practices were immediately and so widely accepted by the American industry that the U.S. Congress had to investigate labor unions' contention that "the scientific management" dehumanized the workers to machines (and robots, which were yet to be discovered). Yet the American industry continued to adopt Taylor's methods because they facilitated mass production and growth—the strategic needs of that era. And the "scientification" of the production process into simplified tasks helped manufacturers to use low-skilled immigrant labor.

2. Deming Frontier: Statistically Process Controlled Quality. Initially, many industrial managers considered quality of produced goods indirectly, as a component of effective supervision issue. As industrialization proceeded, variability in the quality of produced goods demanded management's more direct attention. It was quickly recognized that a worker was responsible for only a limited fraction of the total defects related to product quality. A larger number of product defects came from production process variations and other constraints not directly determined by the workers.

In efforts towards elaborating such a seminal discovery, statistics was the first field of scientific knowledge which contributed towards a systematic development of quality-related practices. The role of statistics in quality was popularized by Dr. W. Edwards Deming, a statistician with a Ph.D. in mathematical physics from Yale, who worked in the 1920s at Western Electric's Hawthorne plant in Chicago. In the 1930s, Deming moved to the U.S. Department of Agriculture and was influenced by a Bell Laboratories' statistician, Walter A. Shewart. Shewart's seminal 1931 book, *The Economic Control of Quality of Manufactured Product*, converted Deming to Shewart's preachings about the critical role statistics could play in controlling the quality of manufactured goods. In 1940, Deming worked at the U.S. Bureau of Census developing sampling techniques. During World War II he helped U.S. defense contractors use statistical quality control to discover "systematic" defects in defense production operations.

After the atomic defeat and destruction of Japanese industries in World War II, Deming was invited to Japan by the U.S. Secretary of War to help conduct a population census of Japan. In 1950, the Union of Japanese Scientists and Engineers (JUSE) invited Deming to present his principles of statistical quality control to Japanese business leaders. Deming became so committed to improving the standards of Japanese goods, and he was so impressed by the sincerity of the Japanese to learn about quality, that he donated his royalties from a book based on his lectures in Japan, to fund prizes for quality improvement. This resulted in the establishment of (a) the Deming Prize for individuals and (b) a Deming Application Prize for the company with the greatest improvements in quality of all phases of its business during the past year. Over the past three decades, Deming award ceremonies contributed very significantly, by providing highly visible incentives for organizations and managers to contribute towards Japan's national quest for quality leadership.

Deming's paradigm for quality was based on statistical process control and reduction of process variations. Deming repeatedly demonstrated this explicitly with his red-bead experiment. In this highly popularized experiment, he mixed a few red beads (representing defective goods), into a bowl full of white beads (representing quality goods). He then asked different supervisors to perform the impossible task of picking a spoonful of white beads without getting any red beads in it. Deming demonstrated that workers cannot be held responsible for poor-quality performance if they were not permitted to monitor and control their manufacturing process. And, in order to control the process, Deming suggested the use of statistics in production planning and control. The product quality was thus integrated to the production process control, and the Deming Frontier can be classified as *process control integrated quality.*

Subsequently, Deming also recommended that to reduce variations in process operations the participation and leadership of management is essential. He proposed his "14 points" to elaborate his management philosophy (Note 5.7). These points included a stress on institutionalized statistical training, a constancy of purpose for continuous improvement, organizational building of pride and self-esteem in workmanship, and so on. One of his key points, was to drive fear out of the workplace. These subsequent amendments were attempts to build a paradigm for *people management integrated quality.*

Unfortunately, while Deming was being heard across the Pacific Ocean by the Japanese managers, supervisors, and workers in different industries, most of the U.S. organizations in their postwar economic boom overlooked the potential of Deming's work and philosophy. During the 1960s and the 1970s, the U.S. industries continued to produce goods in large volumes with low quality and sold them to customers with limited choices. After a lag of about 30 years, in June 1980, an NBC white paper documentary entitled "If Japan Can . . . , Why Can't We?" introduced W. Edward Deming who was in his late seventies, to the wider American population for the first time. The documentary described Deming's pioneering work with the Japanese, and it resulted in many invitations for him from organizations such as GM, Ford, Western Electric, and others. Until a ripe age past 90, Dr. Deming operated very actively from his modest home near Washington, D.C., delivering lectures to a wide variety of interested groups.

3. Juran Quality Frontier: Company-wide Quality. Joseph Juran, a U.S.-based consultant, approached quality as cross-functional integration for company-wide management of quality by all (horizontal and vertical), segments of an organization. In early 1950, Juran looked at quality as an integration issue, rejecting the more traditional Tayloristic practice of specialization, differentiation, and delegation of responsibility for quality to a quality control department only. Juran distinguished between the control/inspection-type status-quo management from the managerial breakthrough with continuous improvement. He demanded "hands on" leadership and involvement by senior management. Juran divided quality management into a trilogy: quality planning, quality control, and quality improvements. These were recommended to be integrated in a highly structured manner. He recommended an ongoing rather than piecemeal improvement in quality. He stressed that quality breakthroughs cannot be brought about by an operational work force, but can only be ushered by the upper echelons of management. The upper managers and their decisions accounted for more than 90% of quality-related problems.

Juran was also invited to Japan in 1954 by JUSE, 4 years after Deming, and contributed significantly to Japan's quality-based global success. JUSE asked Juran to teach Japanese executives the secrets of managing quality. Juran proposed that quality improvement should be an integral and essential part of the business plan of an organization and should be linked to each manager's annual performance review.

During the 1950s, 1960s, and 1970s, while the Japanese were eagerly embracing the ideas of their foreign gurus like Deming and Juran, very few U.S. companies paid much attention to their native countrymen preaching quality. In the quality management movement in the United States, AT&T, Corning Glass, Hewlett-Packard, IBM, Eastman Kodak, and Westinghouse are often cited as the traditional leaders due to their early quality consciousness. Managers at Eastman Kodak of Rochester, New York, initially sought out both Deming and Juran, but eventually hired Juran as their preferred consultant. Compared with Deming's general philosophy, Juran had a higher appeal because he focused on Kodak's specific problems and helped Kodak Park managers develop tangible solutions. According to an Eastman Kodak spokesperson, "Deming is more the practitioner, and Juran is the theologian."

4. Feigenbaum Frontier: Total Quality Control. While Deming and Juran were busy in Japan, Armand V. Feigenbaum, while still a doctoral student at the Massachusetts Institute of Technology, originated the concept of total quality control (TQC). In 1951 he completed editing the first edition of his treatise entitled *Total Quality Control.* He elaborated on his concept of TQC as follows:

> Total quality control's organization-wide impact involves the managerial and technical implementation of customer-oriented quality activities as a prime responsibility of general management and the mainline operations of marketing, engineering, production, industrial relations, finance, and service as well as of the quality-control function itself.

Feigenbaum argued that the traditional quality control programs were limited and narrowly defined in one functional area only. He stressed that the total quality control must include "most importantly the interdependent quality activities throughout the organization." He identified the eight stages of an industrial cycle: (1) marketing, (2) engineering, (3) purchasing, (4) manufacturing engineering, (5) manufacturing supervision and shop operations, (6) mechanical inspection and functional test, (7) shipping, (8) installation and service. (These correspond to primary activities in value-chain framework proposed in this chapter.) Feigenbaum considered quality as determined by the customer and not by an engineer, a marketer, or a general manager. He supported this by pointing out that "much of the quality improvement lies outside the work of the traditional inspection and test-oriented quality control function, such as in product design, basic production process, and scope of service, etc." (Note 5.8).

5. Ishikawa Frontier: Prevention Quality. Kaoru Ishikawa, the late president of Musashi Institute of Technology, an institute located on the outskirts of Tokyo, shifted Japan's quality focus away from monitoring and control to prevention of defects in production operations. He did so because of Japan's excessive dependence on imports for most of its raw materials. Ishikawa targeted his focus towards prevention of production of defective goods in the first place. According to Ishikawa, "Quality begins with education and ends with education." He urged that in order to prevent production of defective goods in the first place (and save misuse of materials and manpower therein), there was a need to continuously collect more information and to develop a better understanding about processes and their outcomes.

To implement and map critical issues related to "total quality control," Ishikawa popularized the use of a cause-and-effect diagram or fishbone diagram, now often referred to as the *Ishikawa diagram.* See Exhibit 5.2 on page 222. In a typical cause-and-effect diagram or Ishikawa fishbone, all the major and supporting causes can be identified systematically for each quality-related problem or effect. The effect is written in the head of the fishbone on the right-hand side. The backbone is drawn from the head to the tail on the left. Then main causes are identified by brainstorming "why" the quality-related problem occurred. This identifies the main causes. These are written on inclined side bones connected to the backbone. Subsequently more "why" questions result in successive levels of causes.

Ishikawa developed this technique in 1943 at the University of Tokyo, to educate engineers from the Kawasaki Steel Works. He used this technique in conjunction with a Pareto diagram, which classifies relative magnitudes of the effects of different factors contributing to a quality problem. The fishbone and Pareto diagrams, when used together, can help to identify relative significance of various major and minor factors in a quality problem. According to the quality value-chain framework proposed here, Ishikawa's causality-based efforts can be classified as *prevention integrated quality.*

Ishikawa focused further on preventive aspects of quality through his "Total Quality Control" approach. He also shifted his focus from process orientation to customer orientation. He expanded definition of the word "customer" by including internal as well as external customers, and he suggested that "the next process is your customer." By developing such a chain of customers, Ishikawa wanted to bring down barriers of communication across the different functional areas and thereby integrate one functional department with another functional department in the same organization. These contributions by Ishikawa can be termed *horizontal cross-functional integration of quality.*

6. Taguchi Frontier: Design Integrated Quality. While Deming and Ishikawa were stressing statistical quality control of production processes in the Japanese industry, a new frontier of quality management was opened by a Japanese design and development engineer.

Genichi Taguchi, a Tokyo-based Japanese consultant, challenged the prevailing quality paradigm of American industry based on strict process control. He instead stressed a focus on the "robust quality" of design. Taguchi integrated product quality with design and functional deployment. The robustness of design implied that with normal variations in process operations, the product so designed was less likely than before to fail any criteria for acceptable quality. In other words, a robust process consistently produced high-quality goods even with variations in raw materials.

Taguchi did his early work on new product development at the Electrical Communications Laboratories of the Government of Japan. There he realized that theoretically there were a large number of possible permutations and combinations for operating even a simple process with few process variables. For example, for a simple process in semiconductor production at Bell Labs with nine variables, if only one process variable was changed at a time, then over 6000 sets of alternate experiments were required to compile all the necessary experiments to develop information needed to arrive at the optimum processing conditions. By using more sophisticated statistical procedures, Taguchi designed experiments by changing multiple variables simultaneously. Thus, with his approach, a much smaller number of experiments resulted in an optimal process. For example, using the Taguchi method the nine-variable semiconductor production operation at Bell Labs required only 18 experiments to arrive at an efficient process (Note 5.9). Whereas the earlier process of optimization took months of experimentation, Taguchi's design of experimental methods achieved the same result in a much shorter period of few weeks.

With the speeding of process optimization, such information could be quickly fed back and integrated with the design department to make the product designs more robust, with a much reduced sensitivity to normal process variations. With additional technological developments in powerful workstations and desktop computers such as computer-aided engineering (CAE) systems, the Taguchi method integrated design robustness with process manufacturability to produce higher-quality goods. For example, in 1983, using the Taguchi method, American Telephone & Telegraph (AT&T)

revamped its age-old basic approach to manufacturing by breaking down the walls of cultural barriers between designers and manufacturing engineers within an organization.

According to Edward Fuchs, the director of AT&T's Quality Assurance Center, this led to a new focus on "an entire product-realization process that views everything from design and systems engineering through to manufacturing and marketing—as totally integrated" (for details see Note 5.9).

In the 1980s, the Taguchi method appealed to the U.S. managers managing increasingly complex manufacturing processes who could not accommodate any further tightening of process control parameters without exponentially increasing their costs of doing so. Taguchi's *design integrated quality* thus became far more popular across the Pacific Ocean in the United States than in his home country Japan. This was a sort of reverse "Deming effect."

7. Crosby Frontier: Cost of Quality. With an increasing involvement of senior managers in quality movement, as urged by the previous quality champions, the cost of quality became a frequently debated topic among managers. Philip Crosby, the author of a 1979 best-seller, *Quality Is Free*, popularized the concept of cost of quality, and the "price of nonconformance." Crosby worked at ITT and rose from a line inspector to a Corporate Vice President for Worldwide Quality. Using Crosby's ideas, ITT claimed to have saved $720 million in 1 year in quality-related costs. Crosby proposed and instituted cost-of-quality programs at ITT by taking into account costs for rework, scrap, warranty, inspection, and testing of defective goods. These costs were distinguished from the cost of running a quality assurance department. Crosby proposed that managers should evaluate these costs of quality so that they could demonstrate to their senior executives the cost-related benefits of quality efforts. By reducing the cost of quality, their profits could be raised without increasing their sales or by investing additional funds in new equipment or by hiring of more people. In the context of the quality value-chain framework proposed here, these contributions by Crosby are classified as *cost integrated quality*.

Crosby proposed five stages of maturity in the involvement of management in quality. These spanned from the stage of uncertainty to the stage of certainty about knowledge of potential of quality as an effective management tool. Other intermediate stages proposed by Crosby were: awakening, enlightenment and wisdom, with successive increases in the awareness of potential of quality in the overall management of an organization.

Crosby believed that the only effective performance measurement was the cost of quality, and that it was always cheaper to pay attention to quality to do the job right the first time. Crosby also reiterated his belief in three other "absolutes." These attributes stressed quality as conformance to requirements; the prevention of quality defects (similar to the Ishikawa viewpoint); and the zero defect performance standards (a performance integrated quality viewpoint).

8. Kearns/Xerox Frontier: Customer Quality. Crosby's paradigm requiring conformance of performance to specification developed cracks as the domestic competition in the United States grew increasingly intense and turned global. In the 1970s and the 1980s, the competitive domain of the U.S. corporations crossed their domestic national borders. The major competitors were not the other domestic companies, but instead the foreign companies. Under such economic context, the next frontier of quality movement could be attributed to an industrial manager whose organization was getting a beating at the hands of its global competitors.

In the 1970s, David Kearns, President of Xerox Corporation and an ex-IBM marketing man, was facing major organizational hurdles in his attempts to rejuvenate his organization. Xerox, which launched the xerography technology in March 1960, had held dominant and undisputed market share for over 15 years. As a result of this market control and confidence, Xerox also became complacent regarding customers' requirements and complaints and thus neglected the potential threats of entry by new competitors. Xerox's reluctant entry into the quality arena was finally forced by a sharp erosion of its long-held market share in the field of xerography, which was steadily poached by new entrants from Japan.

In 1975, as a result of an inquiry determination made by the Federal Trade Commission, Xerox was forced to open and share its technological know-how and patents for its closely held xerography technology with its competitors and others. The federal consent order required Xerox to cross-license its patents with their competitors, and it also forced Xerox to revise its large account discounting practices. As a result of this court order, in the late 1970s and early 1980s Xerox saw a steady loss of its competitive edge to more aggressive marketers eager to satisfy wishes of their customers. In 1981, *Businessweek* magazine reported that Xerox's share of the U.S. copier revenues had plummeted from 96% in 1970 to about 46% in 1980 and was still falling.

David Kearns joined Xerox in 1977, and he tried many ways to revitalize Xerox's marketing organization. After trying a few initiatives, finally in 1982, as the CEO of Xerox Corporation, he initiated a "Leadership Through Quality" program to turn around the ailing company. Xerox's strategic alliance with Fuji Xerox provided Kearns an insider view of the prevailing quality-related practices in Japan. Kearns defined quality as "meeting the requirements of customers" and urged the Xerox employees to conduct extensive conversations with its unhappy customers. He believed that "customers are the reason why Xerox exists—the reason Xerox men and women come to work each day" (Note 5.10).

Xerox's "Leadership Through Quality" program was a management system that depended heavily on employee involvement. Kearns initiated company-wide training programs beginning with himself; he cascaded this training initiative down the organization through work groups, to Xerox's over 100,000 employees worldwide. According to chairman Kearns, the phrase "team Xerox" must become more than just a phrase and ought to permeate through thousands of quality improvement teams across the entire Xerox organization. Kearns emulated Japanese obsession with customer integrated quality, by integrating product quality with the big "C" customers outside the organization, as well as with the small "c" customers inside the organization.

In its search for competitive quality improvement, Xerox also embarked on an extensive "benchmarking" exercise. To improve performance of its different operations, Xerox sought the best practices of some competing and noncompeting organizations. Xerox managers systematically identified the best performers and their best practices, and they considered how these practices could be adopted by Xerox. For example, in this search for "best of the breed," Xerox managers "benchmarked" L. L. Bean and Nordstorm in the area of customer service; for product development they studied Motorola, 3M, Sony, and Digital; and for training they borrowed practices from Ford, General Electric, and Polaroid. Competitive "benchmarking" is used in an iterative manner for continuous improvement. Kearns's efforts at Xerox can be classified under our quality value-chain framework, as *market competition* and *customer integrated quality.*

In 1989, after persistent hard work for 7 years, Kearns' commitment to market competition integrated quality paid dividends and Xerox Business Products & Systems, based in Rochester, New York, became one of the two winners of the Malcolm Baldrige National Quality Award in their "large company" category (Note 5.11). To Kearns personally and to the employees of the Xerox unit at Rochester, New York, the Baldrige award was an endorsement and encouragement to their long-held beliefs in quality management. According to Kearns, "We are probably the first American company in an industry targeted by the Japanese to regain market share without the aid of tariffs or other governmental help."

When using market competition integrated quality as the driving force for competitive survival, the pressure to benchmark manufacturing and other business operations of the organization depends on the intensity of competition faced by the organization in its existing markets. Other market competitive forces such as barriers to entry, negotiating power of buyers, negotiating power of suppliers, and so on (Note 5.12), may also have significant influence on market competition integrated quality.

9. The Next Quality Frontier: Innovation and Market Creation Quality.
In the 1990s, a new strategic flaw became visible in the previous market competition integrated paradigm for quality. Edward Biernat, manager of quality engineering and the facilitator for quality programs at Bausch & Lomb in Rochester, New York, in a lecture at Rochester Institute of Technology in March 1991, stressed that "If you try to catch the tail of your leading competitors, you will always remain behind them. Instead, you have to focus on the direction in which their head is turning. Only then can you ever think of overtaking them."

In 1990, in a cover story on quality in automobiles, *Business Week* noted that "Just as the U.S. car makers are getting their quality up to par, the Japanese (competitors) are redefining and expanding the term (quality)" (Note 5.13). With the search for "admirable quality" (as noted earlier), Japanese designers such as Mazda's Toshihiko Hirai wanted to make cars that go beyond what their U.S. competitors offered—that is, they wanted to make the cars that were more than reliable and that "fascinate, bewitch, and delight" the owner. This philosophy had already led to the birth and success of Mazda's Miata. Similarly, Nissan Motor Company made its sporty 240SX and 300ZX models more visible and appealing based on the observations made by anthropologists noting the expressions and behaviors of auto buyers while they were browsing in Mazda showrooms. According to Richard D. Rechia, executive vice-president at Mitsubishi Motor Sales of America Inc., "Now it's the personality of the product that dictates quality."

Japanese global competitors also started going beyond competitors' offerings or current customers' expectations, by exploring what the customer's customers wanted. They made a clear distinction between the expectations of two roles embodied in a customer—that of a buyer and that of a consumer. For defining product quality, instead of asking and relying on what the customers (buyers or consumers) were able to articulate verbally, Japanese car makers increasingly depended on their own observations of how the consumers behaved in conjunction with their products. In the context of the quality value-chain framework proposed here, this implied an integration of quality with first the buyer part of a customer and then, particularly more so, with the consumer side of the customer. In the case of quality of automobiles, Japanese car makers used cameras fitted inside a car to record how the buyer–driver and buyer–passenger consciously or subconsciously interfaced with their automobiles under different circumstances. In our quality value-chain model this can be classified as *consumer integrated quality* (a subset of *customer integrated quality*).

Furthermore, Japanese competitors also recognized that, due to their chronic trade surplus with most industrial nations of the world during the past few decades, in the years ahead they are likely to face growing tariff or nontariff hurdles against import of their products. In the future, this was expected to get emotional and acute as Japanese imports increasingly competed with similar products produced locally with native jobs. With growing trade barriers, more Japanese manufacturers quickly grasped that whereas in the past, competitive quality based on their customers' expectations enabled them to match their competitors' offerings, this approach will not help them in the future to gain global markets. Furthermore, with growing pressure from their low-wage competitors in South Korea, Taiwan, and Singapore, the Japanese managers were forced to look for market creating products like Walkman, innovated by Sony Corporation, which could not have been conceived through traditional customer satisfaction surveys. The global markets are therefore expected to see more Japanese products embodying innovation-based value addition, resulting in the creation of new markets. In our quality value-chain framework, this paradigm can be classified as the market creating quality.

Quality Value-Chain and Enterprise-wide Integration

With these successive contributions of quality champions, we illustrated how the different quality frontiers are related to an enterprise-wide integration of quality. The contributions of different quality champions have significant implications on either the primary value-adding activities or the secondary value-adding activities of an enterprise. For the sake of discussion here, an enterprise will be considered as comprising of an organization and all its allies—such as suppliers, distributors, and so on. We have shown here how the different quality champions embarked on different challenges based on either (a) horizontal cross-sectional integration along the primary value chain or (b) vertical hierarchical integration integrating production shop floor with upper echelons of an organization. The quality value-chain framework can also help to illustrate how constituencies outside the traditional boundaries of a production organization are increasingly contributing towards definition and determination of product quality. For example, for discrete goods such as consumer electronic goods, the strategic alliances with upstream suppliers and their process-related practices have a critical impact on the real and perceived quality of finished products at the downstream end of the quality value-chain. We expect that the consumers and consumer-based innovations will drive the next frontier of innovation integrated quality. The innovative entrepreneurs will offer products that will delight customers beyond their expectations. In other words, in the years ahead, the producers of goods and services will need to listen to the voice of customer, but they must innovate beyond their expectations.

Caveat Against Universality

In the quality management movement, there has often been a common tendency of supporters of different quality champions to offer their guru's perspective as the universal panacea for all organizational difficulties and diseases. Based on a multi-industry research study, Joshua Hammond, president of the New York City–based American Quality Foundation, proposed that there was no such thing as "a generic model" for quality in all organizations. This study indicated that the practice of "benchmarking" for quality was most useful to the high-performing firms, but it had little significance for the low-performing firms (Note 5.14). It was therefore proposed that the low-performance firms were likely to be better off by imitating not "the best of the breed," but their competitors who had only marginally superior quality.

The quality value-chain framework was proposed here as a descriptive model attempting to metasynthesize different quality-related paradigms known to managers. It does not intend to recommend that a quality paradigm proposed by one champion was superior to quality paradigm proposed by another champion. Such a cross-sectional comparison of effectiveness should be the subject of future studies. In a similar spirit and to make this chapter relevant to practicing managers, we also stayed away from discussing empirical results of quality-based research.

The De-layering of Quality Mosaic

Thus, the enigmatic quality as seen by a practicing manager today can be understood as a mosaic of multiple superimposed layers of different quality paradigms proposed over many years. Our understanding of quality has evolved a long way away from its original conceptualization as inspection of the manufacturing function. Over these years, these different layers of quality, involving integration of increasingly larger or distant domains of business operations, within and outside an organization, can be quite conveniently understood using the quality value-chain model proposed in this chapter.

To the managers whose organizations have joined the bandwagon of quality movement only recently, the quality value-chain model developed here will hopefully help grasp the overall perspective of quality management movement. This can help by identifying the underlying multiple frontiers of quality individually. Managers of quality in these organizations can thereby seek different individual frontiers of quality, using different components of the quality value-chain. This selection should depend on a strategic assessment of the organization and its external and competitive environment.

In this regard it is interesting to recall that about two decades ago Phil Crosby (Note 5.15) noted that

> . . . quality has much in common with sex. Everyone is for it. (Under certain conditions, of course.) Everyone feels they understand it. (Even though they would not want to explain it.) Everyone thinks execution is only a matter of following natural inclinations. (After all, we do get along somehow.) And, of course, most people feel that all problems in these areas are caused by other people. (If only they would take time to do their things right.)

Based on these observations made in 1979, Crosby concluded that ". . . it is difficult to have a meaningful, real-life, factual discussion on quality (or sex)." It is hoped that our understanding of quality has progressed much during the past two decades. Our investigation of contributions made by different quality champions identified a common core dimension of cross-functional integration in an enterprise. Hopefully, the quality value-chain analysis and the discussion in this chapter regarding metasynthesis of different frontiers of quality pioneered at different times, based on this core dimension of enterprise-wide integration, will somewhat help in understanding the otherwise complex and enigmatic mosaic of quality.

QUALITY TOOLS AND CONTINUOUS IMPROVEMENTS

A primary goal of quality management is to achieve a continuous improvement. Achieving quality is not a destination, but a journey. With learning, participation, and motivation, every successive day should usher in a level of quality better than the previous one. In order to achieve this continuous improvement, a number of quality tools have been used. We will discuss a select group of these.

Quality Control Circles

The quality control circles were small groups of five to twelve employees working together on a quality problem of common interest to them. They were either working together in the same department, or they were drawn from different interrelated departments. They met regularly, say once a week, and voluntarily took up pressing problems related to poor quality, low productivity, or work safety.

These quality circles were the backbone of Japan's continuous improvement movement. However, they never became very popular or effective in the American workplaces. There were other differences too. In Japan, the employees in quality circles met on their own time, outside the usual work hours. On the other hand, in the American factories, quality circles met only during the regular work hours. The employees were not willing to get together and discuss work-related problems during their private time.

The success of quality circles depended on certain other prerequisites.

1. Open communication between quality circle members, as well as between these members and the management.
2. Access to all relevant data related to the quality circle's work.
3. Management commitment and recognition for the quality circle activities.
4. Quality circle members must receive (a) necessary training in different problem-solving techniques and (b) the tools needed to continuously improve quality.

Given below are some of the quality tools that different employees can use to solve quality-related problems or for continuous improvement.

Fishbone or Cause-and-Effect Diagram

As discussed before, a cause-and-effect diagram is also often referred to as a fishbone diagram or Ishikawa diagram, named after the pioneer who spread its use (see Exhibit 5.2). These diagrams represent, in an easy to understand graphic format, the relationship between a quality characteristic (the effect) and the causes which may have influenced the same. These

EXHIBIT 5.2. Ishikawa Cause-and-Effect Fishbone Diagram

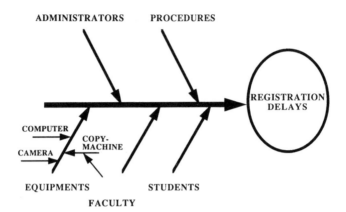

potential causes are identified and listed by a brainstorming session. The listed causes are then classified under different headings and subheadings.

Example: Registration Delays. Let us consider the cause-and-effect diagram for the student registration delays caused at the beginning of a college semester. (A similar approach can be used for the patient registration delays at a hospital, or the guest registration delays at a hotel check-in counter.)

Usually, the effect, or the problem to be resolved, is put in the head of the fishbone on the right side. In our case, we will write "registration delays" on the fish head. See Exhibit 5.2.

Draw a thick "backbone" arrow from the tail on the left side to this fish head. The primary causes (from four to eight) are put on slanting "sidebone" arrows pointing to the backbone. Let us say that the brainstorming session raised the question; Why do registration delays take place? This generated five primary causes as the sources of such delays: (1) procedures, (2) administrators, (3) equipment and technical difficulties, (4) the student's own actions causing delays, and (5) the availability of faculty members who must sign their approvals for registering some courses.

The secondary causes for each primary cause are generated by second-level "whys." These are put on smaller arrows pointing to the slanting sidebones. For example, the primary cause of equipments and technical difficulties may include secondary causes such as (1) camera for picture I.D. not working, (2) computer breakdown or slow obsolete computers, or (3) copying machine waiting for technician's service.

The third-level "whys" could produce more detailed causes for the slow technician's response. This could be caused by a lousy contract with a poorly equipped service vendor, instead of using Xerox or a more reliable service vendor.

Once the cause-and-effect diagram is ready, the comprehensive graphic map can be reviewed carefully to decide the next steps to be taken for improving the quality promise.

Histogram

A histogram plots the frequencies of occurrence against the quality values of output. The data values are grouped in meaningful intervals. The number of intervals, the range, and the center point values are used to characterize the distribution of values. Usually when there are less than 50 output values, then five to seven intervals are used to plot a histogram. For 100 output values, six to ten intervals may be used.

Let us say that in the case of our earlier example of student registration delays, a survey was taken and 120 students responses were received. Of these total responses, 48 surveys complained of the procedures required, another 36 were due to students' own actions, 24 were due to the equipment and technical difficulties, 6 were due to administrators' availability, and the remaining 6 were due to the faculty's approvals. A histogram will show these frequencies against each cause.

Pareto Analysis

This is an important and frequently used quality tool. The goal of this quality tool is to identify the most significant causes of the quality defect or to identify the solution that will provide the most effective solution. This technique is based on the Pareto principle or the 80/20 rule. According to this principle, 80% of a problem can be attributed to 20% of its causes. These are also called the vital few causes, compared to the trivial many.

Let us consider a different example: dissatisfied customers' complaints for a hotel's services. Let us say that the top five sources of complaints in descending order are room air quality, children's noise, check-in, check-out, and other factors. Say they generated 48, 36, 30, 12, and 24 complaints, respectively, or a total of 150 complaints. The frequency for the "other" category is listed at the end, and usually should not exceed half of the frequency for the largest category. These frequency values in descending order are written in column 2 in front of each category listed in column 1 in a table. See Exhibit 5.3. In the next column 3 the cumulative numbers are added. These values will be 48, 84, 114, 126, and 150, respectively. Then in column 4, the percentages of overall total are listed. These will be 32%, 24%, 20%, 8%, 16%, respectively. These should add to 100%. Finally, in column 5, cumulative percentages are written down. These values for our example will be 32%, 56%, 76%, 84%, and 100%. This verifies our calculations. All these values are shown in Exhibit 5.3.

The Pareto graph plots the five primary causes of complaints along the horizontal axis. Two vertical axes are used. One vertical axis ranges from 0 to the maximum number of observations. This has the value of 150 in our example. One plot is of the frequencies for each cause. The first two causes have frequencies of 48 and 36, respectively.

The other vertical axis uses the cumulative percentage, from 0% to 100%. Here, the values from column 4 in the previously developed table are plotted. The first cause of room air quality has a 32% share of customer

EXHIBIT 5.3. Pareto Analysis: Dissatisfied Hotel Guests

Customer Complaints	Number of Complaints	Cumulative	Percent	Cumulative
1. Room air quality	48	48	32	32
2. Children's noise	36	84	24	56
3. Check-in delay	30	114	20	76
4. Check-out delay	12	126	8	84
5. Others	24	150	16	100
Total	150		100	

dissatisfaction. The second largest cause of children's noise has a share of 24%. The cumulative percentage value for the second largest cause will be 56%. The third largest cause of check-in delays accounts for 20% of customers' dissatisfaction, with a cumulative value of 76%. The Pareto plots are shown in Exhibit 5.4.

Scatter or Correlation Diagram

The scatter diagrams plot the relationship between an independent variable and a dependent variable. The quality effect can be the dependent variable and is plotted along the Y axis, while the causes or the process variables can be the independent variables and are plotted along the X axis. For example, we may be testing the quality of a tire tube by measuring its bursting pressure in pounds per square inch. This is the dependent variable, to be plotted along the Y axis. The corresponding independent process variable may be

EXHIBIT 5.4. Pareto Graph

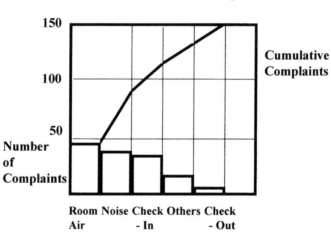

Sources of Customer Complaints

the rubber cure time in minutes. The paired data are collected for each value of the tube bursting pressure, along with the corresponding value of the rubber curing time.

The scatter diagrams may also be plotted between two related quality parameters, or two process conditions related to a common quality parameter. If such relations are highly correlated, the measurement of one quality parameter or a particular process variable may be monitored, assuming that the other variable would also follow a similar pattern. For example, the tube bursting pressure may be correlated with the life of the tube, and the curing time may be correlated to the curing pressure.

When an increase in bursting pressure occurs with an increase in the rubber curing time, we say that the two variables are correlated positively. On the other hand, if the bursting pressure decreases as the curing time increases, the two variables are correlated negatively.

Not all relationships are perfectly correlated. Thus the strength of correlation can be measured quantitatively by the correlation coefficient. The correlation coefficient is 1.0 for a perfectly correlated set of data, where a variation in variable X corresponds to a predictable variation in variable Y. The data points lie along a straight line. The correlation coefficient is 0.0 for a perfectly uncorrelated data. In this case, a value of variable X cannot help make any estimate of the corresponding value of variable Y. The correlation coefficient thus gives us a measure of how far the data points are deviating from a perfectly correlated straight-line relationship (see Exhibit 5.5).

EXHIBIT 5.5. Scatter or Correlation Diagram

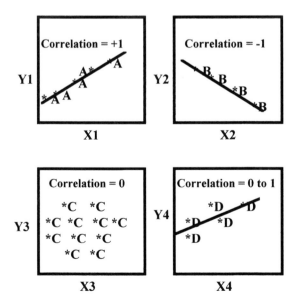

Regression Analysis

The correlation between two variables is represented by a simple straight line relationship, represented by the following equation:

$$Y = A + B * X$$

This is called a *regression line.* Y is the response or dependent variable, say a quality parameter. X is the explanatory or independent variable, which may be a process condition. A is a constant, or an intercept on the Y axis, when the value of X is zero. B is called the *regression coefficient.* This regression line is developed based on a mathematical procedure called the "least-squares" fitting method. Given a set of scattered paired data points obtained by experimentation, alternate straight lines are fitted through the scatter and their cumulative total of a sum of their distances squared is calculated. Finally, the best regression line is selected when the sum of squared distances from data points to the lines attains the minimum value possible. The regression coefficient B is the slope of such a best-fitting regression line (see Exhibit 5.6).

Statistical Process Control (SPC)

All processes, no matter how good they are, have some variation. Even if a production process is very tightly controlled, there are variations in inputs and in environmental changes. This variation is caused by a number of different types of causes. Some causes are chance causes and are called *common causes.* They occur randomly and cannot be humanly controlled. For example, daytime and nighttime temperatures and sunlight vary because of uncontrollable natural forces.

EXHIBIT 5.6. Regression Analysis

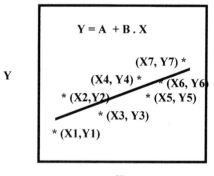

There are other causes which can be assigned, and they are called *special causes*. These special causes can be assigned to the variations in some process elements, such as a machine, the second-shift operators, and raw materials procured from vendor X. These causes can be controlled provided that they are identified first. By controlling the special causes, the overall process variation can be reduced.

The use of principles of statistics to control quality of products and processes has provided many advantages. In many production operations, only a small sample of output can be measured and tested. Based on these sample measurements and using the principles of sample statistics, we are able to make inferences for the entire production/operation process. Based on the sample statistics, such as the mean, median, standard deviation, or variances of such data points, corrective actions can be made. Thus inputs and/or processes are modified to bring about the desired corrective actions. However, the first step to understanding a complex multioperation process is to map its flow.

Process Mapping and Flow-Charting

Many production operations involve a large number of steps. These steps are interdependent. A first step to improving quality is to map the various steps involved in a production process or operation. Different steps in a process are designated by different symbols. Ellipses indicate the starting and ending steps. Diamond shapes represent decision points. Rectangles represent processing steps.

Mapping helps identify the precedence relationships. That is, what steps must be done before a subsequent step can be undertaken. A simple advertisement by a process mapping consultant shows the picture of an executive wearing his clothes in wrong sequence. Thus, the underpants end up on top of the trousers. Even though the executive does all the operations and wears all the designer clothes, the wrong process sequence can make a big difference.

A step-by-step process mapping can help streamline the production operations. Mapping of material flow in auto manufacturing at Volvo plants in Sweden revealed how a simpler flow can save significant amounts of time and cost.

Control Charts

Walter Shewart at Bell Telephone Laboratories pioneered the use of control charts to control the quality of telephone communication equipment used by AT&T. Dr. J. Edwards Deming spread the use of control charts to the Japanese manufacturers who were being prepared to supply military goods to the U.S. Army during the Korean War.

A control chart is like a run chart, where quality values are plotted as a production operation proceeds. Different control limits are marked on the

EXHIBIT 5.7. Process Control Chart

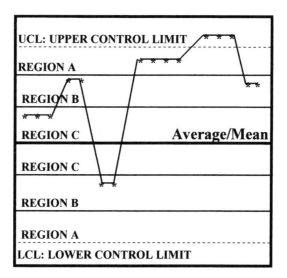

run charts. Above and below mean center lines, upper control limits, and lower control limits are indicated (see Exhibit 5.7).

The values for upper and lower control limits are calculated mathematically using the process data and assuming a normal distribution. These control limits represent the range of the variation in a process due to the different causes. The normal statistical distribution is represented by a bell-shaped curve, which is symmetrical around a mean value. The upper and lower control limits are marked at plus and minus three standard deviation values around the mean value. The area under the normal distribution curve is represented as 100%, or 1.0. With this normalization, the area under a specific interval of variation represents the probability of getting a process value within that interval.

The output values from a process are plotted on a control chart to identify and differentiate assignable special causes from the uncontrollable common causes. When all the values lie within the specified control limits, the process is considered in control. There are certain rules available to distinguish variations due to the special and nonassignable common causes.

Additional Approaches to Continuous Improvement

The use of the above-mentioned quality tools can facilitate a continuous improvement of quality in the production and operations of an organization. For a continuous improvement in the total quality of an enterprise there are other assessment tools available. The Malcolm Baldrige Award has certain criteria which are used by many U.S.-based companies to assess the state of quality in their organizations. Similarly the international, national, and industry-based standards can be used to benchmark one's own quality against that of the competitors' quality. The international standards, such as the ISO 9000 standards series, have evolved to become the minimum requirements for doing business with the members of the European Union market. In a similar manner, the standards adopted by the Big Three U.S. auto makers must be met by a supplier such as Goodyear to do business with these major buyers of parts, supplies, and components. These and similar other issues will be discussed next.

THE MALCOLM BALDRIGE NATIONAL QUALITY AWARD

The Origins of America's National Quality Award

An ongoing assessment and conformance to world standards can help a technology-driven enterprise to establish quality targets for its continuous improvement. These targets and measurable achievements are the milestones of a company's quality journey. The Malcolm Baldrige National Quality Award in the United States, the Deming Prize Audit of Japan, and the ISO 9000 standards of the European Community Union can be of significant help in this regard. So can some state-level awards, such as the Excelsior Award in New York State. Certain industries and companies have also established their own standards. The Big Three auto makers have their own quality standards, while Motorola's Customer Satisfaction Award achieves a similar effect within that enterprise.

The Deming Prize Link. In 1987, the U.S. Congress passed a legislation to create the Malcolm Baldrige National Quality Award (hereafter referred to as the Baldrige National Quality Award or BNQA). The Baldrige Quality Award was patterned after Japan's Deming Prize for quality, the highest recognition for an industrial organization in Japan. The Deming Prize was established in Japan in December 1950, more than three decades earlier than the Baldrige National Quality Award. Japan's highest quality award was named in recognition of the contributions of W. Edwards Deming, an American. Dr. Deming visited Japan and introduced the concepts of statistical process control to the war devastated Japanese. He helped the Japanese managers apply the principles of statistics to improve the quality of goods produced by the war-devastated Japanese industries. These goods were required urgently by the U.S. forces deployed in the Far East to combat the spread of communism in Asia. Over the past four decades, the Deming Prize has played a major role in helping many Japanese companies gain competitive advantage in international markets.

The Baldrige National Quality Award, like the Deming Prize, seeks a total and an integrated involvement of a company in improving its quality of goods and services. Both the American and the Japanese awards share their concern for continuous improvement of quality of products and services. The Baldrige National Quality Award differs from the Deming Prize in that the U.S. Award puts a much higher weight on quality defined by customers than does Japan's Deming Prize. The U.S. quality award is also more specific and less ambiguous than its Japanese counterpart. The assessment for the Deming Award depends more on the subjective interpretation of the examining judges. The Deming Prize, because of the influence of the person after whom it was named, puts a higher weight on the statistical principles used to implement the statistical quality control of processes. The Deming Prize is administered by the Union of Japanese Scientists and Engineers (JUSE), whereas the Baldrige National Quality Award is administered by a U.S. federal agency. The guidelines for the Deming Prize are given in Exhibit 5.8.

What was the need to create a national award for quality in the United States? During the 1980s, one American industry after another faced stiff competition from the highly reliable quality goods produced by their Japanese counterparts. Thus, compact and fuel-efficient German and Japanese cars requiring less maintenance quickly replaced the big, gas-guzzling American cars earlier liked by the American consumers. America's semiconductor manufacturers, who had earlier pioneered the birth of transistor, the integrated circuits, and the large-scale integrated circuits, conceded their technological leadership in the production of dynamic random access memory (DRAM) to the Japanese producers. Some critics claim that most of these gains were made by the Japanese companies because of their closer ties with the Japanese government.

With this background setting, America's national-level quality award was established with the intent to be the cornerstone of the U.S. Federal Administration's initiative for encouraging the U.S. companies to improve their quality and global competitiveness. The Baldrige National Quality Award was intended as a visible gesture of the newly emerging partnership in the United States between the U.S. government and the U.S. businesses. This was a strategic change considering that in the past the two sectors had often taken adversarial and confrontationary positions.

The American award was named after Malcolm Baldrige, a much admired former Secretary of Commerce in the Ronald Reagan Administration. Malcolm Baldrige served as the Secretary of Commerce from 1981 until his sudden and tragic death in 1987 in a rodeo accident. He was described by President George Bush, at the 1990 Baldrige National Quality Award ceremony for the year's winners, as follows.

> Mac Baldrige . . . prior to becoming secretary of commerce, was a true leader in the business. When it came to business, he really did understand that quality cannot be assured with some slogan or an ad campaign. And he knew it began with winning and keeping business (Note 5.16).

EXHIBIT 5.8. Criteria for Japan's Deming Prize

1.0 Policy
1.1 Management's policies for quality and quality control
1.2 Methods used to establish policies
1.3 Constancy and justifiability of policies
1.4 Use of statistical methods
1.5 Communication, transmission, and diffusion of policies
1.6 Review of policies and the results achieved
1.7 Relationship between short-term and long-term planning

2.0 Organization and Its Management
2.1 How explicit are the scopes and responsibilities of authorities?
2.2 How appropriately is the authority delegated?
2.3 Cooperation across different divisions
2.4 Committees and their activities
2.5 Effectiveness in the utilization of staff
2.6 Use of quality-control circle activities
2.7 Quality control and diagnosis

3.0 Education (and Training)
3.1 Education programs and their results
3.2 Extent of understanding of quality control by the employees
3.3 Teach statistical methods and concepts
3.4 Effective grasp of quality control
3.5 Educate related companies, such as suppliers, distributors, etc.

4.0 Collection, Dissemination, and Use of Quality-Related Information
4.1 Compilation of internal information
4.2 Transmit information among divisions
4.3 Speed of transmission of information
4.4 Statistical analysis of information and utilization of the results

5.0 Analysis
5.1 Selection of key problems and themes
5.2 Propriety of analytical approach
5.3 Linkage with proper technology
5.4 Quality analysis, process analysis
5.5 Use of statistical methods
5.6 Use of analytical results
5.7 Assertiveness of suggestions for improvement

6.0 Standardization
6.1 Systemization of standards
6.2 Method for establishing, revising, and abolishing standards
6.3 Outcome of the establishment, revision, or abolition of standards
6.4 Contents of standards
6.5 Use of statistical methods
6.6 Accumulation of technology
6.7 Use of standards

7.0 Control
7.1 Systems for quality control, and related matters of cost and quantity
7.2 Control items and control points
7.3 Use of statistical control methods
7.4 Contribution of performance and activities of quality-control circle
7.5 Actual conditions of control activities
7.6 State of matters under control

EXHIBIT 5.8. (*Continued*)

8.0 Quality Assurance
8.1 Procedure for the development of new products and services, including analysis and improvement of quality, design, reliability, etc.
8.2 Safety and immunity from product liability
8.3 Process design, process analysis, process control, and improvement
8.4 Process capability
8.5 Instrumentation, gauging, testing, and inspecting
8.6 Maintenance of equipment and control of subcontractors, purchasing, and services
8.7 Quality assurance system and its audit
8.8 Use of statistical methods
8.9 Evaluation and audit of quality
8.10 Actual state of quality assurance
9.0 Results
9.1 Measurement of results
9.2 Substantive results in quality, services, delivery time, cost, profits, safety, environment, etc.
9.3 Intangible results
9.4 Measures for overcoming defects
10.0 Planning for the Future
10.1 Grasp the present state of affairs and the concreteness of the future plans
10.2 Measures for overcoming defects
10.3 Plans for further advances
10.4 Linkage with long-term plans

Source: Adapted from *The Deming Prize Guide for Overseas Companies*. Union of Japanese Scientists and Engineers (JUSE), Tokyo.

The primary responsibility for administering the Baldrige Quality Award was given to the National Institute of Standards and Technology (NIST), of the Department of Commerce. They are headquartered at Gaithersburg, Maryland—on the outskirts of Washington D.C. Subsequently, it was assigned to the U.S. Department of Commerce Technology Administration to manage, and the awards were to be administered by the American Society for Quality Control (ASQC). In the mid-1990s, to spread the message of the Baldrige National Quality Award winners, the Award Foundation sponsored the Annual Quest for Excellence Conference at several major cities in different parts of the United States.

Most of the cost of soliciting, receiving, processing, and judging of applications was proposed to be covered by a $10.4 million endowment sponsored by the U.S. industry. Compared to the impact of the Baldrige National Quality Awards, the U.S. government did not spend much money.

The Objectives of the Baldrige National Quality Awards

The Baldrige Quality Awards are awarded annually. These are awarded for the quality-related practices of a company division, and are not meant as endorsements for the products or services of the entire company. Each year, a number of these awards are awarded in different categories to recognize those U.S. companies that excel in their achievement of high-quality results and that follow exemplary quality management practices.

The Baldrige National Quality Awards are aimed at promoting following three main objectives:

1. Spreading an awareness of quality as an important and integral element of gaining sustainable competitiveness.
2. A good and clear understanding of the prerequirements for achieving excellence in quality.
3. Sharing of information on successful and reproducible quality strategies, as well as disseminating the information about the benefits from implementation of these strategies.

The winners of Baldrige National Quality Awards are expected to share their understanding of quality-related practices, strategies, and the benefits of successfully implementing these with other organizations in the United States.

For example, after Motorola won the Baldrige Quality Award in the manufacturing category in 1988, many of its senior executives spoke at numerous conferences, and presented their findings to other corporations and organizations. Motorola answered over 1000 outside inquiries about their quality practices within a few years of winning the Baldrige National Quality Award. For a long time, every month Motorola conducted a 5-hour-long quality briefing for managers and executives from other companies interested in learning how Motorola achieved excellence in quality.

Westinghouse, another Baldrige Quality Award winner in 1995, within a few months of winning the award, published and distributed over 6000 copies of their Annual Report that chronicled its award-winning journey.

The Evaluation Process

With the twin goal to protect the confidentiality of the applicants and to improve the validity of evaluation for the Baldrige National Quality Award, a quite elaborately designed robust process of evaluation was established for judging the Malcolm Baldrige Award applicants.

The Applicaton. The applicants for the Baldrige National Quality Award have to follow a 42-page-long set of instructions, and they must include a mass of quality-related data regarding seven criteria of their quality management system. The large companies must keep their documentation to less than 75 pages, and the small business applicants must not exceed beyond the limit of more than 50 pages. The application for the Baldrige National Quality Award must be accompanied by a $2500 application fee for the large manufacturing and service companies, and a $1000 application fee for the small businesses with less than 500 full-time employees. For large organizations a corporate division can apply, but only one division per company can apply in a year.

Since the inception of the Baldrige National Quality Award in 1988, over 2000 applications for the Baldrige National Quality Award have been received, but a much greater number of companies have requested the applications. Even if a company is not applying for the Baldrige National Quality Award, the managers in the company can use the award criteria for self-assessment and continuous quality improvement.

The Award Examiners. The applications for Malcolm Baldrige National Quality Award are examined by volunteer examiners. The administrators of the award at the National Institute of Standards and Technology (NIST) pay a great amount of attention to (a) the selection of volunteer examiners and (b) any potential conflict-of-interest issues they may have. All of the examiners must disclose their business affiliations, and they are barred from judging their affiliates as well as their applicants' competitors. While examining the award applicants, the examiners are not allowed to accept even a souvenir pen. However, many examiners, with the insight into how the Baldrige National Quality Award process works, have built their consulting practice by helping other contestants for the award.

The Judging System. An elaborate judging system has been developed to pick the winners of the Baldrige National Quality Awards each year. This involves multiple phases of evaluation.

1. The first phase of the evaluation for the award includes the determination of "eligibility" of applicants by the volunteer examiners. This establishes that the company is a U.S. company.
2. This phase is followed by the completion of a two-page "site listing and descriptor" form.
3. The eligible applicants then pursue the next stage of the assessment process by submitting the appropriate application fee, along with an appropriate packet of documents by following the 42-page-long set of instructions in the application.

4. The volunteer examiners for the Baldrige National Quality Award continue the examining process by evaluating the applicant's answers to 133 detailed questions. The applicant company describes what efforts it has taken in the areas of the seven criteria that constitute a "total quality management system."
5. The award examiners then recommend a site visit for an applicant if the applicant's assessment gets 750 points or more out of a maximum of 1000 points.
6. The announcement of winners of the Baldrige National Quality Awards is usually made by the President of the United States, at a gala.

The Categories. A maximum of six Malcolm Baldrige National Quality Awards may be offered every year to business units in three different categories. These are:

Two awards for manufacturing companies
Two awards for service companies
Two awards for small businesses with less than 500 employees

In a particular year, if sufficiently high-quality standards are not achieved by any of the applicants, then fewer awards may be given in a category.

The Assessment Criteria. In the documents and the information submitted by the applicants for the Baldrige National Quality Award, the applicant company must describe its quality-related practices and strategies under seven different categories. See Exhibit 5.9 for the assessment system. These

EXHIBIT 5.9. Quality Management System and Malcolm Baldrige National Quality Award

		1.0 Senior Executive Leadership	
2.0 Information and Analysis	3.0 Strategic Quality Planning	4.0 Human Resource Development and Management	5.0 Management of Process Control
	6.0 Quality and Operational Results		7.0 Customer Focus and Satisfaction

seven categories combined together reflect the total quality management system of the company. These seven criteria are as follows:

1. Leadership
2. Information and analysis
3. Strategic quality planning
4. Human resource utilization
5. Quality assurance of products and services
6. Quality assurance results
7. Customer satisfaction

A maximum of 1000 points are awarded to all the seven criteria. Each of these seven criteria carry different weights. The weights for the different award criteria have changed slightly over the years, and these are listed in Exhibit 5.10. Each Award criterion is divided into two or more key examination items. These items identify the critical elements that require direct management attention. Finally, each of these items contains several specific areas that each applicant must address. The information provided by the applicants must help the examiners understand the quality-related intents of the company as well as its specific practices.

Given below are the specific ways these seven criteria, and their subelements, were identified in the 1995 application for the Malcolm Baldrige National Quality Award.

EXHIBIT 5.10. Changing Criteria for Baldrige Quality Award

| | Maximum Criterion Points in: | | | | | |
	1991	Percent	1992	Percent	1995	Percent
1. Leadership	100	10	90	9	90	9
2. Information and analysis	60	6	80	8	75	7.5
3. Strategic quality planning	90	9	60	6	55	5.5
4. Human resource development and management	150	15	150	15	140	14
5. Management of process quality	150	15	140	14	140	14
6. Quality and operational business results	150	15	180	18	250	25
7. Customer focus and satisfaction	300	30	300	30	250	25
Total	1000	100	1000	100	1000	100

1. Leadership. The Baldrige Quality Award recognizes that the senior and top-level executives provide personal leadership for quality. They should also be involved in creating and sustaining a customer focus and visible, clear quality-related values. The examiners also examine how a company's quality values are intimately integrated into the company's management system. They should therefore be reflected in the way the company addresses its public citizenship.

In 1995, this criterion carried 90 points, or a 9% share out of a total of 1000 points, and looked at whether and how the management support system guided all the company activities towards quality excellence. This criteria examined the involvement of senior executives and the company's quality leaders and leadership in the external community. The way these companies integrated their public responsibility with their quality values and practices were also examined. This criterion was divided into three subelements.

1.1 Senior executive leadership   4.5%
1.2 Leadership system and organization, including assessments   2.5%
1.3 Public responsibility and corporate citizenship   2.0%

2. Information and Analysis. This criterion of the Baldrige National Quality Award in 1988 carried 60 points or 6% out of a maximum of 1000 points. In 1995 the weight of this criterion was increased to 75 points or 7.5% of the total. Under this criterion the examiners assessed the scope, validity, use, and management of raw data and derived information that underlie and drive quality excellence and improve its operational and competitive performance—in other words, the role that information played in the company's total quality management system. Under this criterion the applicant company is evaluated for the adequacy and appropriateness of its data, information, and analysis system used to improve the quality of its products or services, operations, and customer focus. Examiners assess the adequacy and effectiveness of collecting and analyzing data and information. They check if this will lead to the support of a responsive and prevention-based approach—to quality based on "management by fact." The following three subelements were considered under this category:

2.1 Management of (quality and performance) information and data   1.5%
2.1 Competitive comparisons and benchmarking   2.0%
2.3 Analysis and uses of company-level data   4.0%

3. Strategic Planning. In the beginning, this criterion carried 90 points, or a weight of 9% out of a total of 1000 points. The weight was reduced in 1995 to 55 points, or 5.5% of the total. The examiners assessed the planning process for retaining and achieving quality leadership, and they also assessed how the quality improvement planning was integrated into the overall business planning of the organization. This criterion also examined the short-term and long-term priorities used by a company to achieve and sustain a quality leadership position for the organization in its relevant business sector. The focus was turned around a little bit in the 1995 criteria. The focus was shifted more to (a) the overall business strategy and planning process and (b) how all the key quality efforts were integrated into the same. In this criterion, specific attention is paid to the following two elements.

3.1 Development of strategy   3.5%
3.2 Deployment of strategy   2.0%

4. Human Resource Development, Utilization, and Management. In 1995, this criterion carried a weight of 140 points, or 14% share of the total maximum points. Earlier, it used to carry slightly more—that is, 150 points, or 15% out of a maximum of 1000 points. This criterion was used to examine the efforts and initiatives undertaken by a company to involve all employees and work force of the applicant company. How were they kept informed? How were they kept motivated? Under this criterion, the judges for the Baldrige National Quality Award also examined the effectiveness of the organization's efforts to develop and realize the full potential of the work force, including those in management. This criterion also looked at how the company maintained an environment that promoted excellence and one that was conducive to full participation, along with personal as well as organizational growth. These efforts were assessed under five subcriteria.

4.1 Human resource planning and evaluation   2.0%
4.2 High-performance work systems   4.0%
4.3 Employee Education and Training   4.0%
4.4 Employee Performance and Recognition   2.0%
4.5 Employee well-being & Morale   2.0%

5. Management of Quality by Process Assurance of Products and Services. In the past, this criterion carried 150 points, or 15% share in a maximum of 1000 points. This was reduced in 1995 to a slightly lower 14% weight. The focus of this criterion was on the quality processes implemented by a company to achieve, sustain, and continuously improve quality. Using this criterion, examiners assessed the systematic processes used by the company for its continuous improvement of quality and performance. This included a total quality control system for goods and services based on the management of processes involved in the process design and control, research and development, and the management of quality of procured materials, parts, and services from suppliers. This criterion was split into the following four subelements.

5.1 Design and introduction of products/services   4.0%
5.2 Process management, including product/service production
     and delivery   4.0%

| 5.3 Process management and support services | 3.0% |
| 5.4 Management of supplier performance | 3.0% |

6. Business Results. Originally, this criterion was focused on quality assurance and carried 150 points, or 15% out of a total of 1000 points. Examiners assessed the quality levels and quality improvements achieved, as measured by objective measures derived from the analysis of internal business operations as well as the external customer requirements and expectations. This was one criterion that actually looked at quality of the applicant's products and services in relation to those of its competing firms.

In the 1995 application, the focus was significantly changed to business results, and the weight was increased to 250 points, or 25% of the total maximum points. The company's performance was also compared with the performances of its competitors. These achievements were assessed based on the following three subcriteria.

6.1 Results related to product and service quality	7.5%
6.2 Company and operational results	13.0%
6.3 Supplier performance results	4.5%

7. Customer Satisfaction. Under this criterion, which accounted for 300 points or as much as 30% out of a total of 1000 points, the examiners examined the organization's knowledge of the customer, its overall customer service, responsiveness, and ability to meet the customers' requirements and expectations. This criterion also covered the current levels and trends in the organization's customer satisfaction level.

In 1995, the relative weight of this criterion was reduced to 25% of the total maximum points. The criterion included the company's understanding and use of the customers' requirements and other key factors driving their competitiveness in the marketplace. The examiners assessed the methods by which the company determined customer satisfaction, current trends, and levels of customer satisfaction and retention. These results were compared to those of the company's competitors. The company answered questions related to how well their customers liked their products, and how did the company measure that. The customer focus criterion was further subdivided into the five subcriteria listed below.

7.1 Customer and market knowledge	3.0%
7.2 Customer relationships management	3.0%
7.3 Determination of customer satisfaction	3.0%
7.4 Results of customer satisfaction	10.0%
7.5 Comparison of customer satisfaction	6.0%

These seven criteria, together with their subelements and weights in the application for the 1995 Baldrige National Quality Award are summarized in Exhibit 5.11.

EXHIBIT 5.11. Summary of Seven Major Criteria and Their Subcriteria in 1995 Application for Baldrige Quality Award

Criteria	Weights
1. Leadership	**9.0%**
1.1 Senior executive leadership	4.5%
1.2 Leadership system, organization, including assessments	2.5%
1.3 Public responsibility and corporate citizenship	2.0%
2. Information and Analysis	**7.5%**
2.1 Management of (quality and performance) information and data	1.5%
2.2 Competitive comparisons and benchmarking	2.0%
2.3 Analysis and uses of company-level data	4.0%
3. Strategic Planning	**5.5%**
3.1 Development of strategy	3.5%
3.2 Development of strategy	2.0%
4. Human Resource Development, Utilization, and Management	**14.0%**
4.1 Human resource planning and evaluation	2.0%
4.2 High-performance work systems	4.0%
4.3 Employee education and training	4.0%
4.4 Employee performance and recognition	2.0%
4.5 Employee well-being and morale	2.0%
5. Management of Quality by Process Assurance of Products and Services	**14.0%**
5.1 Design and introduction of products/services	4.0%
5.2 Process management, including production and delivery of products and services	4.0%
5.3 Process management and support services	3.0%
5.4 Management of supplier performance	3.0%
6. Business Results	**25.0%**
6.1 Results related to product and service quality	7.5%
6.2 Company and operational results	13.0%
6.3 Supplier performance results	4.5%
7. Customer Satisfaction	**25.0%**
7.1 Customer and market knowledge	3.0%
7.2 Customer relationships management	3.0%
7.3 Determination of customer satisfaction	3.0%
7.4 Results of customer satisfaction	10.0%
7.5 Comparison of customer satisfaction	6.0%

Baldridge National Quality Award Winners for 1996

Let us next illustrate the Baldrige assessment criteria by considering the Baldrige Quality Award winners in different categories in a particular year and then reviewing their major achievements.

In October 1996 (October is the month that is annually celebrated throughout the United States as Quality Month), the U.S. President Bill Clinton and Commerce Secretary Mickey Kantor announced the winners of the 1996 Malcolm Baldrige National Quality Awards. After 4 years, since 1992, there were once again winners in every one of the three award categories: ADAC Laboratories of Milpitas, California won the Award in manufacturing, Dana Commercial Credit Corporation of Toledo, Ohio won in service category, and two winners were announced in the small business category. These were Custom Research of Minneapolis, Minnesota, and Trident Precision Manufacturing of Webster, New York. These winners were selected out of 29 applicants, the lowest number of applicants since the award's inception in 1988.

Manufacturing Category. ADAC Laboratories was in the business of designing, manufacturing, marketing, and supporting products for nuclear medicine and radiation therapy segment of the health care industry. This company of 710 employees was commended for their excellent practices in the area of customer satisfaction. All their executives were expected to spend at least 25% of their time with customers. They took customers' calls and invited them to attend weekly quality meetings. As a result of these efforts, the company improved its customer retention from 70% to 90%. They achieved a 10% increase in its customer satisfaction for post-sales technical support. The company was also able to reduce downtime for customers caused by a problem, from 56 hours to 17 hours.

Service Category. Dana Commercial Credit was in the business of leasing and financing services in selected market niches. They earned a reputation for servicing and custom-leasing those transactions for their customers which most of their competitors could not.

Small Business Category. Custom Research Incorporated (CRI), with 105 employees, was one of the two winners of the 1996 Baldrige Quality Award in the small business category. CRI, a Minneapolis-based full-service national marketing research company, designed market research projects and collected information for their clients. Their areas of expertise included market research of consumer behavior, business-to-business marketing, and medical marketing. They designed and conducted market research projects for large multinational companies. By using technology-driven processes, CRI was able to reduce their cycle time for tabulation of data from 2 weeks to 1 day. More than 70% of their clients rated CRI as exceeding their expectations, and they did this for more than 97% of their market research projects.

Finally, Trident Precision Manufacturing of Webster, New York, the second winner of the 1996 Baldrige Quality Award in the small business category, was in the business of contract manufacturing of precision sheet metal components and electromechanical assemblies. They had customers in the office equipment, banking, medical supplies, computer, and defense industries. They also produced custom products. They measured the quality of their custom products by defects per 100 machines produced, and they found zero defects for the past 2 years. For its major customers, Trident's quality performance rating was over 99.8%. Since 1990, Trident improved its on-time delivery from 87% to 99.9%, and reduced the direct labor hours for rework from 8.7% to 1.1%.(Note 5.16).

Exhibit 5.12 shows a list of the winners of the Malcolm Baldrige National Quality Award winners since its inception.

Promises of the Baldrige National Quality Award

From the foregoing discussion it is quite clear that the Malcolm Baldrige National Quality Award is based on the assessment of the quality processes used in the management of total quality system in an organization. In 1990 the Cadillac Division of General Motors won the Baldrige Award in the manufacturing category. When General Motors extensively reported their achievement of this prestigious national award, many pointed out that winning the Baldrige Award did not mean that the Cadillac division in particular, and General Motors in general, made the best-quality cars during the Award period. As stated earlier, the Baldrige Quality Award does not endorse the quality of the winner's products. Thus, when Cadillac advertised that the Commerce Department had praised its 4.9-liter V-8 engine, the Texas attorney general pressed charges of false advertising (Note 5.17). General Motors also claimed that its division won the award, out of 167,000 considered. This was the number of applications mailed out, often in large lots to individual companies. The number of applicants for that year's awards were only 97.

Joshua Hammond, president of the New York-based nonprofit American Quality Foundation, has warned business executives not to consider Baldrige Award as a cure-all for their businesses. He has pointed out to *Fortune* that it "does not address key elements of business success—innovation, financial performance, and long-term planning" (see Note 5.17).

The Baldrige National Quality Award is not a panacea, but is instead a quality award. Winning the national quality prize does not guarantee superior competitive performance in the market place. The processes examined in the Malcolm Baldrige Award criteria cannot guarantee, but may help, improving the organization's market share, growth, and long-term profits. A 1990 survey by the General Accounting Office of 20 top scorers in Baldrige competition concluded that most of these companies had improved operations such as delivery of products on time, number of defects in products, customer satisfaction, and so on. They also corresponded with significant improvements in overall performance measures such as market share, sales per employee, and return on assets (see Note 5.17).

EXHIBIT 5.12. Baldrige Quality Award Winners Since Its Inception

Year	Award Category and Winning Business Organization
1988	**Manufacturing** 1. Motorola Inc., Illinois 2. Commercial Nuclear Fuel Division of Westinghouse Electric Corporation, Pennsylvania
1988	**Service** None awarded
1988	**Small Business** 1. Globe Mettalurgical Inc., Ohio, a metal alloys maker
1989	**Manufacturing** 1. Milliken & Company, South Carolina 2. Xerox Business Products & Systems, Connecticut
1989	**Service** None awarded
1989	**Small Business** None awarded
1990	**Manufacturing** 1. Cadillac Division of General Motors, Michigan 2. IBM's Rochester, Minnesota facility
1990	**Service** 1. Federal Express Corp., Tennessee (first awardee in this category)
1990	**Small Business** 1. Wallace Co. Inc., Texas, a distributor of pipes and fittings
1991	**Manufacturing** 1. Solectron Corp., California 2. Zytec Corp., Minnesota
1991	**Small Business** 1. Marlow Industries, Texas
1992	**Manufacturing** 1. AT&T Network Systems Group, New Jersey, Transmission Systems Business Unit 2. The Texas Instruments, Texas, Defense Systems & Electronics Group
1992	**Service** 1. AT&T Universal Card Services, Florida 2. The Ritz–Carlton Hotel Co., Georgia
1992	**Small Business** 1. Granite Rock Co., Watsonville, California 2. Zytec Corp., Minnesota
1993	**Manufacturing** 1. Eastman Chemical Company 2. Ames Rubber Company
1994	**Manufacturing** 1. AT&T 2. GTE Directories
1994	**Small Business** 1. Wainwright Industries

EXHIBIT 5.12. (*Continued*)

Year	Award Category and Winning Business Organization
1995	**Manufacturing**
	1. Armstrong World Industries' Building Products Operations (BPO)
	2. Corning's Telecommunications Products Division (TPD)
1995	**Service**
1996	**Manufacturing**
	1. ADAC Laboratories, Milipitas, California
1996	**Service**
	1. Dana Commercial Credit Corp., Toledo, Ohio
1996	**Small Business**
	1. Custom Research, Minneapolis, Minnesota
	2. Trident Precision Manufacturing, Webster, New York

Private Versus Public Good

We must also consider the trade-off of benefits in the private and public good that a Baldrige Award winner gains. A company winning the Baldrige Award may accrue private good in conjunction with the impact of winning the award on the "public good" to the society. One of the obligations of winning the Baldrige National Quality Award is to communicate and propagate the lessons learned by the winners during their quality journey to others interested in learning the same quality lessons. For example, Motorola's top executives, after winning the Baldrige Award, have been speaking from time to time at conferences and making presentations to individual corporations. Motorola, since winning its award, has answered over 1000 outside inquiries. Motorola managers conducted 5-hour-long briefings every month for executives from other companies. Similarly, Westinghouse published and distributed over 6000 copies of their annual report that chronicled its Baldrige Award winning story.

It is therefore quite likely that the Baldrige Quality Award had a significant impact on the "public good," by improving the overall competitiveness of the U.S. industries in the global markets (see Note 5.18).

Some Criticisms of the Baldrige Quality Award Assessment

The Baldrige National Quality Award also attracted some criticisms. A major criticism against some of the winners of the Baldrige Quality Award was that the winners exaggerated what the Award actually represented. The Baldrige Award only assessed the quality-related practices of the winning company, whereas a winning company may be tempted to claim its overall superiority. Given below are some other criticisms against the Baldrige National Quality Award.

1. Incremental Improvements. Baldrige Quality Award assessment depends on incremental annual improvements. Hypothetically, some applicant companies may be tempted to first report the results of a bad year, and then show dramatic incremental improvements in the subsequent year in which they apply for the Baldrige Award.

2. Narrow Focus. Some companies may work hard only on the categories and the numbers that the evaluation procedure for the Baldrige Quality Award evaluates. They may thereby overlook other critical aspects that define their competitiveness in the global markets.

3. The Baldrige Award Is Division-Based and Not Organization-wide. In the categories related to the large organizations, the Baldrige Quality Awards were awarded based on the assessment of the quality practices and achievements of a particular division of the large organization. The Baldrige Award therefore reflected the practices and achievements of that particular division only. This implied that an award-winning organization can have an island of excellence in one part of the large organization. Other divisions of the award-winning large organization may have much lower or even poor quality standards. The winning organization as a whole may even have an overall poor performance, and many dissatisfied customers.

4. Overadvertisement. One expectation (and almost a condition for winning) from the award winners of the Baldrige National Quality Award has been that they were expected to spread the information on how they improved their quality. While doing so, some companies could use the winning of the award in their advertisements and promotion materials. However, they should do so in a manner that may not suggest that their entire organization practices and achieves excellent award-worthy performance. For example, the Cadillac Division of General Motors was the winner of the 1990 Malcolm Baldrige National Quality Award in the manufacturing category. Some critics considered their promotion of winning the award as overadvertising.

5. Applying for the Award Is Too Expensive. The assessment process used for Baldrige Quality Award is quite lengthy. Some critics have argued that the assessment process required full-time involvement of many senior-level employees and managers for a considerably long time. According to some estimates, the award procedures could cost up to $250,000. The critics of the Baldrige Quality Award considered this as a nonproductive investment that did not add value to the bottom line.

6. Use the Award's Assessment Process and Not the Award. Some companies considered the Baldrige National Quality Award assessment by outside examiners to be an expensive process. They therefore embraced the award's assessment criteria internally, rather than actually applying for the Baldrige Award. This is why every year the number of requests for the Baldrige Award applications has been much larger than the actual number of applicants for the award.

Xerox's Lessons from Winning the Baldrige Quality Award

Xerox underwent a grueling scrutiny involving more than 400 hours of analysis and on-site inspections by the Baldrige Award examiners. As a result of this careful examination, Xerox picked up a few pointers for winning the Baldrige National Quality Award. These are listed below.

1. *Malcolm Baldrige National Quality Award Has a "Process" Orientation.* Baldrige examiners are primarily interested in the "processes" pursued by the applicant organization in achieving high-quality standards.
2. *Applying for the Malcolm Baldrige Award Is a Win–Win Situation.* According to the organizers of the Malcolm Baldrige Award, if an organization applies for Malcolm Baldrige National Quality Award, it cannot lose. The application and the applicant's examination helps the organization and its managers identify their strengths as well as weaknesses—areas which can benefit from further improvements. While applying for the Malcolm Baldrige Award, "companies that do not win the award get nearly as much benefit as the winners of the award" (see Note 5.17).
3. *Listen to the Customers' Voice.* Look at your organization through the eyes of your customer—even if it hurts. You have to let the customers define your quality for you. Many large companies become too inward-looking.
4. *Get Management Commitment.* To achieve high-quality goals, you have to have commitment from the top, from the CEO down to the mail room.
5. *Set "Stretch" Goals.* You have to establish benchmarks that really "stretch" you. Set impossible goals. Then do your level best to meet them.
6. *Get Your Suppliers Involved.* In 1983, when Xerox began their "Leadership Through Quality" process, 92% of the parts received from their suppliers were defect-free. This was considered a fair standard for the office equipment industry. After Xerox developed their quality orientation, they attained a level of 99.97% defect-free parts from their suppliers.

In summary, one can see that the Malcolm Baldrige National Quality Award can be used for self-assessment as well as for spotlighting a company's quality-related achievements to the customers, suppliers, and other competitors. Like any other initiative at the national level, the Baldrige National Award means different things to different companies. It is also continually evolving with time. Therefore, until a company has developed a better assessment tool to guide its quality-related efforts, this is a good assessment tool to begin with. Next, we will look at the role that standards have played as an alternate assessment tool for achieving a high level of quality performance.

ISO 9000 STANDARDS OF QUALITY

To Certify or Not to Certify?

Most U.S. managers in technology-driven organizations wish to have some presence in the international markets. Most of them may have also heard that ISO 9000 certifications are the tickets, and sometimes the minimum requirements, to enter into the European markets. Some managers fear that the ISO certification costs a huge amount of money, takes a lot of time, and does not directly help in improving their firm's competitiveness. Depending on the remaining number of years until their retirement, and the uncertainty associated with their own jobs, some managers may not be too inclined to invest their time and efforts in the ISO certification. They may even successfully argue that most of their business is in the domestic U.S. markets, so why bother. So, the question that haunts many managers is very similar to the one that haunted Hamlet, the hero of a Shakespearean play. To certify, or not to certify, that is the question.

To answer this question, the following six questions must be answered first.

1. Is a significant amount of revenue of the firm generated in international or European markets?
2. Will the firm like to grow the international revenue stream in the firm's top line?
3. Are some of the customers asking or demanding an ISO 9000 registration?
4. Will the firm like to expand a similar international customer base further?
5. Do the firm's competitors have ISO certification, and do they repeatedly claim the same in their promotions to your customers?
6. Is the firm losing some of its current customers because of such claims by the competitors?

If the answer is yes to any one of these six questions, then the answer to the earlier haunting question is also yes. The firm must then try to certify for ISO 9000 registrations.

This is a good news—bad news thing. The bad news is that this decision to seek ISO 9000 certification will require a significant attention of a large number of senior managers and their subordinates. And it may still cause a lot of confusion. The good news is that seeking ISO 9000 registration may also help streamline and reengineer the production operations and lead to higher productivity.

Military Origins of Industrial Quality Standards

The industrial quality standards, such as the ISO 9000 series, are based on the rough experiences of military. In every war, the military services are clearly made aware of the high number of casualties on the battlefield due to the poor quality of their supplies. During World War I, the military forces of many countries were shocked by the lack of reliability and consistency of quality of their ammunitions and weapons. These were procured in a hurry from the civilian suppliers. It is estimated that during World War I, about 200,000 soldiers might have died primarily because of the poor quality of military supplies. To avoid such losses in the future, the U.S. military adopted certain military quality standards. Their suppliers and manufacturers were expected to conform to these standards as a prerequirement to doing business with the military.

In 1945, after the end of World War II, many civilian industries and government agencies also adopted similar standards. With time and increasing international trade, other specialized quality-related organizations emerged. These were chartered primarily for the development and administration of industrial quality standards. For example, in 1975 the British Standards Organization developed the British Standard (BS) 5750. This has been popularly considered to be the "mother document" for the International Standards Organization (ISO).

Sponsors of the ISO 9000 Series

The ISO 9000 Standards Series is sponsored by the International Organization for Standardization. It is an international body chartered to administer the standards, and it uses task forces and technical committees to do so. The ISO/Technical Committee (TC) 176 was chartered to review and revise these standards. More than 90 national standards bodies around the world are members of ISO. ISO consists of approximately 180 technical committees, each responsible for a specific area of specialization, spanning from abrasives to zirconia. Their technical work is published periodically. For example, in 1987 the ISO 9000 Series was published, and in 1989 it was adopted by the European Community. By 1991, compliance to the ISO 9000 series was integrated into the European laws related to product safety of goods, such as the telecommunications equipment, the electronic appliances, and the medical devices.

The American National Standards Institute (ASI) is the representative body of the United States in the International Standards Organization. In 1987 the ISO 9000 Series was adopted by the United States as the ANSI/ASQC Q-90 Series.

European Community Union of 1992 and Standards

In 1992 the European Community (EC) was united together in a historical manner. Twelve major European countries, including Belgium, Denmark, France, Germany, Greece, Ireland, Italy, Luxembourg, the Netherlands, Portugal, Spain and the United Kingdom, agreed to break down some of the barriers of their national borders for trade across these major markets. Soon thereafter, many technology-driven organizations in the United States and other parts of the world (such as Du Pont, Union Carbide, and others) quickly realized that as a result of this unification, a system of standards proposed by the International Standards Organization (ISO) had become a prerequisite to doing business in this market. The ISO 9000 certification became a ticket required to enter into this big and highly prosperous unified market. This marked the birth and the rise of strategic significance of ISO 9000 Series. The ISO series consisted of a set of five standards. Unlike the product standards used for decades by organizations such as American Standards for Testing Materials (ASTM), the ISO 9000 standards are standards for quality management systems, and not endorsements for a specific product or service.

The ISO 9000 series is an elaborate system of framework intended for assuring and achieving excellent quality. The objective of the ISO 9000 series is that after a company has received ISO 9000 certification by an elaborate system of internal and external audits, that company's customers are assured that the published quality standards of the company will be delivered in a consistent and repeatable manner.

ISO 9000 Global Phenomenon

Within 5 years of the first publication of ISO 9000 standards in 1987, the series had an enormous impact on the way the business was conducted around the world. Many other countries, besides the European Community, adopted their own standards based on the ISO 9000 series. For doing business in the unified Europe, tens of thousands of suppliers around the world were expected to get these third-party registrations, as a requirement for taking part in these markets. Europe-based technology-driven firms, such as Siemens, which procured equipment, parts, and components from technology-driven suppliers in the United States, started requiring its suppliers to make efforts to receive ISO 9000 certification as a condition for doing business with Siemens. The buying company argued that this saved them time, effort, and money to test and inspect the quality of the parts procured from their external suppliers.

It was estimated that by the start of the 21st century at least 100,000 U.S. companies would have registered, or applied for registration, for ISO 9000 standards. An increasingly larger number of companies came under pressure to become ISO 9000 certified as early as possible because their customers demanded it or because their competitors had already certified ISO 9000 registration. The companies which did not have ISO 9000 registrations were likely to lose some of their business. There was, however, one question that many companies worried about. Where do they start? The section below will describe the ISO 9000 series, and what each standard in the series stands for.

The ISO 9000 Series

The ISO 9000 series consists of five standards, numbered from ISO 9000 to 9004 (see Exhibit 5.13). This includes three standards, one set of guidelines, and one management framework for the entire series. The ISO 9000 series of standards was first published in 1987 and revised in 1994. These are standards and not awards. They are not prescriptive recommendations. They do not imply or suggest how a company must manage its product and process quality. But their compliance mandates that the certified company must clearly and specifically define its quality-related standards and describe in detail the processes used to assure those standards. The company must also prove that it consistently adheres to the claimed standards and procedures. The objective of the ISO 9000 series of standards is to reassure the customers that a holistic quality system is in place and that the company has the capability to provide goods and services of the promised quality.

The ISO 9000 series of standards applies to all industries, in manufacturing or service sectors. And these standards apply to all types of companies in different industries. The ISO 9000 registration ensures that the different

EXHIBIT 5.13. Conceptual Organization of ISO 9000 Series

	ISO 9004 Quality management and quality system	
ISO 9001 Quality of design, development, production, installation, and servicing	**ISO 9002** Quality assurance in production and installation	**ISO 9003** Quality assurance in test and inspection
	ISO 9000 Definitions and selections	

companies doing business with one another have robust quality management systems put in place. The specific levels of acceptable quality levels are left to the negotiations between the individual customers and their suppliers.

Together the five standards in the ISO 9000 form a quality system. Unlike the inspection-based quality pursued by many companies in the past, the ISO 9000-based quality system is a prevention-based quality assurance system. The goal of this company-wide quality system is to ensure a consistent quality of products and services produced by robust and reliable processes. The objective is to assure customers that the promised quality goods and services will be delivered in a dependable manner each and every time.

1. ISO 9000 Standard. This standard provides the definitions of the underlying concepts and principles governing the application of other standards (ISO 9001, ISO 9002, ISO 9003, ISO 9004) for a specific business firm. This standard applies to all industries, including the newly emerging software industry. Good cross-references of the different elements used in the ISO 9000 standard series are available under this standard.

2. ISO 9001 Standard. This is the most comprehensive standard in the ISO 9000 series. The ISO 9001 standard covers the most comprehensive quality system, with the maximum number of elements in its quality system. This standard is to help conformance to quality and quality assurance for the design, development, production, installation, and servicing parts of an organization. This standard is particularly applicable to construction and engineering firms, designers, and manufacturing firms which design, develop, install, and service products. The ISO 9001 standard recommends a three-tier documentation of quality system. These three tiers are as follows:

Level 1. A quality manual that defines the overall management of the quality system. The quality manual answers the why question. The quality manual is the governing document (Note 5.19). It documents the quality policy and objectives that the organization pursues. It lists the responsibilities and the degrees of empowerment of different individuals involved.

Level 2. Procedures that guide interdepartment coordination. They answer the what, when, and who questions. The organization-wide procedures for each system element are developed from the quality manual. These interdepartmental procedures should be developed from the quality manual with the consent and consensus of the concerned departments.

Level 3. Work instructions used within each department. The instructions explain and document how each task is accomplished. More documentation is required as proof and evidence of the practices followed by the organization. The instructions for individual departments are step-by-step task-level instructions derived from the interdepartmental procedures.

3. ISO 9002 Standard. This standard documents and records quality assurance during the production, assembly, and installation stages only. This may be relevant to companies, such as polymer injection molders, who are themselves not actively involved in the design of their products or its after-sales service. The structure and documentation of the quality system for ISO 9002 standard is very similar to that of ISO 9001 standard.

4. ISO 9003 Standard. This standard covers quality assurance in the test and inspection stages of an industrial enterprise. This standard may be applied to small test houses, or divisions of a large manufacturing organization, or distributors, who inspect and test products procured from their manufacturers.

5. ISO 9004 Standard. Finally, this standard deals with the quality management and with the elements of the overall quality system. This standard can be applied to many different types of firms in different industries.

For a company to be certified ISO 9001, ISO 9002, ISO 9003, or ISO 9004, it must have an on-site audit of its operations by an accredited, third-party, independent examiner. This examiner will verify that the company is in compliance with the requirements of the standard that the company seeks to register for.

Registration Examples

Du Pont's ISO Staircase and Grace's Success Story. The managers at Du Pont, the world's largest multinational chemical company, recognized very early the advantages of receiving the ISO 9000 registration. They could foresee that ISO 9000 registration was likely to quickly become a minimum expectation for doing business in the international markets. After all, inside Du Pont there was an early believer in the ISO 9000 system. One of the original members of the Task Group Committee (TC) 176, which was chartered to write the ISO 9000 standards, was Donald Marquardt. He was also the founder of Du Pont's Quality Management & Technology Center. With Du Pont's many worldwide operations along with active guidance from Donald Marquardt, Du Pont earned its first ISO registration in 1987 in Europe. In the next 7 years, the number of Du Pont operations with ISO registrations swelled to over 225. About 100 of these were in Europe and the United States, and the rest with the exception of four were in Asia Pacific. Du Pont's South American operations had four ISO 9000 registrations (see Note 5.20).

Cost and Benefits of ISO. Du Pont also learned that if ISO certification is not done systematically, attempts to gain ISO 9000 registration can get very chaotic as well as costly. On the other hand, by proceeding in a systematic manner, Du Pont was able to save significant amounts of costs as a result of fulfilling the ISO 9000 requirements. From Du Pont's record of savings directly accrued as a result of ISO 9000 registration, given below are some of the benefits.

1. A Du Pont electronics plant saved $300,000 cost in its molding section.
2. Another electronics plant was forced to calibrate its process control equipment, resulting in savings of $440,000. More savings were made by replicating these improvements in other plants.
3. As a result of the internal audits for ISO 9000 registration, one Du Pont site was able to consolidate more than 3000 tests to 1100 tests.

A Road Map to ISO 9000 Registration

Based on these and many more savings, Du Pont's Quality Management & Technology Center developed a "Road Map to ISO 9000 Registration." Note that this road map did not recommend rushing to a registrar. On the other hand, it required a considerable amount of preparation and soul-searching before starting the journey. The proposed certification project included many steps with clearly marked milestones. The milestones in the original Du Pont road map have been modified somewhat. We have included an additional step for rewards and recognition. Furthermore, each step is illustrated with the actual actions taken by Grace Specialty Polymers (GSP), based in Lexington, Massachusetts, which successfully received its ISO 9001 registration in September 1994. Roger Benson and Richard Sherman shared GSP's successful journey in an article in *Quality Progress,* with their belief that the ISO registered companies should share their experiences with other U.S. companies interested in similar registration (see Note 5.21).

Step 1. Acquiring Management Sponsorship. The journey for ISO 9000 registration must begin with a clear and committed sponsorship by the senior management of the registering organization. The process generally required significant resources that the senior managers must be willing to allocate. The project also puts a heavy demand on the senior managers' time and direct involvement. To get senior managers' commitment, they first must be educated about what ISO 9000 stands for. Du Pont's experience showed that a half-a-day presentation to the senior management by an experienced facilitator could generate the required buy-in.

At Grace Specialty Polymers, management established a goal in early 1993 to become ISO 9000 registered by the end of 1994. The vice president/general manager had envisioned the benefits of total quality improvement (TQI) many years earlier. The objective of the project was clearly defined. GSP would seek ISO 9001 certification for four neighboring sites, which included the headquarters, the research & development facility, and two manufacturing units. All the four sites were located within 30 miles of one another.

Step 2. Planning for ISO 9000 Registration. Once the senior management is convinced, then a working plan must be developed with their help. The working plan included an outline of the goals and scope of ISO registration project, an initial assessment of the organization, desired state of the organization for registration, resources available for such efforts, and the time schedule. Du Pont's experience showed that an 18-month-long time horizon was common.

Grace Specialty Polymers had several factors in its favor. The company had been using military standards (MIL-I-45208A) and had earlier achieved Ford's Q1 automotive quality requirements. The company also had a strong TQI system already in place. Therefore pursuing ISO 9000 registration seemed a natural extension of these activities. The steering committee attended a one-day management awareness training session. The session discussed the basic requirements of the ISO 9000 registration process, and explained the responsibilities for the management.

Step 3. Developing Internal Resource Infrastructure. The working plan could help establish a basic infrastructure of people who are assigned the overall responsibility. This included the members of the steering and implementation committees. Du Pont recommended that a senior manager must set aside over 75% of his or her time for ISO 9000 efforts in its early stage. Other members could spend from half to full time for the registration. Their time commitment usually tapered down with the progress of the ISO registration project. The steering committee coordinated the efforts and progress of coordinators appointed in different departments. All the steering committee members and the coordinators must be educated in the requirements of the ISO 9000 registration. In order to do this, Du Pont recommended a 2-day seminar.

Grace Specialty Polymers chartered an ISO steering committee. This consisted of the general manager and those who reported directly to him. They were to provide commitment, direction, and the needed resources. A seven-member ISO Implementation Team (IIT) was also chartered by the steering committee. A TQI director was appointed as the leader and liaison member of the IIT. The cross-functional implementation team consisted of department managers from four sites and three major departments: quality, R&D, and manufacturing. Each member was asked to coordinate the ISO registration-related activities in his or her jurisdiction.

All the members of the ISO Implementation Team attended an in-house lead assessor training program. This was put together with the help of other business units of W. R. Grace & Co. The IIT decided to meet every week, adopted Du Pont's stair-step approach, and outlined a road map. This was reviewed and adopted at the first meeting of the implementation team. The team members also instructed their staff members to undergo company-wide awareness training using commercial videotape presentations. These presentations were supplemented with site-specific information.

Step 4. Initial Self-Audit. Before any changes are made, the initial condition of the organization must be carefully defined. This could be done with the help of experienced and specially trained auditors. Some internal auditors may be trained by the company, but they must not audit their own places of work. The initial self-audit should document the current state of the organization's quality system. During this self-audit, the prevailing policies, actual procedures, practices, and paperwork must be carefully documented.

In this step, the members of the ISO Implementation Team at Grace Specialty Polymers started networking with other ISO registered companies. They soon discovered that a highly qualified consultant was not available at that time, and therefore they decided to do in-house preaudit and gap analysis themselves. The Implementation Team split into three audit teams to systematically audit each site. They diligently followed the different sections of the Noncompliance Report (NCR) of the ISO-9001 standard. The audit members classified each noncompliance either under a major or a minor category. Each subteam presented the findings to all the members of the ISO Implementation team, and they discussed the same at their weekly meetings.

Step 5. Improved Standard Procedures and Policies. Next, the initial state of the organization's quality system was compared with the expectations of the desired ISO 9000 registration. The senior managers must be involved in prioritizing the proposed changes. New and improved standard policies and procedures must then be carefully documented.

Cultural change is an intimidating experience for many workers and managers alike. Over time, people get complacent and set in their old ways. Involvement of workplace and multifunctional teams in the process of change can help relieve some of the real and imaginary anxieties.

One of the primary documents used to document all changes is the quality manual. The quality manual lists the major quality-related policies and procedures. This is generally about 20 to 40 pages long. Usually, a draft is first written by the employees, and then an expert well aware of ISO 9000 requirements is chartered to put it in a form that meets the ISO requirements.

As the ISO Implementation Team auditors at Grace Specialty Polymers reviewed their old quality system, they also proposed some possible solutions to eliminate the major and minor discrepancies in their NCR. The ultimate responsibility for modifying procedures in a particular area was assigned to the IIT member with the most knowledge for that area. The major noncompliances that were identified required changes in the company's management philosophy or some corrective actions beyond the ability of a single team member. To handle these major noncompliances the ISO Implementation team appointed subcommittees of two to four members. These subcommittees studied the causes of such noncompliances. They proposed corrective actions and estimated the associated costs. These were given to the relevant functional manager who made the final decisions in accordance with the organization's prevailing management philosophy. These decisions were reported to the ISO Implementation Team. The leader of the Implementation Team communicated these to the steering committee for ISO registration.

At W. R. Grace & Co., benchmarking was used wherever possible. Other business units within W. R. Grace & Co., as well as the suppliers and some customers, provided useful baseline information. For example, a major concern was the degree of details for documenting the task instructions, procedures, and policies in the three-tier documentation. Instead of building a quality manual from scratch, the ISO Implementation Team decided to use GSP's existing quality assurance manual as the starting point. The IIT members reviewed, critiqued, and revised the procedures listed in the quality assurance manual. Gradually, a new ISO 9000 manual emerged. This manual followed the standard ISO registration format, which later helped the external auditors in reviewing the procedures more efficiently.

Most of the documentation control was housed in the existing quality assurance department. Exception was made for product design documentation. This documentation was created and controlled by the research and development department. Later, this evolved into GSP's new-product development manual.

Step 6. Selecting a Certification Registrar. At this midpoint in the registration road map, the company seeking ISO 9000 registration must select their third-party registrar. The choice must depend on the credibility of the registrar with the firm's major customers. There are some variations in the practices followed by different registrars. An initial half-a-day informational meeting is of immense help, whereby both the parties can assess each other, ask questions, and present their own philosophies.

At Grace Specialty Polymers, the leader of the ISO Implementation Team decided the registrar. This was done after contacting different registrars directly and after benchmarking other similar organizations.

Step 7. Implementing Quality System. As the improved quality-related procedures and policies are implemented, the documentation for the quality system must be validated and updated. The efficiency as well as the effectiveness of the firm's quality system must be carefully measured and documented. Additional information such as customers' response to the improvements in product and process quality must be tabulated over time to show the trends. The new quality system's stability must be checked for at least 6 weeks. When the new quality system is about 80% implemented, the top management of the organization may be invited to review the improvements.

Management of change is not easy. People have their egos involved in their work. They are driven by their need for self-preservation and territorial ownership. Some people are also prone to act with biased or insufficient information. People must be told how the changes associated with ISO registration are going to affect or not affect their positions in the workplace.

To facilitate the implementation of the improved quality assurance system at Grace Specialty Polymers, corrective action teams (CATs) were appointed. These teams helped to resolve specific noncompliances discovered during the internal assessment audits. For example, an elaborate training program was developed for developing policies and procedures. This was not a direct requirement for the ISO 9000 registration, but it helped. The training of each employee's job function, however, is covered specifically by a section in the standard. A training system was developed and implemented. It included identifying and developing the minimum training requirements for each job, the training procedure, and requirement forms. A matrix of training requirements for each employee and the training certificates were also developed. The training system documented how each employee met the training requirement, or had sufficient on-the-job experience instead of the required training. Five years on the job experience was considered a cutoff point for meeting such training requirements.

Step 8. Preassessment Rehearsal. As an organization gained significant confidence in its new quality system, a full-scale self-assessment must be conducted with the help of the third-party auditors. Such external assessment ensures that the new quality system is functioning smoothly. The experienced third-party auditors can help identify the remaining weaknesses. They would also check whether the policies and practices listed in the quality manual and other documents are actually practiced on a day-to-day basis. The third-party auditors should not be the certifying registrars, who cannot help in removing the discrepancies while auditing for registration. The selected third-party auditor should be the one who is familiar with the practices of the chosen registrar. The identified discrepancies should be carefully removed in a timely manner, and the documents should be updated accordingly.

Grace Specialty Polymers originally made plans to do preassessment by outside auditors. However, they later decided to skip this step, because they believed and discovered that they had a strong quality assurance system in place. They also knew that if there were noncompliances identified by the auditing registrar, they will have about 40 days to focus and correct those noncompliances specifically.

Step 9. Registration Assessment Audit. Finally, the chosen registrar is requested by the registering organization to arrange a scheduled audit visit. This audit visit takes place in three phases. There is an opening meeting, then the investigation and interviews take place, and finally there is a closing meeting where the registrar announces the results of the assessment. The assessment is either a recommendation for registration or its refusal.

The assessment audit for Grace Specialty Polymers took 3 days by two auditors. One of them was designated as the lead auditor. They audited the R&D facility and headquarters for half a day each, and they audited each of the two manufacturing facilities for a full day. At each of the facilities the organization assigned two guides of their own, who were the members of the ISO Implementation Team and who were also experts in the audited area. These guides were to correct the auditors' "observations" as they arose. Such observations are not considered noncompliances by the auditors, but are expected to be corrected in a timely manner.

One of the guides was the quality assurance manager. He accompanied the auditors throughout the entire audit. The leader of the ISO Implementation Team was one of the two guides at the company's headquarters. From there he also coordinated the assessments at other sites as the assessments proceeded. He attended the closing meetings at all the sites. He started addressing corrective actions for noncompliances as soon as these were noted by the auditors. Thus, of the 11 minor noncompliances pointed out by the auditors, Grace Specialty Polymer was able to solve seven noncompliances before the end of the assessment audit. No major noncompliances were pointed out by the auditors.

Step 10. Acquiring the ISO Registration. If the registrar's assessment is favorable, then the organization receives a certification of registration in 1 or 2 months. If some minor and nonsystemic discrepancies are pointed out by the registrar, then the organization has 6 to 8 weeks to fix them. Systemic deficiencies result in rejection, and they call for a new audit of the quality system.

The auditors recommended Grace Specialty Polymers for ISO 9001 registration, and the company received its registration on September 16, 1994. This was 3 months before the planned 20 month deadline.

Step 11. Celebration, Rewards, and Recognition. Seeking ISO 9000 registration is a long and arduous journey. Any excuse to help participating people keep up their progress is a good idea. Recognition of extraordinary efforts by managers leads to further reinforcement of such efforts by the participants. Sometimes a good word can make a big difference. It helps to celebrate completion of milestones. Usually, by the time registration comes, some key people are extremely exhausted by the demands of the registration process. It may be therefore worthwhile to recognize and reward them in an appropriate manner.

It was not clear how Grace Specialty Polymers celebrated its ISO 9001 registration, but everyone seemed relieved and proud of their achievements.

Step 12. The Certification Journey Continues. The registration for ISO 9000 series of standards is just the beginning of the ISO certification process. The ISO 9000 certification only signifies that a reasonably good and reliable quality assurance system has been installed and is being implemented. It must be continually upgraded, and the changes must be carefully documented.

Grace Specialty Polymers was able to receive its ISO registration in a timely and efficient manner because of its clear goals, well-defined milestones, and widely shared status reports. Dividing the overall project into smaller manageable segments helped the implementation of the registration project. By involving key actors in the various training programs, the company was able to build in-house expertise and avoid the use of expensive outside consultants. The company also benefited from benchmarking and networking with other divisions in their parent company. This kept them on target, and they were able to achieve their goal ahead of time.

Limitations of ISO 9000 Standards

From the foregoing explanation of the ISO 9000 series of standards, it is clear that the ISO 9000 series of standards do not aim to help directly in finding out how to run a technology-driven company. The standards merely provide a framework for a robust quality system with improved reliability. Therefore, the pursuit of ISO 9000 registrations has produced mixed results for different organizations. Some companies benefited immensely and survived in the market primarily because of their pursuit of ISO registration. Other companies were so frustrated by their efforts to gain ISO registration that they eventually abandoned their mission for ISO certification. In some cases the failure to achieve ISO 9000 registration resulted from the fact that they started very far from having any form of a quality assurance system in place. They had to invent the whole quality system from scratch. Some other companies were frustrated because they had unrealistic high expectations from their ISO registration efforts. If a company has a good quality assurance system already in place, then preparing for ISO 9000 registration will require only a few incremental changes.

The ISO 9000 framework can help managers identify gaps and weaknesses in the different segments of their organization. However, filling those gaps may only produce isolated islands of disconnected quality-related practices. The companies' diverse practices still need to be integrated together to work in unison and produce outstanding results in the market.

Integration of Islands of Practices. A major criticism of pursuing ISO 9000 series registration or its audit has been that it results in a large set of diverse procedures for the different business activities. These activities are like islands which are hard to manage and control. They often fail to synergize and not act in unison. The lack of company-wide integration does not help in improving the competitive position of the company in the marketplace.

Use MRP to Integrate Islands of ISO 9000 Requirements. Some integration of quality system can be achieved by using a computer-based operation system that many technology-driven companies are very familiar with. For example, the traditional materials requirements planning (MRP) or manufacturing resources planning (MRP II) system can be used for integration. These systems have been commonly used for materials purchasing and planning, inventory control, production scheduling, and so on. With the use of a common system in different parts of a company, the MRP or MRP II can be modified for the quality assurance system of a technology-driven organization. By doing so, a significant amount of repeat efforts and paperwork is eliminated in different departments of the organization. Using MRP or MRP II, the data entered once can be used by the different departments in the organization. The purchasing department can monitor and expedite procurement of urgently needed parts and supplies, while the production scheduling department can use the same data to program the availability of different materials. The same system helps the human resource department hire the needed work force, along with the accounting department to arrange for their payments (see Note 5.22).

The MRP and MRP II systems can be further adapted to satisfy some of the quality-related requirements of ISO 9000 certification, while at the same time they are integrating and improving the business performance of the technology-driven organization. For example, the use of MRP and MRP II can meet the ISO 9000 requirements for purchasing, capacity planning, testing of raw materials and supplies, and so on. For the ISO 9000 registration, the company managers will have to show how MRP or MRP II systems are used by the company on a day-to-day basis to coordinate the activities required to maintain and certify the high-quality standards expected under ISO 9000 requirements.

The ISO 9000 standards insist on detailed and specific documentation for making changes in procedures, test results of incoming parts and supplies, statistical analysis and results of process capability studies, inspection reports of the outgoing finished goods and so on. The reports thus generated should provide reliable information about the specifications of the products produced, and variations in the processes used to produce these products. Producing these reports manually is a hard and tedious job. Use of computer automation and software, integrated with various elements of the technology value-adding management system, can generate these required reports with more reliability and less time.

In some companies, different departments acquire and operate different software programs. These departments do not coordinate with each other before ordering their softwares. These programs are therefore often not compatible with each other and do not "talk" to each other. This happens when the different departments operate as independent "silos" with very rigid boundaries. These departments operate in a mutually confrontationary manner, and they communicate only when they are forced to. As the required information flows from one department to another, because of the differences in the software programs, most of the data must be reentered manually each time. This consumes a lot of unproductive time and effort. Furthermore, this practice increases the possibility of errors in repeated data entry. The decisions about acquisition of appropriate software programs should be based on their capability to integrate across different departments. This decision has significant implications on the efficiency and effectiveness of different departments and the competitiveness of the company.

In general, the ISO 9000 standards do not specify requirements for selection of the software programs. However, in practice most auditors for ISO 9000 certification tend to look for a validation of the software used by an organization to validate product specifications and process capability. Like in the case of many other operations, the ISO 9000 auditors also tend to look for the documentation regarding changes in the different versions of a software. They want to see the documentation manuals, as well as the procedures used for the control of the master copies and documentation manuals of the software programs used by the company. The users of a software should obtain validation certificates from their vendors, or they should themselves validate their softwares in-house before the scheduled visit of an ISO auditor.

Postregistration Surveillance

First Fix the Minor Findings. Once a company is audited, unless the firm was in total compliance with their proposed quality system, the auditing registrar would typically issue a number of minor findings. These are deficiencies which are nonsystemic. But they require some quick attention. These must be fixed if the firm does not wish to risk its registration in future audits. The management of the firm must carefully assign responsibilities and deadlines for carrying out these corrective tasks. Proper resources must be allocated. The consequences of these changes on other parts of the organization must be carefully evaluated and communicated appropriately to the relevant people concerned. Some internal auditors must be appointed to verify that the proposed deficiencies are fixed prior to the registrar's revisit.

Even after a firm is recommended for ISO 9000 registration, there is still more work to do. Most of the accredited registrars conduct periodic surveillance audits to ensure that the registered firm is maintaining its compliance. The surveillance systems are not standardized for all registrars. Some registrars issue certification with no expiration date. Their registration is valid provided that the registrar sees evidence in follow-up inspections that the registered firm has maintained its quality system properly. Other registrars, such as Det Norske Veritas Quality Assurance use a 4-year periodic cycle (see Note 5.23). Lloyds Register, the German Association for the Certification of Quality Systems, and Yardley use a 3-year cycle for their periodic assessment.

Some auditing registrars may even insist on a continuous rigorous re-assessment. For example the British Standards Institution (BSI) considers the postregistration surveillance as a continuous assessment process. They issue their registration as a permanent certificate, provided that the registered firm stays in compliance. Typically, during the continuous assessment audit some minor discrepancies are found and pointed out. These are fixed by the registered firm as early as possible, and the registration continues (see Note 5.23).

Continuous Improvement of Critical Processes. Finally, with time and changes in the market's competitive environment, sometimes critical business processes need to be changed after an ISO 9000 registration is acquired. For example, the acquisition or development of a new technology may warrant such changes. Mergers and alliances may also impose the need for such changes. Such continuous improvements must be made in a systematic and planned manner. A Plan–Do–Check–Act cycle, explained below, may be used to guide such efforts.

1. Plan and Assign Ownership. The need for changing the critical processes must be clearly defined. The responsibilities, accountability, and ownership must be specifically assigned. Adequate resources should be allocated, and controls should be put in place for budget and time overflows.

2. Do the Change. Each group should be empowered to carry out and schedule their assigned changes. These groups are more intimately aware of the new developments in technology and other practices in their area of expertise. Their cycle-time goals, however, must be monitored. These groups should be encouraged to use flowcharts. Changes from one review to the next review should be diligently noted on these flowcharts. Professional writers may then be brought in to write the modified processes.

3. Management Checks. The changes in critical business processes should be reviewed by the senior managers. Management reviews are recommended by ISO 9001 and ISO 9002 standards. The management is also likely to be more aware of the shifts in the firm's competitive position. Maybe market research has indicated the existence of a price war or a demand saturation. These changes demand changes in the company's procedures. Impact of these external changes must be integrated with the internal modifications of critical processes. Management must also review the cost of quality and ensure that it is held well within its desirable limits.

4. Act Beyond the Minimum ISO Expectations. Finally it must be clearly understood that the ISO 9000 requirements are only minimum guidelines for doing business in a globalized market. To win the competition, the firm must exceed far beyond these minimum requirements. Sustainable competitive advantage is gained by developing core-unique competencies. These competencies are rare and are hard to imitate, and they add significant value to the firm's success in the eyes of its customers.

In conclusion, the ISO 9000 series of standards can be deployed not only to gain access to the markets in the European Union, but also to improve competitiveness by the deployment of a robust quality assurance system.

In the next two sections we will discuss two forms of industry-level standards: the ISO 14000 environmental standards and the QS 9000 industrial standards adopted by the U.S. auto makers.

ISO 14000 SERIES STANDARDS

The ISO 14000 standards are the standards for the management of environmental impacts of an industrial organization. Many technology-driven organizations such as the ones from semiconductors, chemicals, polymers, electronic appliances, office equipment industries and others fall under this purview. Like the ISO 9000 series, the ISO 14000 standards are also administered by the International Standards Organization. They were jointly developed by the U.S. Technical Advisory Group (TAG) for Subcommittee 1.

The primary objective of the ISO 14000 standards was to draw worldwide attention to the environmental management issues, and urge industrial organizations to manage their environment beyond complying with the basic requirements specified by national regulations. The ISO 14000 standards hoped to promote a voluntary approach to developing consensus environmental standards. This was expected to promote international trade and to ensure consistency and predictability across nations without damaging the global environment.

The Universal Environmental Standards

The ISO 14000 standards were developed as the universal environmental standards. They were universal in that they were meant for all industries in all countries. They provided a framework for managing environmental compliance for all aspects of an organization's operations pertaining to environment.

The ISO 14000 series standards consisted of the following five categories:

1. Environmental management systems
2. Environmental auditing
3. Environmental labeling
4. Environmental performance evaluation
5. Life-cycle assessments

These standards were not expected to have a single governing document or manual, nor did they have a prescribed rate of performance improvement. There was no mandated governmental or organizational policy requirement.

The ISO 14001 Series

The ISO 14001 standard defined an environmental aspect as any "element of an organization's activities, products or services that can interact with the environment" (see Note 5.24). The environmental standard implicitly wanted organizations to go beyond the minimum mandated requirements by their respective governments. For example, all U.S. companies are expected to comply with regulations such as filing a Toxic Release Inventory report for the U.S. Environmental Protection Agency (EPA). Most industrial organizations in the United States prepared some kind of documentation to comply with environment, health, and safety requirements. This involved monitoring and reporting of their company's air emissions, wastewater drainage and other hazardous effluents. When the industrial operations of a company spanned beyond the United States into many other countries, such information often got scattered in different departments of the organization. A universal environmental sensitivity and standard made managers assign specific responsibilities for such environment-related information.

Besides these regulated operations, the organizations must monitor their unregulated environmental effluents and practices—for example, their consumption of energy, water, and the gaseous effluents such as carbon dioxide, sulfur and nitrous oxides, polluted water, and so on. The ISO 14000 standard also proposed that the industrial organizations should carefully assess disposals at the end-of-the-product life cycle. The end-of-the-product life cycle extended beyond the useful life of a product.

For example, the ISO 14004 standard proposed a systematic multistep audit process. According to their guidelines, organizations should select a process or an operation, define its relevant environmental aspects, and evaluate their impacts on the environment. Finally, the significance of this impact should be evaluated.

The choice and the use of materials in quality industrial operations is closely related to the environmental aspects. A proper understanding of hazardous substances and other chemicals can go a long way in searching for the less harmful substitutes. In a similar manner, quality defects also often correlated with the discharge of pollutants. More defects in a product or service means more rework, and therefore more use of environmentally harmful materials such as the cleaning solvent.

The registration for ISO 14001 standard is done by an accredited third-party audit. This requires evidence of implementation of procedures, along with an environmental management system that ensures compliance to the local, national, and other applicable laws. The environment management system must also show commitment to the prevention of pollution, along with continuous improvement measured by recycling, efficient use of energy, substitution of harmful materials, and so on.

In January 1996, San Diego–based SGS-Thomson Microelectronics became the first U.S. facility to gain certification to the draft ISO 14001 registration. The corporate office had mandated that by 1997 all their worldwide facilities must have a third-party audit for compliance to the European Union's Eco-Management and Audit Scheme regulation. They applied for ISO 14001 registration as "an icing on the cake" (see Note 5.24). The environmental management team listed all the major environmental aspects and estimated their impacts. These were then classified for their severity and actual intensity.

Quality Digest, a trade journal in the area of quality management, reported in its August 1996 issue that after more than a year-long preparation, LeMoyne, Alabama's plant of Akzo Nobel Chemicals received the first joint ISO 14001–ISO 9001 certificate. This was awarded before the final version of the ISO 14001 was out. A team of 14 managers and engineers investigated and prepared Akzo's quality management system (QMS), as well as its environment management system (EMS). They found enough reasons to seek the joint certification. The joint assessment audit was conducted by registrar Bureau Veritas Quality International. To do the audit the registrar sent two lead auditors, one for ISO 9001 assessment audit and the other for the ISO 14001 audit. They worked together where the QMS and EMS overlapped.

Other leading U.S. companies, such as the Procter & Gamble Company (P&G) and Digital Equipment Corporation, also considered using the ISO 14000 standard as a basis for incorporating environmental sensitivity into their management of operations.

Let us next turn to standards for automobiles.

QS-9000 AUTO STANDARDS

Gone are the days when the new auto models could be launched with defective parts. During the mid-1990s the Big Three auto makers—General Motors, Ford, and Chrysler—were able to change the perceptions of their customers significantly. This happened to a great degree because of their partnership with the Tier-1 components and parts suppliers. These suppliers were expected to pass a rigorous launch testing phase and to ensure zero defects thereafter. Supplier Quality Assurance departments at the Big Three auto makers had worked hard to ensure this level of assurance for each one of their Tier-1 suppliers. In 1995 there were about 13,000 Tier-1 suppliers. The auto makers' representatives periodically visited the suppliers to review their progress and suggest corrective actions.

The Big Three auto makers in the U.S.—General Motors, Ford, and Chrysler—chartered a Supplier Quality Requirements Task Force to develop standards for their various suppliers and other associates. The QS-9000 standards were proposed in September 1994 to define the auto makers' expectations for all the internal and external suppliers providing production materials, supplies, heat-treated parts, and service parts. GM quickly announced that it required third-party registration of its new suppliers by January 1, 1996, and of the current suppliers by middle of December 1997. Chrysler similarly expected third-party registration of new as well as current suppliers by July 31, 1997. Ford took a little longer time in deciding its expected dates for suppliers' registrations (Note 5.25). The auto makers seemed serious about their expectations, and started their QS-9000 audits in the first quarter of 1996.

The QS-9000 standard contained the ISO-9001 standard, but its requirements were broader than ISO 9000. Some argued that the requirements of the QS-9000 were more rigorous. An anomaly was caused by the observers counting and arguing over the numbers of requirements in the two standards. The QS-9000 series of standards was divided into three major sections:

1. Common requirements, consisting of exact requirements of ISO 9001, with the addition of requirements for automotive and trucking industries.
2. Additional requirements common to all three auto makers, which are beyond the scope of ISO 9001.
3. Customer-specific requirements unique to General Motors, Ford, and Chrysler.

The Big Three U.S. auto makers felt the need to develop their own standards because of certain inconsistencies in administering ISO 9000 certifications. This included inconsistencies of interpretations by different registrars, as well as a perceived conflict of interest in registrars who offered consulting services also. Furthermore, they also wanted to highlight certain quality-related aspects more than the others. For example, the auto makers wanted their suppliers to spend more time and effort on upstream quality planning. They also urged more robust production systems with a higher weight on prevention rather than on inspection. The auto business also required more emphasis on on-time delivery and authorized parts qualifications.

These expectations by the Big Three U.S. auto makers caused a big dilemma to the ISO 9000 administrators. They had to strike a balance between (a) the needs of a specific and powerful industry and (b) the generic requirements that can be met across all industries (Note 5.26).

At the time of the launch of the environmental standards, the QS-9000 assessments were conducted by registrars accredited by the U.S.'s Registrar Accreditation Board (RAB), the Netherlands' Road Voor de Certificatie (RvC), and the United Kingdom's National Accreditation Council for Certification Bodies (NACCB). By having common acceptable standards, the Big Three auto makers have saved significant cost and overlapping efforts.

The Future Expectations

Soon after their launch, the QS-9000 standards were also adopted by many truck manufacturers, such as Freightliner, Mack Trucks, Navistar, Paccar, Transportation Manufacturing Corp., and Volvo/GM.

The auto makers soon became aware that their own efforts at improving quality of the Tier-1 suppliers were not being translated by the Tier-1 suppliers' guidance to the Tier-2 suppliers. The Automotive Industry Action Group therefore wanted to mandate Tier-2 suppliers to meet QS-9000 section I and section II sector-specific requirements. They must either go through their own self-assessment audits or seek second-party audits by their customers. A database of successful audits by Tier-1 suppliers would avoid the need for repeated audits for the different auto makers.

The QS-9000 Standard by the Big Three auto makers was not without opposition. Major electronics industry suppliers, such as Motorola and Hewlett-Packard, opposed the ISO-9000 and QS-9000 standards because of the cost of plant-by-plant registration required under these systems. They received endorsements from 37 other major electronics firms worldwide for their Supplier Auditor Confirmation (SAC) scheme. This has the support of many registrars, and it would eliminate the need for the plant-by-plant registrations.

The Big Three auto makers rejected the acceptability of the SAC scheme for their suppliers. They believed that the scheme did not provide them enough assurance as customers. However, they were willing to loan some of their auditors to the suppliers for a specified number of days.

In summary, the standards and the quality awards provide important assessment tools to help improve quality on a continuous basis. In the 1990s, the standards have also emerged as the new norms of doing business in a highly globalized world economy.

A BRIEF REVIEW OF BENEFITS FROM THIS CHAPTER

In this chapter, we learned to define quality in different ways, including by using the four P parameters. We noted the nine frontiers of quality movement and that continuous improvement was an integral part of the quality promise. To facilitate continuous improvement, the workers and managers alike must learn to use certain quality tools. These tools included histograms, Pareto analysis, Ishikawa diagram, and statistical process control charts. Another approach to continuous quality improvement is by using the assessment tool provided by the seven criteria of the Malcolm Baldrige National Quality Award. The origins and the contents of the ISO 9000 series of standards were discussed. ISO 14000 standards for universal environmental standards, as well as the QS-9000 standards adopted by the U.S. automobile industry, were also explored.

SELF-REVIEW QUESTIONS

1. Define the four P parameters of quality for the following:
   (a) sharp pencil
   (b) CD player
   (c) college education
   (d) health care

2. How have the perspectives of quality changed from the craftsman to the customer?

3. What are the seven criteria for the Malcolm Baldrige National Quality Award?

4. Explain the ISO 9000 series of standards. How will you integrate the islands of excellence resulting from the same?

5. Discuss the following:
   (a) ISO environmental 14000 standards
   (b) QS-9000 auto standards

6. Do international and industry standards hinder or nurture global competitiveness of a technology-driven enterprise?

NOTES, REFERENCES, AND ADDITIONAL READINGS

5.1. Parts of this chapter were published earlier by the author. See Gehani, R. Ray. 1993. Quality value-chain: a meta-synthesis of the frontiers of quality movement. *Academy of Management Executive,* 7(2):29–42.

5.2. For instance see Fuchsberg, Gilbert. 1992. Quality programs show shoddy results. *Wall Street Journal,* May 14: A1.

5.3. See Richard Schonberger's 1992 article: Is strategy strategic? Impact of total quality management on strategy. *Academy of Management Executive,* 6(3):80–87; and see his two books cited therein. *Academy of Management Executive* has recently carried other articles on the Baldrige Award and on the international comparison of quality management practices in America, Japan, and the former Soviet Union.

5.4. For example, see Porter, Michael E. 1985. *Competitive Advantage: Creating and Sustaining Superior Performance.* New York: The Free Press, pp. 33–61. Please note that Porter considers the value chain for business level strategy. The enterprise-wide value chain proposed in this paper integrates the producer's value chain with the upstream supplier's value chain and with the downstream buyer's value chain.

5.5. Taylor, Frederick W. *Principles of Scientific Management.* New York: Harper, 1911.

5.6. For a critical review of Taylor's contributions see Charles D. Wrege and

Amedeo B. Perroni 1974. Taylor's pig-tale: a historical analysis of Frederick W. Taylor's pig-iron experiment. *Academy of Management Journal,* 1974, March: 6–27; and Wrege, Charles D., and Stoka, Ann Marie. 1978. Cooke creates a classic: the story behind Taylor's principles of scientific management. *Academy of Management Review,* October: 736–749. For a more favorable review of Taylor's work, please see Locke, Edwin A. 1982. The ideas of Frederick W. Taylor: an evaluation, *Academy of Management Review,* January: 14–24.

5.7. Many books have appeared on Deming's work. Two recent ones are: Gabor, Andrea. 1991. *The Man Who Discovered Quality,* New York: Random House; and Aguayo, Rafael. 1990. *Dr. Deming: The American Who Taught the Japanese About Quality.* New York: Fireside Books/Simon & Schuster.

5.8. For details see Armand V. Feigenbaum, 1963. *Tatal Quality Control,* 3rd edition, New York: McGraw-Hill.

5.9. Port, Otis, 1987. How to make it right the first time: a revolutionary new approach to quality control. *Business Week,* June 8:142.

5.10. See Placek, Chester. 1990. Milliken and Xerox Garner 1989 Malcolm Baldrige National Quality Awards. *Quality,* January:13. Also see Kearns, David, and Nadler, David. *Prophets in The Dark: How Xerox Reinvented Itself and Beat Back the Japanese.* New York: Harper Business.

5.11. See Placek's review of the 1989 awards (Note 5.10); also seeKearns, David, 1990. Leadership through quality. *Academy of Management Executive,* 4(2):86.

5.12. For details see Michael Porter's five-force model for competitive analysis of an industry and competitors, as described in Porter, Michael. 1980. *Competitive Strategy,* New York: The Free Press, p. 4.

5.13. See Woodruff, David, Miller, Karen Lowrey, and Peterson, Thane. 1990. A new era for auto quality. *Business Week,* October 22:84.

5.14. In this study, the high- and low-performing firms are based on their relative performance from the mean in the industry. Results were announced in National Quality Forum VIII, a teleconference relayed nationwide on October 1, 1992 to mark the beginning of Quality Month. Details are available in a priced report from American Society for Quality Control or American Quality Foundation.

5.15. Crosby, Philip B. 1979. *Quality is Free: The Art of Making Quality Certain,* New York: The New American Library.

5.16. For example see the announcement in Dusharme, Dirk, and Reeves, Cecelia. 1996. President announces 1996 Baldrige Winners. *Quality Digest,* December:8.

5.17. Main, Jeremy. 1991. Is the Baldrige overblown? *Fortune,* July 1:62–65. Also see Main, Jeremy. 1990. How to win the Baldrige Award? *Fortune,* April 23.

5.18. Additional information about Malcolm Baldrige National Quality Award is available in a number of informative sources. See Annual Application Guidelines for the Malcolm Baldrige National Quality Award. http://www.quality.gov/; Hart, Christopher W. L., and Bogan, Christopher E. 1992. *The Baldrige: What It Is, How It's Won, How to Use It to Improve*

Quality in Your Company, New York: McGraw-Hill. The authors are a Baldrige Award examiner and a consultant who helped Federal Express win the Award. They have written a practical book to help a conpany's self-assessment; Easton, George S. 1993. The 1993 State of U.S. Total Quality Management: A Baldrige Award Examiner's Perspective, *California Management Review,* Spring: 32–54; Garvin, D. A. 1991. How the Baldrige Award Really Works, *Harvard Business Review,* November/December; Stevens, Tim. 1994. Management doesn't know what its job is. *Industry Week,* January 17:26. This covers the last interview of Dr. Edwards Deming; Stratton, Brad. 1991. A different look at the Baldrige Award, *Quality Progress,* February 1991:17–20.

5.19. Corrigan, James P. 1994. Is ISO 9000 the path to TQM? *Quality Progress,* May:33–36.

5.20. Hockman, Kymberly, Grenville, Rita, and Jackson, Suzan. 1994. Road map to ISO 9000 registration. *Quality Progress,* May:39–42.

5.21. Benson, Roger S., and Sherman, Richard W. 1995. ISO 9000: a practical step-by-step approach. *Quality Progress,* October:75–78.

5.22. Murphy, John M. 1996. Let the system drive your ISO 9000 effort. *Quality Digest,* December:34–38.

5.23. Dzus, George, and Sykes, Edward. 1993. How to survive ISO 9000 surveillance. *Quality Progress,* October:109–112.

5.24. See Hale, Gregory J., and Caroline G. Hemenway, 1996. Tackling the ISO 14001 task: identifying environmental aspects, *Quality Digest,* June:43–64. Also see Hemenway, Caroline G., and Hale, Gregory J. 1996. The TQEM-ISO 14001 Connection, *Quality Progress,* June:29–32.

5.25. Streubing, Laura. 1996. 9000 Standards? *Quality Progress,* January:23–28. Also see Quality system requirements QS-9000. Chrysler Corporation, Ford Motor Company, and General Motors Corporation, 1994.

5.26. Scicchitano, Paul. 1996. Big hoops. *Quality Digest,* May:21. Also see other sources such as, Jackson, Suzan. 1992. What you should know about ISO 9000. *Training,* May:48; and Miller, Cyndee. 1993. U.S. firms lag in meeting global quality standards. *Marketing News,* February 15:1.

CHAPTER 6

Processing Information and Communication

PREVIEW: WHAT IS IN THIS CHAPTER FOR YOU?

In technology-driven enterprises, the evolution of cross-functional communication and the development of information processing technologies have followed the path of two interacting trajectories. In this chapter, we will learn how communication takes place between professionals from different functional areas in a technology-driven enterprise. We will also see how information technologies in general, and computer-based information processing technologies in particular, have evolved over time. Communication has evolved from a one-way communication between two parties, to a two-way communication between many parties.

Many best-selling authors have stressed that since the 1980s the U.S. economy has steadily progressed towards a postindustrial Information Technology Age. In this new age, the knowledge content and the perceptual value of a product or service, instead of the physical value of the product or service, determines the market price. The products and services are valued for the solutions they provide to the customers, rather than for their product attributes.

Information processing technologies have impact on technology-driven enterprises at the operational, tactical, and strategic levels. They help integrate the different functional areas of a technology value-adding chain. For example, flexible manufacturing system uses computers and information technologies to route materials and parts from one workstation to another. Computer-integrated manufacturing links the flexible manufacturing system with computer-aided design, manufacturing, and inventory control departments. American technology-driven enterprises hold a unique competitive advantage in their well-developed information infrastructure. The employees, suppliers, and customers have easy and quick access to large amounts of valuable information.

Finally, we will discuss (a) the benefits of cross-linked technology-driven enterprises with information networks and (b) the ways to implement new information processing technologies effectively.

A POLYCULTURAL CONCURRENT MODEL OF COMMUNICATION IN TECHNOLOGY-DRIVEN ENTERPRISES

In the multicylinder engine model of technology management suggested in this book, communication and exchange of information play very crucial roles. Communication is the glue that binds the cross-functional interactions and integration in a technology-driven enterprise. In these integrated technology-driven enterprises the employees from different functional areas must be in constant communication and exchange information with one another. They do so to design and develop new products and then diffuse them in markets in the most efficient and effective manner. Communication in these organizations is therefore essentially a "concurrent" and often intercultural process.

A Dual-Trajectory Path

Given the current socioeconomic megatrends, the 21st century is most likely to be the age of highly globalized business organizations with intensive interactions and information exchange between professionals from different functional cultures distributed in distant parts. What should be the values for such global organizations to communicate and interact effectively? Can we forecast these values based on the work values that we have seen evolve since the Industrial Manufacturing Revolution. Can we change communication values to help organizations become more agile? In the past, bringing about a change in values seemed a sticky and slow process. Would it be different in the postindustrial Information Technology Age?

From another perspective, the communication and information technologies have evolved rapidly over the past few hundred years. From papyrus etching for written communication we have come to net-surfing in cyberspace. In the 1990s, we have information technologies to communicate across the chasms of long geographical distances, large age differences, different genders, diverse ethnic backgrounds, and varying physical abilities or disabilities. Will the new cybercommunication technologies bring us together, or will they split us apart?

In a technology-driven organization, these two trajectories of incremental and discontinuous changes in communication values and information processing technologies influence each other significantly. Our preferred medium of communication changes with the availability of new developments in information technologies. The development of new information technologies depends on the changing communication needs.

Where would these two trajectories lead us? Will the two trend lines sweep us off our feet, to new levels of mutual understanding, better productivity, and higher standards of living? Or will they crash into each other and replace the irrational human worker with a 24-hour-a-day knowledge-packed robot machine? With the rampant and indiscriminate downsizings and layoffs we saw in the 1990s, are the new communication and computation technologies out to throw us humans out of the workplace and destroy much of the pride we have derived from our work during the past 100 years?

After a brief study of the past break-points in communication technology, this author believes that we humans will once again meet the new technological challenge (of our own making) and learn to tame the new beast of technology. The bad news is that some of us either cannot or will not adapt to these changes and will pay a price of some sort for being inflexible. The good news is that many things about work will perhaps remain either the same or would change only slightly.

For example, one of the constants will be the trust between co-workers and co-members of a self-directed high-performing team. It is postulated that in the next few years, trust across members from different functional cultures dispersed around the world would define the productivity of the individual people and their polycultural teams in technology-driven enterprises. The pull for prosperity by the entrepreneurial, creative, and continuously re-skilling people, rather than the push of computer-drawn plans and top-down reengineering mandates, will govern the wealth of business organizations and the nations. Our economic and social successes will depend on the organizational management of multidimensional dynamic human players who are flexible and agile in space as well as time.

The ideal organizational model for technology-driven enterprises is a polycultural, diverse, and distributed but closely connected work team. These teams are flexible and multiskilled. They may seem relatively more chaotic at first sight. But they offer the potential of producing the most creative and effective solutions in the long run. The available advanced communication and information processing technologies can be deployed effectively by them to facilitate harmony across such distributed but connected organizations. The different subelements of such distributed organizations are connected by their common strategic intents. These intents help leverage trust-building dynamic interactions across organizational players from different functional areas, even when they are distributed in geographic space.

These concepts and other information related issues will be discussed in this chapter from an organizational theory perspective. We will review how communication and information processing technologies have evolved and transformed technology-driven enterprises over time. We will also review how work values have changed over the years, and what are the major socioeconomic forces defining them in modern organizations. Then, we may be able to postulate how these two trajectories are likely to interact together and influence the competitive advantage of technology-driven organizations.

To illustrate the dynamic interaction between the role of communication and information processing technologies with the work-related values in a technology-driven organization, let us consider a case study where the two trajectories crashed into each other with little constructive synergy.

A Case Study: Shared Values and Communication in a Telecommunication Company   In 1996 the top management of Frontier Corporation, a major telephone service provider in Rochester, New York, and its workers were locked in an extended and bitter contract dispute. The workers were members of the Communication Workers of America (CWA) Local 1170. The CWA workers in Rochester were working under a contract that its national CWA had imposed on them a year ago. In January 1997, the CWA members in Rochester rejected the agreement between their management and the national CWA because it phased out pensions and reduced health benefits for retirees (see Note 6.1).

Developing Shared Values.   While the contract negotiations were going on, the top management of Frontier put high hopes on implementing a training program to change the company's culture and improve its people's productivity and job satisfaction. A key part of this Shared Values program, developed by Lebow Company Inc. of Bellevue, Washington (and delivered to Frontier through a local consultant vendor), was to cultivate trust and honesty in the workplace. The proposed Shared Values program was derived from Rob Lebow's book, *A Journey into the Heroic Environment,* and was based on eight principles. The eight recommendations included: (1) lavish trust on your associates, (2) mentor unselfishly, and (3) take personal risks for the organization's sake.

The training program was initiated 1 year earlier in 1996, after conducting an attitude survey among Frontier workers. About 1800 of the 2800 workers in the Frontier telephone group companies were planned to be trained in ten 1-hour weekly sessions according to Pam Preston, the human resource manager for Frontier. Of these people scheduled for the program, about 500–600 were expected to be from the Rochester Telephone subsidiary, the largest telephone company in the Frontier group of companies.

Shared Values Versus Contract Settlement.   Bob Flavin, president of CWA Local 1170, found the proposed new organizational values laudable. But he considered the management's efforts with union in setting shared goals and settling the contract more important to the local workers. He stressed that if the management and workers had shared work values, then they should have been able to develop consensus for their shared goals without much difficulty.

Jerry Carr, the president of the Frontier Telephone Group and the chief executive officer of Rochester Telephone Company, admitted that his management could "take three or four years to be able to recognize . . . a new culture." The management consultants involved in the training also admitted that "The most important variable in whether a change takes root is not the techniques . . . or the programs, but the emotional environment."

CEO Carr insisted that changing the corporate culture was a different issue, separate from the issue of settling the contract with the union. He believed that "settling the contract doesn't promote trust." To him, "Trust gets built every day in a culture where people are allowed to speak up."

Some people in the company did not feel very comfortable speaking out in training sessions in particular, and in the workplace in general. According to a company-related report (see Note 6.1), three workers were "disciplined" for airing their views during the Shared Values program. CEO Carr sent out an e-mail stating that those who were involved should be tracked down, "not so (that) we can convict . . . (but) so we can educate."

A Cornell Professor from the School of Industrial and Labor Relations, James Gross, wondered "Why not have the union involved in these (value changing) programs?" But Frontier had a hard time getting CWA members involved in launching the Shared Values program. Workers of unions from other member companies participated more willingly. The company officials were frustrated too. They asked, "What is there about truth, trust, and honesty that they (workers) wouldn't want present in their workplace?"

This case illustrates how the work organization at Frontier, a technology-driven telecommunication company, faced a major challenge of managing its communication-related organizational values. Under the intensely competitive conditions facing many technology-driven industries in the 1990s, information and effective information exchange played strategic roles in the competitive advantage of a technology-driven organization. These organizations were forced to compete effectively in intensely competitive globalized markets. The role of information exchange became far more strategic in the postindustrial 1990s than it was a few decades ago.

THE POSTINDUSTRIAL INFORMATION TECHNOLOGY AGE

The Third Wave

In 1980 Alvin Toffler presented a revolutionary idea in his world-famous book, *The Third Wave* (Note 6.2). He suggested that a new wave was sweeping through the world. Toffler proposed that the developments in communication and computer-based information processing technology were creating a revolutionary new economic civilization. There was a change of the same magnitude as the previous two revolutions, namely the agricultural and the industrial revolution. In this new postindustrial civilization, information played an unprecedented important role. Toffler wrote in his book, "An information bomb is exploding in our midst, showering us with shrapnel of images and drastically changing the way each of us perceives and acts our private world" (Note 6.2). He argued that the computer was imparting intelligence to the "dead" inanimate machine environment around living human workers in technology-driven enterprises.

The First Megatrend

In 1982 John Naisbitt, another best-selling author, identified 10 major trends shaping the modern society in the 1980s. He called these *Megatrends* (see Note 6.3). Naisbitt's first "megatrend" was a shift from the industrial society to an information society. Naisbitt wrote, "None is more subtle, yet more explosive (trend)." Yet many Americans continued to resist accepting that their economy was built on information and that the industrial era was fast coming to an end. Naisbitt provided some excellent facts as his evidence. He pointed out that in 1956 the white-collar American workers outnumbered the blue-collar workers. Soon after that milestone the Russian Sputnik launched a satellite communication age around the world. This was followed by the start of a computer age. In 1979, the clerical positions outnumbered those employed as laborers and farmers. The next largest job classification was that of the professionals—also information-based workers. They accounted for 16% of the overall work force of about 90 million.

Almost a decade earlier than the fall of communism in the late 1980s, Naisbitt projected that Marx's obsolete labor theory of industrial value, born at the beginning of the 20th century, should be replaced with a knowledge theory of value. In the information society, the value was increased by knowledge. This connection was confirmed by the economist Edward Denison, who showed that most of the productivity gains and economic growth in the United States from the end of World War II to the first Oil Shock in 1973 was due to the higher education and knowledge level of skilled workers. These included women workers and their access to more knowledge (see Note 6.3).

Interestingly, way before Internet became so popular in the 1990s, Naisbitt pointed out that we were drowning in information but starving for knowledge. Prophetically he pointed out that "unorganized information is . . . the enemy of the information worker." Information processing technology was needed to add value to data that would be of much less value. Until 1980 there were only 1 million computers, of which more than half a million personal computers were sold in the year 1980. The demand was expected to grow at more than 40% per year. In 1980 the computer technology was a catalytic agent for the Information Revolution, the way the powered machine was to the Industrial Revolution. These changes, according to Naisbitt, were leading us to "conceptual space connected by electronics (and computers), rather than physical space connected by the motorcar" (see Note 6.3).

Knowledge-Value Creation. No other society has realized the impact of this change more than Japan. With very limited availability of industrial raw materials and an abundance of human capital, Japan was likely to benefit most from such a megashift from the physical goods to knowledge-based value. Taichi Sakaiya, a former official of Japan's Ministry of International Trade and Industries (MITI), in his 1991 book *The Knowledge-Value Revolution,* proposed that in the years ahead the soft portion of a product's design and function, or its knowledge value, would increasingly determine a larger weight in its actual monetary and social value (see Note 6.4).

In the Japanese language the original term used by Sakaiya was *chi-ka.* In the context of Far-Eastern culture, *chi* in Japanese language refers to "knowledge," "intelligence," "wisdom," and so on. *ka* refers to change and transformation. Sakaiya elaborated that in the years ahead the knowledge value of a product will depend on the subjective perceptions of a group of people—such as the paying customers of a technology-driven enterprise.

These "subjective" perceptions of customers are likely to change rapidly with time and competition. These perceptual changes can take place because the customers change their minds. Or the perceptions change because a new superior technology replaces the old technology. This can make the knowledge value of an old product obsolete and reduce its perceived value.

For example, in the mid-1990s an author friend bought a portable Fujitsu Lifebook laptop computer with Intel's Pentium semiconductor microprocessor chip. This was the state-of-the-art technology available at that time for word processing a long manuscript. Because of this perception, he was willing to pay a few thousand dollars, say $3000. However, within weeks of his purchase, laptop computers were introduced with Intel's MMX Pentium microprocessor with faster processing speed. This reduced many customers' "perception" of the desirability of computers with an old Pentium chip. Overnight, the prices of laptop computers with older Pentium microprocessors fell by 20%, to around $2400. My friend's laptop computer was the same physical product as before. But its perceptual value had fallen significantly because of the new substitutes available in the market. After the introduction of the laptop with MMX Pentium microprocessors, customers like him were willing to pay less for this product than they were willing to do the week before. A 20% fall in prices has a significant impact on the book value of inventories, sales revenue, and the potential earnings of a computer retail store and of the laptop computer's manufacturer.

Sakaiya pointed out that the material content of a semiconductor chip does not change. But its knowledge value does. Similar change is frequently observed with the computer software programs. Information-intensive goods and services may lose their value in less than 2 years, and sometimes even 1 year. This drastically challenges our conventional understanding of the utility functions of commodities based on their physical and tangible values. In the postindustrial world of the information technology age, the knowledge content drives the value of a product, service, and technology. Therefore the knowledge-value-driven organization must be flexible and agile enough to respond quickly to such perceptual as well as physical changes.

ROLE OF COMMUNICATION

For enhancing the information and knowledge-based value of products, services, and technologies, the interaction of active participation—and particularly communication exchange between people with different professional and functional cultures—has critical significance. Communication is the backbone of effective participation and interaction in multifunctional self-directed teams. The importance of communication increases further, when people with different professional cultures are involved and must interact, such as in a polycultural work team. We have to accept that different professionals think and act differently. When their work brings them together, they can contribute their different perspectives for a superior solution.

Defining Communication

Communication involves a process of sharing information and messages transmitted via a variety of means. The English word communication is derived from the Latin word *communis,* meaning common. People communicate with each other to seek a "common" ground of understanding. To develop a common understanding, it is important that the meaning of a particular communication be understood in the same way by the sender as well as by the receiver.

In technology-driven organizations, people communicate to share and exchange information for a number of tasks and activities. They do so to comment about a market research study, to develop an idea about a future new product, or to offer a feeling of support for a new production procedure. A project manager may be asked to periodically "communicate" the status of his or her project to a management committee. A sales representative may be asked to report the number of sales calls he or she made in a particular week. These are different types of communication that are going on all the time in an organization. These exchanges involve senders of messages and receivers of messages. These participants must understand the transmitted messages in the same manner. Unfortunately, the messages and their meanings are influenced by other factors which add noise to the signals. The communication process can be broken down into some basic common components.

Interpersonal Communication Process

In technology-driven enterprises, interpersonal communication is a complex process involving the physical aspects with the perceptual aspects. Briefly, the sender encodes and transmits a meaning into physical messages based on his or her own perceptual understanding. The receiver decodes the physical messages into the meaning that is based on his or her own perceptual understanding.

The communication process can be broken down into the following elemental components:

1. Sender or the source of communication
2. Receiver or destination
3. Message
4. Medium of communication
5. Noise
6. Feedback

Let us look at each one of these subelements individually. See Exhibit 6.1.

1. Sender or Source. The sender may be an individual, such as a chief executive officer, a mechanic, a designer, or a planner. The sender may even be a team such as a steering committee, a quality circle team, or a project planning team. Or the sender may be a communication organization, such as a court, a federal agency, or a mayor's office. The sender has a meaning to be conveyed to an intended receiver. The sender encodes this meaning into a message. The meaning as well as the encoded message are influenced by the functional specialization of the senders and their associated value systems. The marketers look at competition from the point of view of market shares, annual growth rates, and profit margins. The designers, on the other hand, consider that the competitiveness of their corporation is driven by how well-designed their products are. The quality controllers insist on conformance to specifications, whereas accountants want costs under control. These different professionals may view a common task differently (see Exhibit 6.1). They are likely to communicate differently.

2. Receiver or Destination. Like the sender or source, the receiver or destination may be an individual, a team, or a communication organization.

EXHIBIT 6.1. Elements of a Communication Process

Subelement	Role in Communication
1. Sender/Source	Interpret meaning
	Encode
	Transmit message
2. Message	Embodies meaning
3. Medium	Transmits messages
4. Receiver/Destination	Decode
	Interpret meaning
5. Noise	Distorts Meaning
6. Feedback	Confirms meaning

The receivers decode the transmitted message and then interpret its meaning.

In a communication, we often get so obsessed about ourselves, or what we want to say, that we forget about the receiver. For example, the research and development professionals with advanced degrees are sometimes enamored by their scientific findings in their narrow specialized fields. In their communication they often overlook who the receivers of their communication are. They may present their ideas in jargon. When that happens, they cannot convince the accountants why more funds must be provided for successful conclusion of their research. Accountants, on the other hand, view these research projects as sinking holes for money. They may not be able to evaluate the future strategic value of the outcomes of such research projects.

People perceive, filter, and selectively understand messages according to what is consistent with their own professional expertise and expectations. Their own professional culture dictates who they communicate with, and how the communication process takes place or does not take place. The professional background is the foundation of people's work-related communication.

In intercultural interactions, such as in cross-functional teams, effective communication depends on building common experiences for team members with different professional cultural backgrounds. For example, if the scientific researchers are trained to understand the terms "return on investments," and "return on assets," they can communicate better with the senior executives who are driven by these measures. In the same way, when the senior executives at a technology-driven company such as Xerox understand their core technological process of photocopying image transfer, they can communicate better with the technical people in the production area.

Many senders worry whether their messages will reach the right receivers. They wonder if the receivers would understand the ideas or the feelings that their coded messages contain. To complete the communication transaction, the receiver must decode the coded message. The senders often worry whether they will be able to develop a common ground with the receivers as a result of the communication transaction just concluded.

3. Message. The communication message is made of symbols with meanings. The words "fired" and "fast-track" have many different meanings attached to them. To a person who has never been "fired" from a job, or who has not learned about this popular organizational practice in North America, the word "fired" is either meaningless or means something different. Once learned, the word "fired" can help both the sender and the receiver relate to the common experiences associated with this message.

A message can take different forms. The message may be written words on paper, TV screen, audible sounds on a public address system, or visual images. Messages may also be coded, such as the Morse code, electromagnetic waves, or the digital bits and bytes of the positions of electronic on–off switches (in a computer). The coded messages have to be decoded to derive the information that would help build the common ground.

4. Medium (and Multimedia). The medium is also an important part of the communication process. The telegraph medium could only communicate one way at a time, and only in codes of dots and dashes. Before the birth and growth of the telephone, the financial sector relied heavily on the messenger boys to convey information to the potential investors. Near the end of the 19th century, the messenger boys ran from one end of New York City to the other end. Many of them were replaced as telephones became popular and easily available. But in the 1990s, bicycle riders still did a similar job crisscrossing streets of Manhattan, carrying important documents and financial instruments.

In the 1990s the Internet, and its corporate cousin Intranet, started emerging as the new medium of choice. This network of computers communicating with each other, originally developed and designed for military communications, became a thriving superhighway of information exchange.

5. Noises in Communication. As mentioned before, the signaled messages are often accompanied by noises. The noises influence the encoding as well as the decoding of meanings associated with messages.

In cross-functional communication, such as between a marketer and a research scientist, a member with one functional background (in this case, say marketing) sends a message to a member of another functional background (in this case, scientific research). The message contains the meaning grounded and encoded in the principles of marketing. When the message reaches the research scientist, it is decoded based on the principles of science and engineering. The professional backgrounds of both the sender and the receiver influence their respective perceptions. These perceptions are defined by variables such as their thought patterns, spoken and written languages, nonverbal communication, organizational memberships, and others.

Members of one profession tend to stereotype the members of other professions by their distorted perceptions of traits and characteristics. This causes noise and misunderstandings in intercultural communication. A sociotype, on the other hand, is a sincere attempt by members of one professional culture to accurately describe the members of other professional cultures by their common characteristics and traits. This helps communication.

6. Role of Feedback. Another important element of communication transaction and process is the feedback that a sender receives when communicating with a receiver. When a project manager sends a message requesting an extension of time or an increase in the budget for his project, a feedback message such as "Oh sure, that seems reasonable" has a different meaning than the feedback message "No way. You cannot do that."

Feedback is received by the sender in a variety of coded messages. For example, a smile, a frown, and a pat on the back from a boss receiver are different forms of feedback messages that senders seek to catch.

It is important to understand that senders also get feedback from their own messages. Senders revise their written memos again and again. They worry about how their speech may sound to others. Teachers and trainers speed up or slow down their messages based on the feedback responses from the students' facial expressions.

What happens when a sender tries to build a common ground with an intended receiver? What happens to the message, and what role does media play in that process?

The senders and the receivers perform multiple tasks during a communication process. The message begins as an idea or a feeling in the mind of the sender—for example, a product concept or a sense of satisfaction that the achievements of an engineer are recognized and rewarded by her bosses. Such messages residing in the mind of a sender cannot be sent or understood by other receivers unless they are transformed. The sender must encode his message into a form that can be transmitted. For example, the message of an idea or a feeling can be written into a commonly understood language. Or the message can be recorded as a verbal message, in a commonly understood language. Sometimes graphics and visual aids are the more appropriate ways of communicating. Communication with senior executives needs to be encoded in accounting language.

Other Factors in Corporate Communication

Multiple Roles That Communicators Play. In cross-functional communication, such as for management of technology-driven enterprises, the senders and receivers play three roles simultaneously. These are:

1. To act as an encoder of outgoing messages
2. To respond as an interpreter of common experiences and meanings
3. To act as a decoder of incoming messages

Each communicator must play the crucial role of an interpreter while they are also encoding and decoding the messages. A message such as "fired" is interpreted with different meanings when someone else is fired than when you are fired. In a similar manner, the words "recognition" and "rewards" mean different things to different individuals at different points in their careers. To a fresh graduate entering the job market and thinking about starting a family, money may be the best form of recognition and reward, whereas to a middle manager the peer recognition and appreciation may be more effective ways of reward and recognition.

Multichannel Communication. Most communication transactions take place via multiple channels. In any communication a wide variety of messages are being sent and received at the same time. Spoken words travel with tone, volume, pitch, and many other bits of information. Some words are stressed. Other words are deemphasized. Pauses make a big difference. Written words are loaded with caring, authority, rudeness, and other attributes. The verbal communication is accompanied with nonverbal communication.

Communication Messages Can Live Forever. Messages, once coded, have a life of their own. These messages may declare their independence from their producers. The message can transform its form. Some messages have even outlived their senders. Given below are two examples.

A. Hammurabi's Code. King Hammurabi of Sumeria conquered Babylon in 1792 BC. In 1761 BC he started a military campaign to bring law and order to the entire Mesopotamia. In 1750 BC Hammurabi developed a detailed document containing 300 laws. This included punishments for crimes such as false accusation, witchcraft, and so on. These were aimed at protecting the weak from the strong. He also detailed a number of laws for loans and debts, wages and taxes, and the business of land. Because this message of King Hammurabi survived long after him, it was used by many subsequent rulers and people to model their own laws, including the individual human rights as we understand today.

B. Johnson & Johnson Credo. The "credo" statement of Johnson & Johnson, a New Jersey–based producer of healing products, is one such modern example of a message with long-lasting value (see Note 6.5). The credo, developed close to 100 years ago by the founding brothers of Johnson & Johnson, was actively followed during the Tylenol poisoning crisis in the 1980s. A sick man tried to blackmail the company and laced the company's number one selling product Tylenol with cyanide poison, killing eight people in Chicago. Faced with the crisis, the senior- and middle-level executives followed their age-old company credo to choose their actions. The J&J Credo began with, "We believe our first responsibility is to the doctors, nurses and patients, to mothers and fathers and all others who use our products and services." In conformance with the company's Credo, the J&J Chairman James Burke recalled 31 million Tylenol capsules from the marketplace and destroyed them without any hesitation. The benefits outweighed the cost.

Understanding the Communication Process Using the Telephone Model

The cross-functional communication transaction can be understood more clearly by considering a telephone-based communication model. The telephone was originally invented by Alexander Graham Bell so that he would be able to communicate better with his hearing-impaired mother. In a telephone the source encoder is the speakerphone, and the receiver decoder is the earphone. It is amazing that we perform these tasks as senders and receivers so many times in a day, without even realizing it.

There are many interesting implications of this model of communication transaction.

1. The Strength of a Communication Transaction Is Determined by the Weakest Element in the Communication Process.

For example, if there is an unclear "image in the mind" of the sender, or there is poor coding of the message, or there is weak transmission medium for the message, or if the decoding of the message is done inaccurately, then the communication transaction would not be strong. The message is therefore likely to be miscommunicated.

2. The Capacity of the Communication Transaction Depends on the Capacity and Intensities of Its Component Elements.

For example, when the telephone lines get busy, messages cannot go through to the receiver. This is why we often build redundancies in our communication. We repeat ourselves. We say the same thing in different ways. When we perceive that a receiver may have a hard time understanding and decoding our message, we communicate our message in a number of alternate ways. We use gestures to reinforce our verbal messages. We give examples and tell stories to elaborate.

3. The Senders and the Receivers Must be "Tuned" to the Same Frequency.

This tuning of senders and receivers is obvious in the case of communication via a radio broadcast. But this is also true in the case of other communication technologies, such as the television, the cellular phone, and so on. We often do not realize this when communicating face-to-face with professionals from other functional areas.

In the case of cross-functional communication, the tuning is often taken for granted. The sender codes a message and the receiver decodes this message, depending on their respective past experiences and professional "cultures." If you have never learned to speak the Japanese language and your Japanese boss praises your work in the Japanese language, it will be of no use. The same is true when some finance and accounting experts are asked to plan and control researchers in a laboratory environment. The two groups of professionals cannot decode the customs and phrases of their colleagues with different professional backgrounds. Many times the workers do not understand what drives senior managers in their organizations. The senders must anticipate the decoding capabilities of their decoding receivers.

Communication Flows in Technology-Driven Organizations

A. Serial Communication in Organizations. Many technology-driven organizations are driven by their chain-of-command hierarchy. Communication in such organizations takes place in a serial and one-way manner, often top-down.

For example, say president Thomas sends a communication message to vice-president Joan. Vice-President Joan communicates that message, after attaching her interpretations to it to manager Peter. Manager Peter calls his production supervisor Jennifer and sends his interpretation of vice-president Joan's message. Supervisor Jennifer then goes to operator Doe and gives her interpretation of the message she received. Such organizational communication involves many intermediaries. Each time a hand-over takes place, there is decoding of the received message, interpretation of the message for meaning, and then coding of a somewhat different message. Each hand-over takes place in multiple channels. There is the interpreted main message, packaged in gestures, body languages, and other side-messages. Each hand-over increases the probability of a miscommunication.

Directions of Communication. The one-way communication could take place in different directions. Communication could flow either from upper echelons to down below in the trenches of an organization, or from the bottom of an organization to the upper echelons of an organization. The serial communication could also take place laterally—that is, to peers and other team members. Sometimes the organizational communication travels faster through rumors. The formal communication is superseded by the informal versions. Rumors travel through their own, often nonhierarchical routes.

Top-Down Fanning Out. When communication travels down the hierarchical levels, the messages get fanned out or expanded and elaborated.

For example, president Thomas sends a message to vice-president Joan of Widget Division to reduce the operating cost of a business unit by 10%. Vice-president Joan communicates that message to her production manager Peter, after attaching her interpretation that the 10% cost savings will come from reduction in payroll expenses. Production manager Peter calls his production supervisor Jennifer, and he sends his interpretation that the payroll expense should be reduced by not scheduling any more overtime and by firing three full-time employees. Supervisor Jennifer then goes to operator Doe and gives him the modified work schedule for the next few weeks, or the pink-slip. In this example, each intermediary sender/receiver added a new dimension to the message received from the upper-level sender.

Bottom-Up Chunking. On the other hand, when communication flows from the bottom layers of an organization to the upper echelons, the messages are chunked together.

Operator Doe in the above example, with many years of experience in his line of work, may have a suggestion on how to reduce the setup time on the machine he operates. He makes detailed drawings of the new jig fixtures that would reduce the cycle time. He goes to supervisor Jennifer with these details. Let us say supervisor Jennifer has 15 operators working for her, and two other operators have similar suggestions. She collects all these details, compares them, analyzes them, prioritizes them, and recommends one or two of these to her boss production manager Peter, and then she requests additional materials and manpower to implement the selected suggestion. Production manager Peter receives similar requests from five other supervisors beside Jennifer. He compares them for their impact on total cycle time reduction, and he comes up with three programs to recommend to his boss vice-president of Widget Division. Thus, the information going up the hierarchy is chunked together or compressed.

B. Concurrent Communication. As mentioned in earlier chapters, the concurrent model of managing technology is different from the sequential silo model of managing technology where a technology is managed and new products are developed sequentially. In the sequential management of technology, the management is focused on one functional silo at a time. Thus the technology-related efforts go through the research and development (R&D) phase, the automation and engineering design (AED) phase, and the market development and diffusion (MDD) phase in a sequential manner. In the sequential model there is bare minimum interaction between researchers, designers, process engineers, and marketers. The communication among these professionals from different functional areas takes place only at the interface between the two functional silos, when the new technology or a new product is handed over from one phase to another. At the interface between functional silos, a project or a new product is thrown over the walls of a functional silo to the next functional silo.

On the other hand, the concurrent multicylinder model of technology engine involves and requires extensive interactions between professionals with different functional "cultural" backgrounds. We know that the market development and advertising professionals have a somewhat different view of the market than the one that researchers and designers have. These professionals may not have different cultures, in the anthropological sense of having backgrounds with different race, ethnicity, or color. (Though they may vary by these cultural attributes too, but we are not considering them here.) These professionals with different functional backgrounds have different world views, which make them look at the enterprise-wide technology differently. A number of researchers focus their studies on subjects such as the way marketers and manufacturers work and communicate with each other. Often this takes place at the interface of their respective functional silos.

Multicultural Groups Versus Polycultural Teams. From the way intercultural interactions are studied in the larger society, we can draw some understanding of the impact of presence of multiple cultures on work organizations. Let us quickly recall and build on what will be covered in the organization of cross-cultural groups in a different chapter in detail under the topic of organization of employee teams (see Note 6.6).

Multicultural Salad Bowl. In heterogeneous societies such as the United States of America, a lot of effort has already been put into a field called multiculturalism. However, often the popular use of the term multiculturalism has some commonly understood meanings. For example, often multiculturalism implies a coexistence of people with multiple cultures—sort of like a mosaic, or a collage. The diverse different cultures stand alone, holding up their respective ethnic identities, with limited interactions across different cultures. They stay out of "each other's face." Multicultural work groups imply a heterogeneous membership of people with different professional cultural backgrounds.

Polycultural Gumbo Stew. A work group facing highly competitive markets, such as for the technology-driven enterprises, cannot operate like a multicultural community. In an organizational setting, professionals with different cultures share common and shrinking resources among themselves. Their work is interdependent—as in a chain or a ring. Mere "standing alone" with independent identities would require excessive amounts of resources and would delay cycle time. Work groups in America cannot afford to do that when the teams and circles in Japanese and German companies are collectively moving rapidly, introducing more new products faster. We need an alternate organizational unit.

That new organizational unit is a high-performing polycultural team. Unlike in a multicultural group, in polycultural teams the members are interlinked. Their achievements are interdependent. Their successes and failures are jointly determined. Like a train or a polymer macromolecular chain, the different members are closely linked to each other.

The members of polycultural teams have the potential to identify and explore larger range of alternatives. They are thus better prepared to resolve ill-defined strategic issues facing a technology-driven enterprise in the age of globalized markets. Thus, they have the potential to generate more superior solutions. On the other hand, the heterogeneous polycultural teams are less likely to be cohesive and may therefore cause higher dissatisfaction among some team members (see Note 6.7). The difference between (a) a team achieving superior above-average performance with superior creativity and (b) the marginal subaverage performance of a group of people with different professional backgrounds is due to the level of trust they have towards each other and their superiors.

In the low-performing multicultural groups there is a general and often vague recognition of the differences in professional backgrounds of different members. In the superior-performing polycultural teams there is high respect among the members with different backgrounds. The two work groups differ in terms of their (a) trustworthy perceptual influences (TWPIs), and (b) trust-building social interactions (TBSIs) (see Note 6.6). The trustworthy influence is the reliance by one person or a group on the part of another person or group to recognize the interests of all engaged in an interdependent task. This is based on the integrity, open-mindedness, consistency, and competence of the members involved. The trust-building social interactions are influenced by factors such as higher education and urbanization of team members. The social interactions are based on (a) the favorable historical record of the past performance of similar teams and (b) the trust-building formal mechanisms and institutional frameworks. A continual, open, and critical public scrutiny and communication help keep in control the uncontrolled use of power by people in authority over others.

In a broader societal context, some observers have postulated that only those societies with a high degree of mutual trust among its diverse members will be able to create the flexible large organizations needed to gain competitiveness in the worldwide global economy (see Note 6.8). Max Weber noted that with increasing economic activity, the focus of trust must shift from a discretionary and sometimes arbitrary behavior of individuals and their relationships between each other, to a more mutually dependable and assured relationships. Such interdependence is a critical requirement for the superior performance of work groups in technology-driven enterprises (see Note 6.9).

Role of Technology and Values in Enterprise Communication

Communication between one or more senders and one or more receivers is facilitated by information processing technology but is driven by the values of the communicating parties in a technology-driven enterprise.

Communication across senders and receivers, separated by long distances, requires electric wires for telegraphy technology, electromagnetic waves for cellular telephone, or fiber-optic cables for cybercommunication on the Internet. But whether we use an Internet or not, and for what purposes, is determined by the prevailing values in the workplaces involved. There are many countries in Africa and the Middle East where the Internet, even if available, is not likely to be utilized freely to the extent that it is in the United States. The people of these developing countries fear that their governments are closely monitoring what they say on the Net and may hold them accountable for what they share on the Net.

These examples illustrate that the message and the media are also influenced by the choice of communication technology. What we can communicate over the telephone is hard to communicate over the telegraph.

Some would argue that the communication message and the media used by the communicating parties also reflect the values of the humans and organizations. Archaeologists dig up ancient civilizations. They estimate the long-gone cultures by the artifacts they used. An inscription on the inside walls of an Egyptian pyramid for a pharaoh reflects a different value than the clay tablets used for trade by the Mesopotamian merchants with the traders in Mohan-Jo-Daro city in the Indus Valley Civilization.

Values for Polycultural Teams. High-performing polycultural teams develop and innovate a high-performance third culture with work values of their own. Their values are not necessarily the values of the team's dominant group. They are also not an arithmetic mean of the values of all its members. The team values are the values that are most appropriate and effective for the tasks and goals of the polycultural teams (see Note 6.6).

For example, Hewlett-Packard competing in highly dynamic and intensely competitive markets (such as for printers and computers) allows its cross-functional polycultural teams to be as chaotic as they wish to be. The team members choose when to bring closure to their "inclusive discussion." Being "right" is valued more than doing it the "right" way.

When More Communication May Not Help

We must consider that more communication is not a panacea under all circumstances. David Stiebel in his 1997 book, *When Talking Makes Things Worse,* suggested that more communication is not always the best solution. Many communication experts would have us believe that we do not communicate enough. In the case of the Frontier Corporation crisis mentioned earlier, these consultants would make its management and workers communicate more. A new communication technology, such as a new newsletter and more chat rooms on the computer Intranet, may be recommended to tide over the management–worker crisis. But Frontier's challenges, and therefore the effective solutions, lie deeper in their workplace values and in the company's communication intents (see Note 6.10). Does management of Frontier believe that its strategic intent for superior performance includes full participation by workers? Or are workers considered dispensable and not an integral part of such strategic relationships?

Stiebel recommended the following four-step process to resolve similar confrontationary situations. (See Exhibit 6.2).

Step 1: *Determine whether the problem is a misunderstanding (that is, failure to understand each other accurately) or a true disagreement (that is, a failure to agree even when both parties understand each other fully).* Stiebel stressed that "it is absolutely, positively, not possible to resolve a true disagreement by understanding each other better." With disagreements, "proceeding incrementally is the fastest way to get what you want."

EXHIBIT 6.2. Communication Intent and Values

Four-Step Communication Resolution
Step 1: Distinguish disagreements versus misunderstanding.
Step 2: Create their "resistence" communication.
Step 3: Understand their beliefs and values.
Step 4: Predict their "success" move.

Source: Adapted from Stiebel, David. 1997. *When Talking Makes Things Worse*. Dallas: Whitehall and Nolton.

Step 2: *Create the other person's next "resistance" move.* To get closer to a resolution, Stiebel suggested that the senders and the receivers must have common understanding. They may disagree, but they must clearly understand what each side is saying or demanding.

Step 3: *Use the opponents' own perspectives, beliefs, and values to convince them.* Most people decide what to do based on their own perceptions and beliefs, not on other persons' beliefs. And therefore Stiebel recommended that the opponent's value system may be leveraged to bring about a solution.

Step 4: *Predict the other person's "success" response.* And see if there is a win–win fit with what you can afford to offer.

This communication guru stressed again and again that the key to productive communication is in the values of the communicating parties and not just in more communication technology. Many companies erroneously throw more money at the communication technology without developing a good understanding of the underlying values of the different parties.

Communication and Information Processing Technologies for Polycultural Teams. Polycultural teams need their own communication technology. As team members of polycultural teams develop trust in each other, new vocabulary grows and new words are adopted as the "preferred" and "shared" terms of the team's communication. For example, for employees working in Disney, their workplace is a "stage," and every employee is a performing artist. An organization's communication technology reflects and often matches its true values and beliefs. At ABB, a global company, broken English is the common working language.

Communication technologies and languages in the Western hemisphere, such as English, have many terms that help its users express their freedom, individuality, and right to compete in their society. On the other hand, Asian languages such as the Chinese, Japanese, and the Southeast Asian languages have abundance of terms that help their users express their groupism, harmony, and collective kinship relationships. Communication in relatively new nations, such as the United States of America, use expressions that look towards the future. On the other hand, the language of the Old British World is more likely to turn to its historical past. If we listen carefully to the nuances of different communication modes, we can notice these differences. In Europe, 200 miles is a long distance, but it is not so in the continental United States. In the United States, on the other hand, 200 years is a long historical period but it is not so in some of the old European and Asian societies. In the same way, a highly interactive polycultural team that has performed together for a while and has its own shared values is likely to develop its own preferences for words and terms used in its day-to-day communication.

This process of development of appropriate communication technology for information exchange in a polycultural team is similar to the need to adopt "politically correct" language in a diverse United States of America full of the foreign born immigrants. One can no longer call a fellow-worker fat or obese. They are referred to as "weight-challenged." People who tell lies habitually are called "ethically challenged." The sweepers and janitors have emerged as sanitary engineers. Many examples of socially acceptable and politically correct words have been listed in the addendum of the *New Oxford Dictionary*. Similarly, the fast emerging cybersociety uses new words such as flaming, spammimg, and many more words and symbols with their own unique meanings. A message with all capitals implies that the sender is yelling. Net surfers use the keys of a standard keyboard to represent their smiles as :-), their sad faces as :-(, exclamation as :-o, and so on.

Like the changing message, the medium of communication for information exchange between professionals from different functional backgrounds also has evolved significantly over time. In the next section we will discuss how new developments in information technology have played an important role in the management of technology-driven enterprises.

Communication Technology in the Cybercommunity

Is the role of communication technology different in the Cyber Age? In the cybercommunities, communication takes place in multiple ways and often involves multiparty interactions. This is a long way from the traditional one-way command-and-control mode of one person in power influencing many other persons in lower ranks. This change did not take place as abruptly as some "experts" and providers of new technologies would want us to believe. The new communication technology has gradually evolved its communication transactions in four step-by-step stages (see Exhibits 6.3, 6.4, and 6.5).

Stage 1: Telegraphic Communication: One-Way, One-Party Interaction.
When a telegraphic transmitter sends a message to another receiver in a communication transaction, the receiver cannot interact or respond to the message in the same communication transaction. If the receiver has an ability to turn into a transmitter, then in the next communication transaction he transmits a response to the new receiver or to the earlier transmitter.

EXHIBIT 6.3. Evolution of Communication Technology

Communication	Medium	Impression	Message
	A. Written Communication		
Seals	Clay	Wedge	Hieroglyphs
Printing	Stone	Chisel	Kanji
	Papyrus	Quill	Alphabet
	Parchment	Pen	Pictographs
	Silk	Pencil	
	Paper	Ball pen	
Computer	CRT	Keyboard	Bits–bytes
	B. Audio Communication		
Telegraph one-way	Wire	Current	Dot–dash
Telephone two-way	Diaphragm	Current	Vibration
Radio	Air	Current	Waves
	C. Visual Communication		
Photographic camera	Ag-halide	Light	Chemical
Television	CRT	Current	Pixel
Computer	Monitor	scanner	Pixel
	D. Intranet Communication		

Word text +
  data +
  audio +
  video interactivity +
  N-way
	Electromagnetic	Bits–bytes	Bits–bytes

Stage 2: Telephonic Communication: Two-Way, Two-Party Interaction.
Telephones allow two-way communication between two parties. The two-way communication normally takes place between one sender and one receiver, interchangeably and simultaneously transmitting and receiving communication in the same transaction. This was an improvement over telegraphic communication.

Stage 3: Broadcasting Communication: One-Way, Multiparty Interaction. Printed books, work-related operating manuals, publishing newspapers or newsletters, and broadcasting or making announcements are all examples of one-way multiparty communication technology. Communication flows in one direction, from one party transmitter to many receivers.

Stage 4: Net-Casting Communication: Multiway, Multiparty Interactive Interaction. Communication interactions in highly cross-linked networked organizations flow in multiple ways and involve multiple parties in-

EXHIBIT 6.4. Communication Communities

Communication	Sender	Receiver	Message
	1. Interpersonal Communication		
Two-way	One	One to few	Words + gestures Vibes
	2. Printed Communication		
One-way	One to few	Many	Written, images
	3. Telegraphy		
One-way	One	One	Codes
	4. Telephony		
Two-way	One	One +	Words
	5. Radio Broadcasting		
One-way +	One to few	Many	Audio
	6. Television Broadcasting		
One-way +	One to few	Many	Images, some text
	7. Main-Frame Computer		
One-way	One-few	Few	Text, data
	8. Intranet Communication		
Multiple-way	Many	Many	Text + data + images+audio + vibes?

teracting at almost the same time. This is like walking on a Ginza intersection in Tokyo. When the lights go red for the cars, the pedestrians can go from any direction to any direction.

Mankind has arrived at this interactive multiparty communication technology in a gradual manner. The information and its different sources were available before the advent of the Internet. The popularity of the Internet has unfolded their access to more of us. Will that make a difference? More than 80% users of the Internet use it for entertainment purpose rather than to access some useful information. This gets back to values. Even with instant access to enormous information, only a few Internet users use the advanced information processing technology for nonentertaining purposes. The choice depends on the underlying values and beliefs behind people's behavior.

EXHIBIT 6.5. Evolution of Communication Transactions

	Two	Many
Two-/Many-way	Telephone	Cyberspace
One-Way	Telegraph	Broadcasting

Communication Traffic

Number of Communication Parties

It is therefore very critical to understand the values and intents of the different members of a work organization. However, in the 1990s the interacting members in a technology-driven enterprise are geographically dispersed around the world and are cross-linked with each other with new communication technologies. The information processing technologies and their developments will dictate to a great extent the way we will communicate with one another. We must therefore first understand how the new communication and information technologies have evolved to their new forms.

THE INFORMATION PROCESSING TECHNOLOGY AND KNOWLEDGE NETWORKS IN THE IDEA AGE

Information technology and computer networks connect people to people in different parts of a technology-driven enterprise. In a traditional bureaucratic organization the relevant information has to travel through many hierarchical hurdles—that is, from one layer to the next, and then from the next layer to the next, and so on (see Note 6.11). New information technologies eliminated some, if not most, of those hierarchical hurdles. New technologies facilitate faster communication by broadening the direct access to the people who need the information most, and can update the information based on their direct involvement.

At Hewlett-Packard every month, its 97,000 employees exchange 20 million e-mail messages internally and about 70,000 externally. They share more than three trillion characters of data, including project status, product specifications, and other crucial information, which gives the company it's reputation as a global company (see Note 6.12).

In Arthur Anderson, one of the Big Six public accounting and consulting firms, more than 70,000 employees worldwide were linked and trained using its own Internet-like network and dedicated databases (see Note 6.13). Coopers & Lybrand, another Big Six accounting firm, also linked its 13,000 employees via a dedicated knowledge network. The success of a knowledge network depends on the willingness of its members to work as an integrated team, even though they may have little person-to-person contact and are located in widely dispersed geographic areas. Work-group information technologies are used to customize communication networks for specific professions—such as for the lawyers or the doctors and their clients. A fully cross-linked company using networking communication technology would interlink purchase, marketing, finance, production, and accounting departments together. But the keys to success in network-based communication are same as the keys to success in person-to-person communication. Trust and feedback play crucial roles.

The Idea Age

What characterizes the current state of human civilization most? We do not rely as much on manpower or animal power. Otherwise China and India, with higher populations of people and animals, would have been the most powerful countries in the world. Modern civilizations are also not defined by our natural resources. Then the "Miracle of Japan" during the 1980s, as well as the growth of Singapore in the 1990s, would not have happened. Japan's bubble economy during the early 1990s, along with a continued growth of the U.S. economy during this period, indicated that we are out of the Industrial Production Age too. The *Industrial Age,* which started the centralized industrial production in Europe, freed mankind from our primary bondage to nature, hunting, fishing, and agricultural farming. Social observers like Alvin Toffler and John Naisbitt have made us aware of the "Megatrend" of the "Third Wave" of the *Information Technology Age,* when we broke our umbilical chord for primary dependence (and survival), on material and physical goods. More recently, ideas and knowledge have become the primary parameters which best explain North America's economic growth in the last lap of the 20th century.

Companies and countries with fast access to rare, hard to imitate, and valuable information seem to command superior economic growth and returns. We find ourselves on the edge of a new economic age, when information is becoming the primary driving force and resource for our economic wealth. Millions of Americans have almost free access to vast amounts of information, distributed worldwide. The manufacturers and providers of goods and services can communicate directly with their customers worldwide. Therefore, in the globalized and highly cross-linked world economy, what is needed most is an innovative value-adding idea that makes use of extensive and widespread information in a profitable manner. And that will lead to the birth of new enterprises—for example, the birth of the world's largest "cyber" book store located on the Internet at Amazon.com (see Note 6.14). In 1997 this unique book store offered 2.5 million titles, with discounts of up to 40%. Barnes and Noble, the traditional book-seller, too opened an online bookstore, to compete with the web-based Amazon.com.

Like in the worldwide society, inside a business organization, the information and new ideas in the 1990s have been flowing between employees faster than ever before. With Intranet, Internet's corporate cousin in the commercial world, and "groupware" (software for teams and working groups), more employees have started interacting with other employees within the organization than ever before. This higher intensity and greater frequency of information exchange is no longer limited by the geographic location of the interacting employees. They can be located a few cubicles or thousands of miles away from each other.

This enhanced information exchange is also available to a technology-driven enterprise's major outside allies. They include the suppliers on one side, with the customers on the other side, of a value-adding chain. However, as in the case of the idea-based entrepreneurs mentioned earlier, for a corporate resource to become a core-competency and a source of competitive advantage, this information exchange must meet three conditions. It must be rare, hard to imitate, and value-adding. It is therefore postulated that in the years ahead, not just more information, but information packaged in a rare, hard-to-imitate, and valuable (particularly of value to customers) idea, will provide a competitive advantage for the originator of such idea.

We can therefore perhaps best label this period as the *Idea Age*. During the Idea Age, innovative and value-adding ideas, rather than mere accumulation, dissemination or processing of information, will derive a firm's performance and success.

Different observers have referred to this by different names. For example, James Brian Quinn, Philip Anderson, and Sydney Finkelstein pointed out in a 1996 paper published in the *Academy of Management Executive* that the intellectual and information processes (in an organization) create most of the value-added for firms in America's large service sector. This sector includes software, communications, education, and health care services. They quote that these sectors accounted for 79% of American jobs and 76% of the U.S. Gross National Product. By the year 2000 the number of knowledge-based jobs is estimated by McKinsey & Co., a Boston-based worldwide consulting leader, to increase to 85%. The corresponding number for Europe was estimated to be a comparable, but a smaller 80% of all jobs (see Note 6.15).

IMPACT OF INFORMATION PROCESSING TECHNOLOGIES ON THE TECHNOLOGY-DRIVEN ENTERPRISE

The information processing technologies have transformed the modern technology-driven enterprise at the operational as well as the tactical and strategic levels. Information processing technologies play an important role in the day-to-day production operations. At the tactical level in an organization, the information processing technologies help managers manage and control the operations and work of others; and at the strategic level, the information processing technologies provide competitive advantage to the technology-driven enterprises by effectively interfacing with the changes in the external environment, competitors, and the target customers.

A. Operations Level Impact

New developments in information processing technologies can influence (a) the choices of technology-driven enterprises and (b) the way these companies produce goods to respond to customers' more demanding requirements. Given below are some of the innovative tools of information processing technologies used in the production operations.

A1. Computer-Aided Design. Designing of products and processes is an important value-adding competency in technology-driven organizations. The designing activity often involves a large number of professionals from diverse professional backgrounds. From time to time, engineering designs must be changed because of a variety of reasons. These reasons could be (a) the changes in the customers' expectations, (b) a new development of technology, (c) added resource constraints, or (d) simply because an important design attribute was overlooked. When engineering designs are made manually, such engineering changes and involvement of diverse professionals add enormous amounts of rework. Computer-aided design (CAD), whereby computers are used to design interactively, help reduce such rework substantially. The CAD tools and software programs help draft conventional engineering drawings, as well as produce three-dimensional perspectives and drawings. Electronic printing circuit designs or architectural drawings can be prepared using specialized CAD programs and tools. These tools also allow designers to estimate and test the mechanical properties and failure characteristics of a designed part or subassembly. The CAD tools help reduce the cost of engineering changes and repeat designs substantially. The CAD technology uses innovations based on graphic–user interface (GUI).

A2. Design for Manufacturability. A proper design can significantly reduce the overall cost of producing a product. Much of the unit cost of producing a part or a product is determined in the design stage. Design attributes influence the efficiency of production and productivity. The software programs for design for manufacturability (DFM) help link the design parameters with the manufacturing requirements. These programs help reduce the overall cycle time. These information tools help simulate a variety of design changes without incurring much additional cost. This helps pick the most optimum design that meets a variety of constraints.

A3. Computer-Aided Manufacturing. Computer-based designing can be integrated with computer-driven manufacturing equipment to provide computer-aided manufacturing (CAM). Such information links help produce a variety of products using a common database of engineering information. With this access, the CAM tools help designers and manufacturing engineers free themselves from mundane details and focus their efforts on the creative and conceptual aspects of a product or a part. This can substantially improve the performance as well as the quality of the products thus produced.

A4. Automatic Process Control. In the case of technology-driven enterprises based on continuous processes, the processing conditions in different parts of a process are monitored and controlled using information processing technologies. Often these continuous processes, such as for the manufacture of synthetic rubber goods, textile fibers or nuclear energy, involve parts moving at an enormous speed and under harsh physical conditions. Automatic process controls use sensors that convert physical parameters such as temperatures, pressures, moisture contents, and so on, into digital signals. Control devices read and record these digital signals on a regular routine basis, as often as a few times every second. Centralized computers then process these digital signals and control processing conditions according to a prespecified protocol. The automatic process controls can substantially reduce the manpower required to run a large continuous plant. Such controls can also improve the reliability of running a complex process with high reliability and safety.

In the past, sometimes the continuous processes (such as used by the Three Mile Island nuclear power generators) have gone out of control because of co-occurrence of a variety of abnormal conditions. This was often coupled with error in human judgement. Such overloads on the central computer result in unreliable process control. More recently, hazardous processes use parallel, redundant, and excessive computing capacity to cover such abnormal contingencies.

A5. Flexible Manufacturing Systems. Many production processes must produce a variety of products depending on the customers' preferences and demand fluctuations. These production facilities must rapidly change from one production program to the next program with the minimal downtime for changeovers. Such flexible production processes use information processing technologies to integrate various computer-driven material-handling equipments with the different production workstations. These are called flexible manufacturing systems (FMS). With automated work cells, different parts needed for a product change are thus automatically fed to a workstation. Different products can be flexibly produced by giving suitable instructions to the flexible manufacturing system. These flexible systems help economically produce a high variety of products in low volumes. However, they require much higher capital investments than the standardized automated equipments producing few standardized products in large volumes.

A6. Computer-Integrated Manufacturing. In this production process, the flexible manufacturing system is integrated with computer-aided design (CAD) and automated material-handling equipments. The entire system is driven by common databases. This substantially reduces the time required to enter data in different functional departments.

These selected examples illustrate how the different innovations in information processing technologies have improved the flexibility of production operations. These innovations reduced the cycle times and lowered the unit costs of manufacturing a variety of products for the increasingly demanding and diverse customers.

B. Impact of Information Processing Technology at the Tactical Control Level

Given the increasingly easy and affordable access to information technology, many progressive firms are using new developments in information processing technology to closely monitor the subtle changes in their internal and external environments. A management information system (MIS) helps decision-makers and workers collect, sort, analyze, and process information needed to support the value-adding activities of a technology-driven enterprise. The MIS provides reliable and high-quality information in a timely and efficient manner. Some examples of information collected by technology-driven enterprises are listed below.

1. Customers' requirements
2. Customers' satisfaction and delight levels
3. Key parameters of manufacturing processes
4. People's and equipments' productivities
5. Problem identification
6. Analysis of trends—external as well as internal
7. Measuring the changes and their implications

Such measurement of actionable data help managers make fact-based decisions. The information thus collected is shared with those front-line employees who need it most to improve their processes. The managers in technology-driven enterprises must accept that they can improve only what they can measure. This is true even for the service sector. Fred Smith, chairman and executive officer for Federal Express, an overnight delivery service provider, believed that "service quality must be mathematically measured" (see Note 6.16).

Example: AT&T Universal Card Services. In 1990, the Jacksonville, Florida-based AT&T Universal Card Services (AT&T-UCS) with only 35 employees started offering an innovative long-distance calling card combined with general-purpose credit card. The company adopted a mission to delight its customers, and it designed the entire organization around that core concept. It carefully measured nine customer satisfiers. These included parameters such as fee price, annual percentage interest rate, and customer service. The key satisfiers were further subdivided. For example, the customer service satisfier was subdivided into accessibility, professionalism, courtesy, patience, speed of answering, and so on.

Using certain mathematical models, many key variables were related with more than 100 internal process measurements. These measurements were made every day, and they were shared with employees the next day. Thousands of customers were specifically contacted to find out about their level of satisfaction. These efforts helped AT&T-UCS acquire 16 million customers within its first two and a half years of operation, and they generated more than $1 billion in revenue. During this period, the number of employees increased to more than 2500, with more than two-thirds having direct contact with customers (see Note 6.16).

Example: Carrier Heating and Air Conditioning. Carrier is a technological pioneer and is one of the world's leading manufacturers and providers of heating and air-conditioning products. It is part of the Farmington, Connecticut-based United Technologies, a conglomerate. Carrier measures standardized corporate metrics in the areas of (1) customer satisfaction, (2) product quality, and (3) employee satisfaction. The three metrics are linked logically along the technology value-adding chain. Product quality reflects the quality of output of Carrier's manufacturing facilities. The customer satisfaction metric is the lagging indicator, and the employee satisfaction metric is a leading indicator of the company's performance. Every year a standard one-page Customer Satisfaction Index survey is sent to about 7000 customers. The product quality is measured by (1) "dead on arrival" units which are defective within 30 days of installation, (2) factory defects at the assembly process based on a 2% sample, and (3) warranty costs measured as a percent of last year's net sales. The employee satisfaction metric is based on an annual survey of 100 questions in 16 categories and in nine languages, which is sent out to all the Carrier employees worldwide (see Note 6.16).

Information Needed to Facilitate Continuous Improvement. Such measurements, illustrated by the cases mentioned above, help technology-driven companies develop a common language for communication across their enterprises. The employees with different functional backgrounds can relate to a common set of facts and other pieces of information.

Harrisburg, Pennsylvania-based AMP, a world leading manufacturer of electrical and electronic connecting devices, shares the collected information with its more than 27,000 employees in 27 countries via on-line computer terminals. At AMP a Quality Scorecard provides a table of the month-by-month measurements for quality, delivery, value, and service at the supplier(s), at AMP, and at the customer stages in the supply chain. Besides providing the overall values in this table, an employee can access more specific details by clicking on each element in the 4×3 table of the AMP Scorecard. Using this Scorecard, AMP has improved its quality significantly. Over a 4-year period, for example, the on-time delivery was improved from 66% to 95% and the loss of abandoned calls was reduced from 15% to 0.4% (see Note 6.16).

Even for making decisions at the tactical and strategic levels, the task-level information obtained from the measurements of operations must be aggregated. The numbers thus produced are likely to be more reliable than those generated without going through detailed measurement and information collection.

Mistrust and Resistance to Measurements for Monitoring. In general there is a human resistance to extensive work-related measurements. Employees are afraid that the measured information will be used against them by their upper management to criticize their work. Senior managers must therefore reassure their employees that the primary purpose of such measurements is to fix and develop the business processes, rather than to blame the persons. The purpose is to identify and monitor the important aspects of performance of a technology-driven enterprise, and then to identify the causes behind the variations in such performance. In the absence of such reassurance by the upper management, the measured information may be highly "polluted" by the anxieties of employees.

These companies use different quality tools to identify the causes behind the different observed effects. These quality tools are discussed in detail in the previous chapter. The measurements help separate the significant few major causes of poor performance from the many trivial causes. Resources are then allocated to these major causes to improve the business processes to a higher level of excellence. The employees who contribute significantly to achieve these goals are rewarded and recognized for their desirable behavior. This is the new way of deciding and managing by carefully measured facts and information, rather than by decisions made by decision-makers sitting in isolated and often insulated offices. The results as well as the rewards are carefully communicated to all the employees to promote the desirable behavior in other employees as well.

The progressive companies, such as AT&T Universal Card Services and Carrier, use the measured information to determine executives' incentives and bonuses. At Carrier, over 300 executives get bonuses that is 20% based on customer satisfaction measurements, 50% based on product quality improvement, and 30% based on employee satisfaction. These executives may thus earn an annual bonus of $30,000 to $50,000 based on their performance in these areas. Employees at AT&T Universal Card Services also get incentives for meeting daily performance targets, determined collectively. Such incentives could add a bonus of up to 12% of an employee's salary.

C. Strategic Impact of Information Processing Technologies

The recent developments in information processing technology have transformed the rules of market competition. They have influenced the way technology-driven companies align their organizational efforts and core competencies with the megashifts in their external environments. Often information-processing-related core competencies influence the choice of product-market segments in which a technology-driven company competes.

In the traditional industrial economy, companies often competed on economies of scale. They produced standardized goods and delivered them to customers in a restricted region with a significant time lag. With the recent developments in information processing technology, the technology-driven enterprises are heading towards flexibly producing a variety of mass-customized products. These are offered in fast response to requirements of customers located in far more distributed geographic areas than ever before.

Good-Bye Middleman: Getting Closer to Customers. Information processing technologies changed the way companies do business in the 1990s. An instant access to accurate data has changed the way many industries conduct their transactions. In the past they did this according to the classical economic theory of supply and demand. In the future customers will base their decisions on perfect, or almost perfect, information. In the 1990s, customers also demanded speedy response on a worldwide basis. Given below are some of the examples of progressive technology-driven companies competing in global markets using the information processing technologies (see Note 6.17).

C1. Hewlett-Packard's Global Service Network. Hewlett-Packard installed a customer response network with 1900 technical support professionals, mostly engineers. When a customer called or left an electronic message, it was routed to one of the four worldwide hubs. The choice of the hub depended on the time of the day. The operators took down the message and noted its urgency into a database that was linked to 27 centers with teams having special expertise in different areas. The "live" database was updated frequently and was accessible to all employees in the network. As teams worked on a problem the progress was reported into the database. More teams were called in if needed.

C2. Retailing in the Information Age. Wal-Mart Stores pioneered the use of computers in wholesale buying, inventory management, and stocking of the popular items. Suppliers were told when and where they were expected to supply. With huge orders and rapid turnovers, they bought branded products at rock-bottom products. Food wholesalers learned a lesson or two from Wal-Mart. They made the most profits when they could pick discounted merchandise from manufacturers and move them fast to where the demand was likely to be. They did so by keeping track of how the grocery shelves were depleted by a new scheme called the efficient consumer response (ECR).

C3. Compact Discs: What Is Hot and What Is Not? In the compact-disc and music recording industry, for decades *Billboard* magazine conducted a weekly survey, based on phone interviews with retail store owners. Record companies influenced the owner respondents by giving them incentives, free discs, and complimentary vacation trips. These practices distorted the projections about the sales of different products. Radio stations and consumers used these to find out what was hot and what was not.

In 1991 Soundscan Inc., a Hartsdale (New York) company, started tracking actual CD sales by electronically taking data from scans at the checkout counters. In 3 years they widened their net to over 14,000 stores. The entertainment industry and consumers, including the *Billboard* magazine, used these accurate data to identify the latest sales trends. For example, the industry was surprised to find out that country music sold more than they had assumed in the past. The revelation convinced record executives to promote country singers more than they did before.

C4. Auto Insurance Claims. Computer networks changed the way auto claims were processed and paid. When a customer called the 800 number on an insurance policy to make a claim, a "preferred provider" got that information automatically. This helped the insurance provider company such as Allstate Corporation consolidate its payouts to a limited few dependable partners. This eliminated, for example, the need for the insurance company to check and verify the claims of payments made to the tens of other local shops in the business of repairing auto glass. The key ethical question for the customers was whether they should be directed to a few preferred providers. The independent service providers opposed such moves, and the insurance companies fought against it. In the meantime the information processing technology was saving the insurance companies millions of dollars in administrative overhead expenses.

C5. Self-Service Travel. American Airline's SABRE Travel Information Network revolutionized the way air travellers made reservations. This industry-wide network, costing about $300 million, provided availability of seats to thousands of travel agents worldwide. The promoter AMR Corporation had the advantage that its flights were listed at the top of the availability tables. Even though flights by other airlines were also listed, flights by American Airlines were picked more often than others. United Airlines received a similar competitive advantage with their APOLLO reservation system. With the widespread use of computer networks and the Internet, consumers could directly make their own reservations and plan their own vacation trips. The information highway, with the help of software for audio, data, and video images, was likely to usher in interactive videos and virtual reality tours of hotels and hot tourist spots.

What will be the impact of these new developments in information technology on the travel business industry? According to Arthur D. Little, a major consulting firm, the expanded direct access of consumers to travel data could reduce 20–30% of travel agency's travel reservation business by the end of the 1990s. The American Society of Travel Agents, on the other hand, predicted that the new information technologies would boost the bookings through the travel agents because of the added value and expert advice they would provide (see Note 6.17).

AMERICA'S INFRASTRUCTURE ADVANTAGE FOR CROSS-LINKED ORGANIZATIONS

In this information age, the United States is way ahead of other industrial nations by virtue of its information infrastructure. Americans can surf and search strategic information about customers, competitors, and critical environment from one end of the world to another end of the globe and retrieve enormous amounts of instantly accessible information. Millions of employees frequently and regularly take the information superhighway and collect timely and detailed information with a click of their "mouse." There is an explosive growth in demand for the employees who have skills related to collecting useful information and can use networks to support more information driven business decisions.

Information Superhighway

Soon after the U.S. President Bill Clinton and Vice-President Al Gore were elected for the first time in 1992, they started talking about improving America's information infrastructure. Vice-President Gore was for constructing a superhighway as America's bridge to the future. The proposed superhighway was far more rapid than the commonly used Internet network communicating over telephone lines. The proposal involved building a national superhighway for text, data, music, voice messages, medical and moving images, and more. It would require an enormous network of fiber-optic cables, sophisticated switching stations, and zillions of man-hours of construction work for the information bridge to the future. In the 1990s, the information superhighway was a project of national significance like the transcontinental railroad for the flow of goods was in the mid-19th century and like the interstate highway was over 50 years ago (see Note 6.18). The information superhighway could radically improve America's education, scientific research, health care, and entertainment choices, as well as make the American companies compete more aggressively in the global economy. Employees located far away from each other could collaborate on new product development by exchanging their data easily and frequently. Manufactures would closely interface with customers and suppliers.

The cost of the Information Superhighway was estimated in hundreds of billions of dollars. Who would pay for this charge? A strategic decision to make was whether this information highway should be a freeway, freely accessible to every citizen, or should it be a private toll-way, charging a fee for the users. The baby-boomer Vice-President Al Gore, whose father sponsored the interstate roadways earlier, argued that the U.S. Government should build it, and the private sector should operate it. Sort of like the Internet computer network constructed with defense budget. The profit-driven publicly held companies were unlikely to invest such sums on so highly risky venture.

One shortcut for America was to use the Integrated Services Digital Network (ISDN), based on a digital technology and the existing copper phone lines. This technology was already developed and would cost much less than the superhighway proposal based on new fiber-optic cables, new switching terminals, and new software programs to run them. Critics argued that the phone system was not fast enough and that the phone companies were not so friendly with the state utility commissions overseeing their rate schedules.

IMPACT OF INTERNET TECHNOLOGY ON COMMUNICATION IN SOCIETY

Cyberspace Surfing and Car-Cruising

Impact of networking on cyberspace has been compared to a number of things, from a social transformation to the industrial revolution. In the 1990s the key characteristic of networking on the information superhighway may be compared to the development and access to cars in the 1890s, particularly when its popularity after Henry Ford's moving assembly line for the Model T made it affordable to the common man in 1913. Henry Ford, by making his "T" accessible to average Americans, freed their spirit to roam and helped connect the continental country from the New York Islands on the East Coast to the shining sea on the West coast. The information superhighway continues that westward expansion beyond the Pacific Ocean to the rest of the world.

People who have started cruising and living along the new information superhighway are the modern-day "idea" homesteaders. Like their predecessors who tamed the Wild West, the Internet surfers find an idea (like a plot of land), clear it, cultivate it, and start living on and off it. The lure of mining information seems no less attractive than the Gold Rush miners in 1849. By the summer of 1994, over 2 million new Internet users were joining the information superhighway every month (see Note 6.19).

The Internet has become the archetype international network of computers, information sources, and storages distributed around the globe. This network of information is commonly called the information superhighway because it seems like a seamless giant information processing monster mechanism. This links zillions of computers owned by individuals, universities, and big computer communication service providers such as America-on-Line, Prodigy, CompuServe, and so on. According to the Internet Society based in Reston, Virginia, by the summer of 1994, there were 25 million surfers on the cyberspace, which was likely to double every year.

Green Card Info Accident

The information superhighway has its own accidents and pile ups. Lawyers and all kinds of other vendors are already on the information superhighway offering their goods and services. For example, cyberspace lawyers offer attractive compensation for accident injuries from computer terminals instead of by chasing behind ambulances in their cars. In April 1994, an Arizona-based husband-and-wife firm of lawyers promised potential immigrants green cards for permanent residence in the United States. This started a major information pileup. From an ordinary IBM PC-type computer parked somewhere in their spare guest room at home in Scottsdale, Arizona, they wrote a program called Masspost. This software program put their little advertisement for Green Card in almost 5500 bulletin boards, to be seen by millions of cyberspace surfers again and again (see Note 6.19).

In the information superhighway jargon, this practice is called *Spamming* the Internet. Its effect is like dropping a can of Spam into a fan and then filling the room with meat. The impact of Spamming is that it hides the computer owner's crucial electronic mail-messages, bills, and checks in a heap of 10,000 pieces of junk mail that cannot be ignored. This angered many cyberspace surfers, who "flamed" for being freely Spammed by the attorneys. The Net users countered the infringing law firm by sending them thousands of phony requests for information from Australia, including a "16-year-old threatening to visit their crappy law firm and burning it to ground." Within 3 days, Internet Direct of Phoenix, the service provider for the attorney firm, decided to discontinue their customer's account.

This is a typical tussle for developing the discipline for using an emerging technology. Millions of information superhighway commuters, operating from the privacy of their basement offices, have their own unique ways of disciplining their abusers. In such incidents, usually the receivers of such Internet rage are so outraged and intimidated by the angry protesters that they shy away from further active or outrageous display of their presence in the cyberspace.

Information Superhighway Patrol

In the case of the Arizona lawyers, the contrary happened. The lawyers claimed that they had succeeded in generating $100,000 worth of new business. They threatened to sue Internet Direct, the service provider, for cutting off their service, and they vowed to the *New York Times* that they will definitely advertise again on the information superhighway.

This angered the cybersurfers even more. They sent the lawyer couple faxes of hundreds of empty pages. Thousands of bogus subscriptions started arriving at their doorsteps. A Norwegian programmer innovated a canceling robot, or cancelbot, that surfed the information superhighway seeking out and canceling this law firm's Green Card advertisements on bulletin boards before they spread any further.

The Green Card superhighway incident exposed the weaknesses of the new networking technology. The information superhighway is likely to get gridlocked very quickly if its freedom for invasion of other users' privacy is abused by greedy or immoral vendors for commercial purposes. By the mid-1990s, some users were building coded fences around their parts of the information superhighway. Others were hoping that the encryption technology will be developed soon to safeguard the commercial transactions on the electronic frontier. There was no doubt in people's minds that the Internet had changed the way they would communicate with others in the future.

CROSS-LINKING THE CORPORATION: IMPLEMENTATION REQUIREMENTS FOR NETWORKING TECHNOLOGIES

Like the society at large, the networking technologies have significantly influenced the way we work in the technology-driven enterprises. The benefits of information networks listed later, however, do not accrue just by throwing in a large chunk of information processing technology at the employees and then telling every one to use it. To derive benefits from a networked organization requires a lot of careful preparation.

1. Drive Out the Fear of Networks

Many employees, including the senior managers, have a fear of the unknown. They have built their success on the old ways of doing things. Often they do not know how to ease into the new ways demanded by a new information processing technology. Many people therefore sit it out. They miss participating in major decisions. Over time their performance suffers. Some managers try to strangle the genie of network technology by restricting others who can use it. Or the employees are afraid of sharing information with others without clearing it first with their bosses. Wired organizations must provide opportunities and sufficient time to help employees acquire the skills needed to use the new networking technologies.

2. Proactively Create Trustworthiness

The traditional hierarchical organizations eliminate the need to rely on trustworthiness of people. In a hierarchy, each person has a well-defined role and authority. Trust has no role in such command-and-control bureaucracy. Bosses get more information than the subordinates because the bosses are assumed to be more trustworthy than the subordinates. The organizational hierarchy and strict operating manuals substitute the need for trustworthiness or trust-building interactions.

Information networks disturb such well-defined hierarchical order. Information no longer flows in one direction. It is flooding in and out in all directions. This requires new interdependencies on others. This can make some bureaucratic bosses very uncomfortable and resistant to change. Only persistent trust-building interactions between people, sustained over time, can cure that mistrust.

3. Signal-to-Noise Ratio

A major challenge of open networks with no protocols is the flooding of useful signal with useless noise. Signal is the useful information needed by a person for carrying out a job. Noise is that other information that every one gets with no useful purpose. The open structure of networks and lack of hierarchical controls on the flow of information can inundate a busy executive. At many major corporations, some senior executives have been complaining that they receive over 200 messages a day. Even a 1 minute scan per message will eat away their whole morning. And some of the messages contain the 15-page report from the sales representative with additional charts and graphs. Reading everything carefully would provide a lot of useful information. But it would leave not much time for anything else, including taking any actions. Employees in technology-driven enterprises must therefore carefully think through how they are going to address this overload of information. Some ground rules and a distinction between essential signal information from the redundant noise will go a long way in improving people's productivity.

4. High-Tech Communication Needs High Touch

Networks are generally dispersed over large geographical and cognitive distances. A wide variety of people with different expertise come from many different places to interact or communicate on a common opportunity or challenge. Experts believe that the more diverse and dispersed a group, the higher the significance and need for periodic face-to-face social contact. Network experts believe that people cannot have a virtual conversation unless they also have real person-to-person conversations.

5. See the Person, Not Just the Work

In cross-linked organizations some people quickly forget that behind each e-mail there is a live person with a heart ticking. Even though a boss may no longer routinely look over the shoulders of her subordinates because they are working from home or elsewhere, the boss's role as a guide and a mentor does not diminish. Some bosses are realizing that in the new networked work setting, managers must be more careful in hiring and training the right kind of persons. These new hires will be mostly working on their own and can't be supervised the old way by their supervisors. The productivity and performance of the subordinate depend more on the commitment of the subordinate rather than on the intensity of his boss's supervision.

6. Carefully Build Polycultural Teams

The cross-linked technology-driven organizations rely a lot on teams of large numbers of people with diverse professional cultures. Putting them together does not turn them into high-performing teams. They have differences and biases which need to be tenderly ironed out. A shy and quiet person may have the most relevant expertise needed to solve a problem, such as the development of a new feature in a new product. On the other hand, a loud and articulate person may have little value to add to the project on hand. Even though the cross-linked organizations do not rely on the traditional hierarchy, there is need for negotiators, timekeepers, scribes, and other important roles in the self-directed teams. These roles do not emerge by themselves. Somebody must silently conduct the team's orchestra to produce a harmonious and productive melody.

7. Vision Setting Leadership

Even when the cross-linked organizations empower their individuals and their teams, they still need the overall vision and guidance of a leader. Leaders do important tasks. They coordinate the work of different teams, set launch dates for new products, and monitor allocation and utilization of resources. A cross-linked organization cannot afford to turn into an anarchy where everybody acts randomly. Organizations and their leaders have fiduciary responsibilities. There are social responsibilities too. The cross-linked organizations need proactive and visionary leaders who facilitate a free but disciplined flow of information and exchange of ideas. The leadership for technology-driven enterprises will be discussed in detail in a later chapter.

BENEFITS OF A CROSS-LINKED ORGANIZATION

As mentioned before, the Internet and its corporate cousin Intranet networking technologies have significantly changed the way people work and exchange information in technology-driven enterprises. Experts believe that cross-linking an organization with information networks could generate many potential benefits. Some of these potential benefits are discussed below.

1. Save Idle Time

Cross-linking a technology-driven enterprise speeds up flow of information and wipes out idling delays. Critical information that gatekeepers withheld from others in the past is now accessible to all. No more waiting for the Monday morning meetings. Manual transfer of information in an organization often took hours, or days, because an important piece of information sat idly in the in-box of an executive. With e-mail such information is immediately out there for everyone to see.

2. Help Cross-Functional Concurrence

The traditional sequential modes of information transfer and exchange involved one functional department at a time. Vertical hierarchical flow of information took place in a roundabout route. Information traveled vertically up and down in functional silos. Information in an organization is often required laterally—by other individuals or departments in the organization. With the progress of human civilization, most products and services have become increasingly complex and sophisticated with time. Developing new products and technologies requires involvement of larger number of people with different functional expertise, often in simultaneous consultations. Networks allow experts in different functional areas to work on a project simultaneously. They can be located in different parts of the world, and they can report or share their results and their difficulties with others in real time. When a problem arises, a person can simply post a note on the bulletin board, "Can anyone help?" Usually someone in the organization can.

3. Save Money

Reducing time, and involving people who can help, usually results in reduction in cost. Lotus Corporation, for example, could introduce its Japanese version of a new software within three to four weeks of its English-language release. Lotus Notes users, according to a survey, recovered their investment and earned high returns within 3 years (Note 6.11).

4. See Market Clearer

Networks can help technology-driven organizations see their dynamic markets and demanding customers more clearly. Under the traditional management of the organizations, the marketers and sales representatives interpret the market and customer requirements for the rest of the organization. Their managers consolidate the information, and then they screen it according to their bosses' preferences. In a cross-linked organization the managers as well as the managers' bosses get the same raw information from the field about what their customers love and what they hate. The raw information goes in a database which can process it into many meaningful ways—without multilayered human screens. More managers get to watch the market performance from the front row (Note 6.11).

These are significant advantages which can go a long way in gaining competitive advantage for a technology-driven enterprise facing global competition. One can argue that the technology-driven enterprises cannot afford to overlook such benefits.

SOME LESSONS LEARNED IN THIS CHAPTER

Communication and information exchange between people play significant roles in technology-driven enterprises facing dynamic markets and demanding customers. Under the intensely competitive conditions these communication and information exchanges must often take place between a large number of people with different professional backgrounds. These experts are also often distributed in different parts of the world. Recent developments in information processing technology, such as the Internet and the Intranet, can help technology-driven enterprises in many ways. We learned through examples how the information-processing technologies have significantly influenced the technology-driven enterprises at the operational, tactical, and strategic levels. However, to derive the full benefits of information-processing technologies in highly cross-linked technology-driven enterprises, these technologies must be implemented with a heavy dose of human touch.

SELF-REVIEW QUESTIONS

1. Describe the communication process in a technology-driven enterprise. How is this process affected when professionals with different functional backgrounds must work together in self-directed teams?

2. How have communication technologies evolved over time? Illustrate with examples.

3. How are new products and services perceived differently in the Information Technology Age? What is the role of knowledge and customers' perceptions?

4. Discuss the influences of information processing technologies on technology-driven enterprises at the operational, tactical, and strategic levels? Give specific examples.

5. How have information technologies transformed the production process?

6. What are the prerequisites of implementing networking information technology in a technology-driven enterprise? List the potential benefits one will accrue from such effective implementation.

NOTES, REFERENCES, AND SOME ADDITIONAL READINGS

6.1. Driscoll, Kathleen. 1997. Frontier hopes 'Shared Values' builds trust. *Democrat and Chronicle.* March 31. 1,3.

6.2. Toffler, Alvin. 1980. *The Third Wave.* New York: William Morrow & Company.

6.3. Naisbitt, John. 1982. *Megatrends.* New York: Warner Books.

6.4. Sakaiya, Taichi. 1991. *The Knowledge-Value Revolution.* New York: Kodansha International.

6.5. Johnson & Johnson Annual report, 1996, and other company documents.

6.6. Gehani, R. Ray, 1996. Poly-cultural teams: a multicultural construct for superior performing teams in the Age of Global Economies. Paper presented at a conference on *Global Multiculturalism,* held in Rochester, New York, July 11–14, 1996.

6.7. Wagnor, G. W., Pfeffer, J. and O'Reilly, C. A. 1984. Organizational demography and turnover in top-management groups. *Administrative Science Quarterly,* 29:74–92. Also see Watson, W. E., Kumar, K., and Michaelson, L. K. 1993. Cultural diversity's impact on interaction process and performance: comparing homogeneous and diverse task groups, *Academy of Management Journal,* 36:590–602.

6.8. Fukuyama, Francis. 1995. *Trust,* New York: Free Press/Simon & Schuster.

6.9. Eisenstadt, S. N. 1968. *Max Weber on Charisma and Institution Building.* Chicago: Chicago University Press.

6.10. Stiebel, David. 1997. *When Talking Makes Things Worse.* Dallas: Whitehall and Nolton.

6.11. Stewart, Thomas A. 1994. Managing in a wired company. *Fortune,* July 11: 44–56.

6.12. Hewlett-Packard company documents and web page.

6.13. Messmer, Max. 1994. Staffing the information superhighway. *Management Review,* November:37–39.

6.14. See Amazon home page on the Internet at www.amazon.com. Also see Robinson, Kathy. 1997. Book wars. Online sellers offer ever-deeper discounts. *Democrat and Chronicle, Rochester,* October 29:3C.

6.15. Quinn, James Brian, Anderson, Philip, and Finkelstein, Sydney. Leveraging Intellect. *Academy of Management Executive,* 10(3):7–27.

6.16. George, Stephen, and Weimerskirch, Arnold. 1994. *Total Quality Management.* New York: John Wiley & Sons.

6.17. Schiller, Zachary, and Zellner, Wendy. 1994. Making the middleman an endangered species. *Business Week,* June 6:114–115.

6.18. Markoff, John. 1993. Building the electronics superhighway. *New York Times,* January 24, section 3:2.

6.19. Elmer-Dewitt, Philip, 1994. The strange new world of Internet: battles on the frontiers of cyberspace. *Time,* July 25:50.

CHAPTER 7

People Resource and Human Capital: Innovators, Agile Learners, Creators, Decision-Makers, Change Agents, and Entrepreneurs

PREVIEW: WHAT IS IN THIS CHAPTER FOR YOU?

People interface with technology-driven organizations internally at three different levels: individual level, group level, and organization-wide level. Different organizations treat their human resources differently. Some consider people as a fixed overhead expense and therefore a liability. Senior leaders in such organizations are perpetually seeking ways to downsize their human resource. They hope to save on that expense to improve their company's bottom line. Other organizations consider people as the source of revenue and profits and therefore an asset. In this chapter we will review how individuals can contribute significantly to their organizations and play the latter role. In other words, we will review the attributes of the learning innovators who can be considered as the asset, and therefore the sources of competitive advantage for their organization.

We will also try to understand how organizations manage their human resources. How do teams operate in organizations? And how do we focus on the individuals in the technology-driven organizations? For a technology-driven organization that faces frequent and dynamic changes, to be effective and efficient, individuals who can adapt fast to external changes are needed. These productive individuals can proactively bring about their own changes.

In this chapter we will learn about the learners who are driven by their continuous improvement. We will also learn about the different types of agile decision makers, about the systematic problem-solving processes they use, and about their values, ethics, and sensitivity to social responsibility. We will understand the different types of teams and the relationships in each type. Finally, we will consider how employees can participate by giving useful suggestions, and we will discuss how they can be creative innovators as well as change agents. We will also describe the way they seek new opportunities and become risk-taking entrepreneurs in their organizations.

DO INDIVIDUALS MATTER IN DOWNSIZING?

Individual employees make up the backbone of a technology-driven and innovative organization. People are considered either its liability or its asset. During the 1990s, many working individuals were blind-sided by rapid and repeated assaults of downsizing and layoffs (see Note 7.1). Individuals who had worked hard for many years and helped their organizations generate wealth and fame over these years were asked to leave overnight—often by a new CEO who thought that these employees were no longer needed. They were considered to be adding unnecessarily to the cost of running the organization. In the face of intense global competition, senior leaders in such organizations either eliminated many jobs or replaced older experienced employees with younger, nimbler, and less expensive employees. All the knowledge and expertise that the veteran employees had accumulated during their work-life was suddenly not needed, even though some of the accumulation of knowledge was paid for by their organizations. Or worse, these employees were considered a liability rather than an asset to the organization.

In technologically pioneering organizations, such as 3M, Federal Express, Hewlett-Packard, and Microsoft, senior leaders have clearly realized over the past few years that their people make a big difference on their corporate performances. These technology-driven and innovative organizations seem convinced that people drive their profits. These innovative firms do not let their fluctuating profits drive innovative people in and out of their organizations. These organizations give their individual workers and managers a lot of freedom. The employees are encouraged to be themselves and be free of fear so that they can innovate. The senior leaders in these organizations get out of the way of these innovators. The role of senior leaders in these organizations is to facilitate and empower their innovators, rather than to introduce hurdles in the learning innovators' paths.

In the 1990s, there were hardly any organizations not affected by significant changes in technology. Technology brought about external market changes in the form of Schumpeterian "constructive destruction." These changes changed the rules as well as the boundaries of market competition. Because of the impact of technology on people's relationships with each other, the technology-driven innovative organizations also came under a continuous assault of technological changes from within the organization. These technology-driven organizations therefore needed individuals who were willing to adapt quickly to such external changes and innovations. Very often they were themselves agents of proactive change and innovation.

If not, the individuals were likely to obstruct changes and the organization's attempts to adapt to external changes.

The technology-driven innovative organizations must continuously rejuvenate their key human resource practices and policies accordingly. These organizations must hire, nurture, reward, and retain agile and innovation-friendly workers and managers, the individuals who embrace change proactively. We will call them the *learning innovators.*

THE LEARNING ORGANIZATION AND INNOVATORS

Once we accept this premise for human resource strategy for technology-driven enterprises, it raises many more questions. Some of these questions are: Who are the learning innovators, and how do we define a learning organization? What do learning innovators do differently? How can learning innovators be nurtured in organizations?

Here we are using the term *innovator* in a broad sense. The innovators sometimes innovate by incremental improvements. At other times they innovate by ushering in radical changes. The choice depends on (a) the contextual circumstances and (b) the risk taken at a particular time.

The Learning Organization

According to a research study sponsored by *Training & Development,* a publication of the American Society for Training and Development, "all organizations learn, but not always for the better. A learning organization is an organization that has an enhanced capacity to learn, adapt, and change. It's an organization in which learning processes are analyzed, monitored, developed, managed, and aligned with improvement and innovation goals. Its vision, strategy, leaders, values, structures, systems, processes . . . all work to foster people's learning and development." According to them a learning organization accelerates its systems-level learning, whereby the organization synthesizes people's learning (see Note 7.2).

In technology-driven enterprises which must cope with frequent revolutionary or evolutionary changes, organizational learning is needed on an ongoing basis. Therefore, Stephen Covey, the best selling author of *The 7 Habits of Highly Effective People,* urged a strong need to rebuild the trust shaken in many organizations by downsizing. With downsizing and a highly turbulent business environment the personal contract between the people and the organizations seems to be in a jeopardy. The trust can be rebuilt by personal relationships between different people in an organization, as well as by nurturing the relationship between the employees and the organization. Covey proposed 10 ways to repair the damage. These included (1) seeking first to understand rather than be understood, (2) keeping promises, and (3) exhibiting honesty and openness (see Note 7.3).

Texas Instruments (TI), when it was still small in size, institutionalized its management of innovation. TI incorporated this belief while developing its statement of purpose and other policies. The purpose statements of TI reads as follows: "Texas Instruments is here to create, make, and market useful products and services to satisfy the needs of its customers throughout the world." (Note that we have covered issues surrounding creation, making, and marketing in earlier chapters.) Leaders at TI institutionalized their innovation process by developing its OST system, which stood for objectives, strategies, and tactics (see Note 7.4). The OST system enabled Texas Instruments to tap effectively its resources from several divisions. If needed, the entire organization was put into action simultaneously. The OST system was a proven, yet evolving system for self-renewal of Texas Instruments. The system was operationalized by:

1. Basing action plans on quantitative goals
2. Providing feedback by measurement of performance
3. Promoting corporate synergism
4. Segregating operating resources from strategic resources
5. Giving visibility and control to strategic investment of resources

Xerox has gone a step further in the 1990s. According to the company documents, Xerox considers its employees' ability to learn as its core-competence. A strategic source of competitive advantage over rivals is Xerox employees' ability to learn faster than their counterparts working with the competitors (see Note 7.5).

The Learning Innovators

In the groundbreaking book, *The Fifth Discipline,* Professor Peter Senge stressed that we must look not only at the organizational "forest" but also at the individual "trees," (see Note 7.6). If we want the technology-driven enterprise to be a learning organization that can proactively adapt to external competitive changes, then we must also make sure that the individual employees become the learning innovators. Organizations in general, and technology-driven organizations in particular, learn only through what their motivated and inspired individual employees learn. Even though what many individuals learn may not accumulate into significant organizational learning, the learning individuals are a prerequisite for the organizational learning to happen.

Continuous Growth Towards Personal Perfection. During the early reconstruction of Japan devastated by World War II, the Japanese workers and managers were urged to improve incrementally but continuously. They were not driven by major radical changes, but were motivated to make a few minor improvements here and there. Senior managers of Toyota Manufacturing came to the United States to learn from pioneering American manufacturers. They visited the River Rouge plant of the Ford Motor Company and saw managers seeking suggestions from their employees to reduce cost. The Japanese visitors returned home and modified the American suggestion system to the Japanese continuous improvement system. They coupled the suggestion system with the quality circles and developed the *Kaizen* way to improve and innovate continuously.

The Japanese employees used these frameworks to suggest and implement minor improvements that made their jobs easier. Their organizations' leaders and managers respected these suggestions to change processes for the better. These employees knew what they were doing. With their persistence and cumulative innovation, enormous cost savings were realized by the Japanese organizations without setting large and frustrating unattainable targets.

Was the Japanese workers' desire to innovate and improve one's work constrained by their nationality or beliefs? Such urge to innovate resides in all of us. It is also not limited to the fine and performing arts. Peter Senge called it "personal mastery" (see Note 7.6). This involves a continual clarification of where we are, coupled with a continual learning about our specific reality. Personal mastery is a lifelong journey, and not a state or a destination to reach.

The personal mastery can be broken down into its two constituent subelements. These are (1) conscious personal vision for a desired future state and (2) continuous learning about the relevant reality.

The first subelement of personal mastery is to have a conscious personal vision that is intrinsically positive and growth-oriented. It is important to have this personal vision anchored in specific actions and not just elegant words. The personal vision should also be not just some goodies and toys a person would like to have now, but a future state based on the key beliefs and achievements that will make a person happy for a long time. It is based on self-growth rather than on obsession to beat the competitors.

The second subelement is an accurate and continual learning about the relevant reality. This requires a commitment to seek the true reality, not just a faked reality. Then, like a coiled spring, as the gap between the true reality and personal vision increases, there is a creative force that tries to bring the two closer.

It is sad to note that instead of using personal growth and mastery with continuous learning as the motivator for employees, most business organizations and their leaders use crisis and fear to bring about a change.

Role of Mental Maps in Learning. The creative force of the gap between the personal intrinsic vision and the accurate learning about reality should lead to corrective and creative actions. Identification of choices and focus help in this. But before jumping into impulsive actions, it is important to develop certain holistic mental maps about the situation on hand. The starting mental maps may be based on the past experience and the obvious facts. But these tentative maps must be continually validated with specific facts about evolving reality.

Professor Howard Gardner has done a lot of research in the area of how the human mind works (see Note 7.7). Gardner proposed that such mental representation manifests in many behavioral aspects. Chris Argyris noted that people believe and act according to their mental models (or the theory in use in their mind), even when they do not behave according to what they say (see Note 7.8). The mental models make us act the way we do. If an employee believes that his manager is out to get him instead of out to help him, then he is likely to act as if the manager is actually coming after him.

Example: Atomic Physics and Mental Maps. Most of the major developments in atomic physics and nuclear technology were based on the scientists' mental models about what is inside of an atom. When the atomic theory was being developed, many aspects of it could not be empirically tested. The atom bombs which ended World War II in the Far East were developed because of Einstein's mental model about how mass and energy are interrelated in an atom. For example, based on his model of atomic particles, Einstein postulated that a ray of light will bend with a gravitational pull of the earth. This could not be proved at the time Einstein postulated his theory. Many years later, Einstein's theory of relativity was validated empirically by two American scientists. See Manhattan Atom Bomb Project in Chapter 8.

Learning innovators have certain common characteristics. These include (a) agility in decision-making, (b) sensitivity to ethical and social responsibility, (c) teaming, (d) active participation, and (e) creativity. We will try to understand these in the rest of this chapter.

LEARNING INNOVATORS ARE AGILE DECISION MAKERS

Managing technology means making decisions all the time. Fast technological changes require agile decision makers. There are many different types of decisions that individual employees must make about technology. They must also realize the significance of time in decision-making. Decision-making is a highly "individual" thing. It varies from one person to another. There are some managers who thrive on making fast decisions. They try to make decision at the earliest that they can. Even when they have insufficient information. Then there are other managers who are perfectionists. They keep on perfecting until the crunch time when they are forced to make a quick decision. Managers and workers alike should understand their own decision-making styles. Given below is an overview of the different types of decisions in a technology-driven organization. This should help reduce the anxiety associated with fast decision-making required under dynamic competitive conditions.

1. Time-Related Decisions. Time-dependent scheduling decisions are one set of dynamic decisions. When an individual worker or a manager says that the product will not be ready that day, it is a decision to reschedule the delivery of the product.

Many businesses are heavily driven by the time dimension. Federal Express is a business conceived on the basis of the guaranteed timely delivery of overnight packages. Time is Federal Express's primary source of competitive advantage. Pizza delivery within 30 minutes and Lenscrafter's 1-hour delivery of eyeware are some other popular examples. These are some service-sector examples.

There are many similar examples in the manufacturing sector too. Toyota's Just-In-Time (JIT) manufacturing and inventory management of automobiles are classical examples. There are different kinds of time lags associated with different decisions. In a later chapter we will discuss management of projects, wherein the estimated time of earliest completion and the slack times have significant impact on the time of completing the project (see Chapter 8).

2. Production Decisions. There is a different set of decisions which deal with production choices. For example, should a production supervisor make 10,000 units of a Christmas gift toy and starve the market, or should she make 10 million units of the same gift toy to flood the market? Should a purchasing supervisor develop 100 vendors for 1000 parts that go in a particular office product like a photocopier, or should he have 500 vendors for the 1000 components in the same office equipment? What should be the frequency of scheduled maintenance repairs to reduce the machine downtime? Cost-cutting in scheduled maintenance repairs saves a buck, but it leads to more downtime.

Some production decisions are closely related and interdependent. That is, one decision is closely related with another decision. Let us say that the shop-floor supervisor at Acme Sports Company makes a decision to make 1000 pairs of roller blades per week to meet the Christmas demand. This decision implies that the purchase department must procure 1000 pairs of roller-blade shoes and 2000 pairs of wheel assemblies, because each pair of roller blades uses one pair of roller-blade shoes and two pairs of roller-blade wheel assembly. The decision to buy the shoes and roller-blade shoe assemblies may have been influenced by the production planning decisions made elsewhere. Furthermore, if ordering the raw material parts takes 1 week, then these decisions must be made 1 week prior to the week for which planning was done. If the shop-floor supervisor plans to change the production capacity, then the raw material decisions must be adjusted accordingly.

This seems elementary. But many shop-floor supervisors and their managers overlook these interrelationships between decisions. If they forget the time lag, then in a particular week either they have too much inventory resulting in excessive inventory carrying cost or they may have too few parts of one raw material, resulting in idling of plant capacity. Things can get more complicated when managers also include the effect of human resource planning with delays in procuring raw materials.

3. Marketing Decisions. Like the production shop-floor decisions, the marketing executives and learning innovators make some unique sets of decisions. For example, they must decide which segment of the market they should target. Who is their target customer? How should the firm approach different customers? These decisions have significant influence on their other decisions related to the market mix—that is, the type of product, pricing, promotion, and placement of their products.

Many of these decisions are subjective. Before the beginning of the fall fashion season, how should innovators decide whether pastel-colored sweaters would be in high demand or whether the neon-colored sweaters would outsell others? Many of these are stochastic (i.e., probabilistic) decisions. They can be estimated only with a chance probability. For an electronic new product based on radical new semiconductor technology, a manager can only estimate the likelihood of delivering a future consignment, or less likelihood of delivering a future consignment of electronic printer circuit boards.

4. Financial and Resource Allocation Decisions. Decisions regarding resource allocations have unique characteristics. First of all they are extremely visible to outsiders, such as the stockholders, bankers, and investors. To safeguard their investments, these external stakeholders maintain a careful watch on the decision makers within a corporation, particularly its senior managers.

The investment decisions are further complicated by the wishful and unrealistic expectations of investors. Investors would want a technology-driven organization to make the least risky decisions, but earn them maximum returns on their investments. As the outside investors get more vocal in the decision-making of a company's strategic resource allocation decisions, many company employees have a tendency to make few or no changes to avoid a confrontation with the vocal investors. Cutting long-term investment in research and development may improve the bottom line in the near-term, but it jeopardizes the long-term competitive advantage over the competitors.

5. Human Resource Decisions. As we have been discussing in this section, the managers and supervisors in technology-driven organizations differ to a great extent in terms of their choices related to human resources. Should people be laid off because they add cost, or should they be retained to meet the economic upsurge in demand around the corner? When the competition gets rough, shouldn't older employees be re-skilled and retrained rather than fired?

The above examples illustrate the range of decisions that individuals must make on a day-to-day basis in technology-driven organizations. These decisions vary in the financial risks associated with them. Different decisions carry with them different risk impact. They also affect different people with different intensities.

Decision-Making Process and Tools

Often learning innovators in technology-driven enterprises face complex decisions. These decisions involve making a decision based on large amounts of information. This information must be subjected to multiple decision criteria. A systematic decision-making procedure helps in making such decisions faster and more effectively.

Types of Decision-Making Situations. In general there are three types of situations for making decisions. These are given below.

A. Decision-Making Under Uncertainty. In technology-intensive enterprises there are many decisions which must be made under uncertain conditions. For example, for petrochemical companies it is difficult to estimate (a) the price of one barrel of oil in the year 2050 or (b) the state of computer technology in the year 2020. Even though there is enormous uncertainty, the decision maker must still make certain decisions heavily dependent on these future outcomes. They make their decisions by considering the maximum possible outcomes they could get from their different alternatives or by minimizing the maximum potential risk they would incur from their different alternatives. A third option is to assume that the best outcome and the worst outcome are equally likely. Such decision-making frameworks help make decisions when there is uncertainty about the possible outcomes.

B. Decision-Making Under Certainty. In case of some decisions, the decision maker has all the necessary information about the possible outcomes. The decision maker can then systematically consider all the available information and make the best decision that maximizes the desired outcome. For example, for duplicating an existing assembly line, the decision maker can use the past information to get a good idea about the new production level. Given this information, the decision maker can then minimize the cost of commissioning such a production line, and maximize the production output from the same.

C. Decisions Under Risk. Some decisions must be made when the decision maker can only partially estimate some of the information needed to make the best decision. In this situation, some of the missing information can be estimated by the probability of its occurrence. For example, a decision maker may not know what will be the actual sale of lawn mowers in a particular month. But the probability of selling lawn mowers is higher in the summer months than in the winter months. Many toy retail companies do from 30% to 40% of their annual businesses during the Christmas holiday season. They can therefore use this information to estimate the demand they are likely to generate during that period, and therefore estimate the production capacity needed to meet that demand.

Decision-Making Models and Tools. We noted earlier that individual managers and workers must make decisions. They must pick and choose from their alternatives. These decisions make the backbone of the management of a technology-driven enterprise. To facilitate these individuals, some decision making tools have been developed. Many additional tools have been discussed in other sections of this book—for example, the quality tools in the chapter on quality promise (Chapter 5). Decision-making tools supported by computer and information processing technology are discussed in the chapter on communication and information (Chapter 6). Given below are some of the tools which can be used for making a variety of decisions.

1. Decision Table. Decision tables help a decision maker define different alternatives clearly. For each potential alternative, the decision-maker faces multiple states of uncontrollable nature (or the context defining the outcome). Depending on the state of nature (for example, a growing or a recessionary economy), the potential outcome is different. All the potential outcome values are arranged in a payoff matrix. The decisions may be then made depending whether there is complete uncertainty, partial certainty, or certainty of information needed to make these decisions discussed earlier.

2. Decision Tree. This is another model commonly used to help decision-makers make the best possible decisions. In this case the available information is arranged in the form of a decision tree. The tree starts from the root where the decision to be made is noted, and it grows right to the branches. The decision-making process involves filling information from left to right, from the roots towards the branches where the potential outcomes are noted. The decision tree is then "solved" by moving backward from right to left, while calculating the expected monetary value at each nodal point.

Decision-Making Process. Decisions may be made intuitively by "gut" feeling. Many managers and entrepreneurs do so. They may even succeed in making good decisions most of the time. But this carries a high risk of failure. A better way is to use a systematic process to make decisions. Many business organizations, therefore, insist that their managers and workers use a systematic and multistep process to arrive at their decisions. Given below is a brief description of a multistep decision-making process.

Step 1. Critical Milestones and Gathering of Major Relevant Facts. In this first step the decision maker collects all the relevant facts and the major milestones up to the point of the decision. Caution must be exercised in compiling facts from the internal 'company' sources as well as from external sources such as the information provided by the analysts and other outside observers. In the same manner, the decision maker must consider the old documents and sources as well as the new sources of information. This will help establish the dynamics of change over an extended period of time.

Step 2. Defining the Overall Issues Facing the Decision-Maker. In this step the decision-maker tries to define the problem as clearly as possible. The decision maker must look at the overall issues involved in the decision on hand. It is important to set the goal by defining the problem clearly. If the goal is not clear then it is hard to measure whether one is making progress towards the goal, or even whether one has reached one's goal.

Step 3. Measures and Mission for Success. The quantitative and qualitative measures help a decision-maker to objectively judge the current situation and the progress toward a desired future state. Measures help the decision-maker understand the priorities underlying a decision. Sometimes the mission or vision of an organization (or a task) can help guide in this direction.

For example, the market success for an organization can be measured by a high market share, or by a high annual growth, or by a high profit margin. During the 1970s and 1980s, many Japanese companies pursued their success in the U.S. markets and monitored their progress by measuring their market shares. They were satisfied with market penetration even if the profit margins they earned were low. American companies, on the other hand, when they enter foreign markets such as in Japan, measure their success by the profit margin these efforts earn. Some senior managers in the American companies expect the emerging international operations to earn as much return on investment (ROI) or return on assets (ROA) as their well-established domestic operations do. As a result of these measures used to judge their success in the international markets, many early efforts in international markets do not succeed and are aborted.

Step 4. Performance in the Past and Present. Using the above-mentioned measures of success, the decision-maker can then objectively define the past and the present levels of performances. This definition of initial state is important. It provides a sense of progress and continuous improvement.

Step 5. Analysis of the Alternatives Available to the Decision-Maker. In this step the decision maker considers all the potential alternatives he or she has, and then he or she analyzes them using the analytical tools and models available to do so. For example, in the case of decisions related to new product development, tools such as the product portfolio matrix and the product life cycle stages can be used to analyze the alternate marketing mixes (four Ps) for the new product under different situations. For decisions related to product and process innovations, the "pattern of innovation" model proposed in the earlier chapter on proprietary know-how may be used. For financial decisions, ratio analysis can be used.

Step 6. Recommended Alternate Action Plans. The analysis done in the previous step should lead to the development of short-term and long-term recommendations for the decision on hand. The short-term decisions are generally those which can be implemented in 1 to 2 years. The long-term recommendations require 3 to 5 years.

Step 7. Execution of Recommended Actions. Finally the ways to implement and execute the recommended actions are proposed and evaluated. A decision-maker may also include the results available from the evaluation of preliminary trials—for example, the information gained by test-marketing.

The seven-step process involves dividing up the task of making a major decision into steps which are focused on the critical milestones and facts, overall decision definition, measures and mission, past and present performance, analysis, recommendations, and execution. Using the first letters of each of these, the seven-step process can be remembered as C-O-M-P-A-R-E. C-O-M-P-A-R-E can be used, for example, to evaluate the performance and health of a company in which an investor wants to invest substantial amounts of money.

The P-D-C-A Cycle. This is another four-step decision-making process that is often used for continuous improvement. The four steps are Plan, Do, Check, and Act. The Plan step corresponds to the first two steps in the seven-step process described earlier. Planning involves defining the problem or the decision facing a decision-maker carefully. The Do step involves steps 3 to 4, developing the measures/mission and analyzing the past and present performance. The Check step corresponds to step 5 of the earlier process, testing the different alternatives available to the decision-makers. The Act step includes proposing the recommendations and the execution of recommendations in the seven-step decision-making process described earlier. Once a problem is solved or a decision is made, then the P-D-C-A cycle begins all over again.

The P-D-C-A cycle can be more easily remembered as the Define, Analyze, Test, and Execute cycle, or the D-A-T-E cycle. It is a date with decision-making.

The four-step problem-solving cycle (or decision-making process) was popularized in the United States by Dr. W. Edwards Deming. It is also quite popular with Japanese managers and engineers. This model throws interesting light on the way the Japanese and the American decision makers decide. On the American side, sometimes there is a tendency to quickly rush through the planning (P) stage and get down to the task of doing (D) step. The Japanese, on the other hand, spend much more time in defining the problem on hand in the planning (P) stage. During this stage they may try to understand a problem from multiple functional perspectives. For example, for a decision to start a joint venture with an American company, in the planning stage the Japanese will look at the decision from the production, financial, marketing, and legal standpoint before moving on to the next stage of doing the contract and committing the resources.

Types of Decision Makers. Different decision makers have different styles. These styles depend on two attributes of decision-makers.

1. *The Number of Thinking Channels a Decision-Maker Uses in Arriving at a Decision.* Either one channel or multiple thinking channels may be regularly used by a decision maker. This could also be a measure of whether the decision maker makes a decision independently, or consults with others before arriving at a decision.
2. *The Amount of Information Required by the Decision Maker to Make a Decision.* The decision maker could use either low or high amounts of information to make a decision. Some decision-makers must wait for the last bit of information to arrive before they feel ready to make a decision. Others are ready to "shoot" a decision without waiting long to "aim."

Based on these two attributes there are four types of decision-makers. These are briefly described below (see Exhibit 7.1).

1. Autocrats. These decision makers have a single line of thinking, and they require low levels of information to arrive at a decision. They make decisions on their own. They do not need to consult others. And, they believe that they know it all. So they do not need much additional information either.

2. Scatters. They use multiple channels of information. They feel that they must contact many people and collect a variety of opinions and information. Sometimes, these are the diverse constituencies which they must satisfy. Yet, they have patience for only low levels of information. Thus, the limited information they gathered is spread thinly in many different directions.

EXHIBIT 7.1. Decision-Maker Types

	Single	Multiple
High	Programmer	Synthesist
Low	Autocrat	Scatterer

Information Required

Single   Multiple

Multidimensional Thinking Channels

3. Programmers. They require large amounts of information. But they can only use linear unidimensional thinking. They must have algorithms or technical procedures to solve problems and make decisions. They are good at making complex but structured decisions.

4. Synthesists. These decision makers synthesize by thinking in multiple channels using large amounts of relevant information. They integrate the information they gather and arrive at creative solutions. The synthesist decision maker is transformational. He or she is able to see and make connections which others have a hard time comprehending. Like a chemist, they take raw ingredients, mix and process them in different ways, and turn them into useful new compounds.

Different Times Need Different Decision Strokes. One must understand that different circumstances call for different types of decision makers. Under turbulent circumstances, or major turning points in a technology or a market, an organization may benefit from the integrating ability of the synthesist. No one person knows the answer to the turbulence. The synthesist gives everybody a chance to contribute to the decision on hand. The synthesist then pools all the information together and comes up with a creative solution.

A corporation close to bankruptcy, on the other hand, may be better off with an autocrat decision maker. Such a decision-maker uses the limited time and information available, along with a limited few channels accessible, to make a quick decision. Controlling damage is what the circumstances on hand demand most.

An advertising agency may benefit from a scatterer, who must essentially consider a limited amount of information and subject it to multiple channels of considerations. The advertiser has many constituencies to satisfy. The advertisement campaign must be good from the client's as well as the customer's point of view.

Finally, circumstances such as those facing the manager of a mutual fund may require the use of a programmer decision maker. Actuaries' decisions also require meeting a few important criteria. These decisions must satisfy the available thumb rules and their interdependent relationships.

We have so far focused our attention on the stages of a decision-making process, the alternate models available to arrive at a decision under different circumstances, and the different types of decision makers. These are visible aspects of a decision-making process. But a decision is like the tip of an iceberg. The decision is the visible part of a complex process, most of which is submerged and not visible to a casual observer—for example, the values and beliefs that mold a decision-maker's actions. Let us next consider the decision-making processes under the surface, which are not visible to a casual observer. The decision is the visible part, whereas the values and beliefs that go behind framing that decision make the invisible part of an iceberg.

INNOVATORS' VALUES AND BELIEFS

An individual's decisions are often guided by the values of the individual as well as the values and the vision of the organization he or she is a member of. Values are intimately related to the decision choices individuals must make. Attitudes reveal our values by how we feel about others. They "color" the way we perceive people, objects, organizations, and societies. The set of choices we are likely to make tend to reflect our value system. Every now and then individuals must make decisions without specific policies. Sometimes these forced choices are due to time constraints. Under these circumstances, individuals act according to their own values or the values promoted by the senior leaders in their organizations.

Take for example, Johnson & Johnson's Tylenol crisis with toxic poisoning (see Note 7.9). Should the individual decision-makers in the field hearing about the Tylenol poisoning seek approvals from their bosses up in the hierarchy and wait for their answers? Or should they ignore the problem because it will add enormous cost, and hope that it will go away by itself? Or should they jump into action and recall the entire supply to save the life of a single customer even if it would add enormous expense? In the case of

Johnson & Johnson, the senior leaders clearly valued the trust of their customers, including parents and physicians, far more than they valued the profits in their bottom line. In the absence of similar clear organizational values, decision makers face a lot of anxiety when they must make fast decisions in a crisis situation.

The progressive companies, such as Johnson & Johnson, understand this dilemma. Johnson & Johnson has therefore clearly articulated their priorities and values in their "Credo" and mission statement. These clearly defined priorities act as the guidelines for managers making decisions under uncertain circumstances. In the case of Johnson & Johnson, safeguarding the health of customers was always given a higher priority over increasing the bottom-line profits.

Protestant Work Ethic and Capitalist Values

During the 1990s, in different parts of the world, and in many technology-driven enterprises in these parts, managers saw the state-controlled communist political structure crumble under its own oppressive control. Communism was replaced with free-market capitalism. This made many people revisit the question of the capitalist value system in the United States. With persistent trade deficits with other nations (such as Japan for decades, and China in more recent years), some observers asked if the American corporations and managers can ever have fair trade with these distant global competitors competing under very different sets of rules. Were these rivals winning the global competition because of the support, encouragement, and sometimes the subsidies they received from their governments? Should the American companies imitate similar value systems? Or should these eastern economies emulate the United States?

In 1904–1905, Max Weber provided a classic insight into the value system underlying the Protestant ethic and the spirit of the Western capitalism. In the footsteps of gaining some freedom from the constraints of a centralized Church, Weber saw that the capitalistic way of seeking wealth was very different from the pursuit of gain by "any and all" means (see Note 7.10). He argued that in different parts of the world—for example, in Babylon and Ancient Egypt, in India and China, and in medieval Europe—the pursuit of gain had existed in one form or another. But only in the West, this pursuit of gain was accomplished by a rational organization of free labor, along with a regulated investment of capital that was accessible to all.

Market Values in the New World Order

In *Head to Head,* Professor Lester Thurow of the Massachusetts Institute of Technology pointed out the different forms and flavors of capitalism practiced in the United States, Japan, and Germany (see Note 7.11). In Japan the visible guiding hand of the government, particularly the Ministry of International Trade and Industries (MITI), held a tight stronghold on the birth and growth of new technology-driven industries. In the United States the "invisible hand" of the marketplace decides the growth and survival of businesses. In unified Germany, on the other hand, the social responsibilities of corporations towards workers' welfare superseded their profit maximization motives.

China under the leadership of Deng Xiaoping and beyond saw the introduction of a "social-free-market" economy. In 1997, after Deng Xiaoping's death, whereas the Chinese economy was still under strict control, the Chinese companies in distant areas "far away from the Emperor" sometimes got away with much liberties. Thus Shanghai, as well as Free Economic Zones off the coast of Taiwan and across the border from Hong Kong, seemed to practice economic rules different from those in the other parts of China, around and near Beijing.

Thus there are major differences in values which are likely to come to interface in global economic encounters between corporations from different parts of the world. How should individuals and managers from the U.S. corporations, driven under the value system of Protestant ethic and free-market profit maximization, interact with their counterparts from the state-controlled Chinese corporation following the value system proposed by Confucius? Are the Japanese corporations, actively guided by their bureaucrats, breaking the rules of the competitive game? Or should the technology-driven American corporations follow the past success of the Japanese corporations? Should the American companies too seek guidance from an empowered U.S. Department of Commerce and Office of Technology Investment? These are difficult choices that individual managers in technology-driven organizations such as Motorola and Eastman Kodak must face when competing with their global competitors.

Social Responsibility and Ethical Values

Social responsibility and ethical values add complexity to decision-making in technology-driven enterprises. Should corporations act to improve the welfare of the society? Or are the corporations only in the business of maximizing their own profits and their investors' returns? Therefore, is there anything wrong in corporate managers selling as many cigarettes and silicone breast implants as they can? Or should they, at the expense of their profits, consider the welfare of their customers too?

Nestlé's infant-formula controversy, whereby the company was accused of being a "baby killer" in the developing countries, highlighted the struggle between (a) the company trying to sell its product in the most proficient manner and (b) the buyers wanting the corporation to be much more concerned with the consequences of its actions. In this case consumers were adding untreated water to dilute expensive infant milk formula. Similarly, whose decisions are to be blamed in the case of Dow Corning's breast-implant cases? The women customers who decided to pursue cosmetic surgery for better appearances, the doctors who recommended and performed the operations, or the corporation promoting and distributing its goods in the fastest possible manner? These issues require careful ethical analysis from the people as well as the profit perspective.

HIGH-PERFORMANCE TEAMS

In the previous section, decision-making was viewed as a stand-alone activity. Very often in technology-driven enterprises, complex and risky decisions must be made in groups, in consultation with others. In this section we will examine how learning innovators can participate effectively in high-performing and self-directing teams.

What Is a Team?

Most of us belong to one form of group or the other. At one time or another, many of us have been members of a sports team, a community group, a religious group, and so on. Technology-driven organizations are also increasingly organizing their work force into different teams to carry out different objectives. The underlying assumption is that in complex and risky situations, more heads are better than one head. Managers use teams and groups to coordinate their human resources into more productive units. Thus most of us have already worked with teams in many different flavors. Teams and groups may also go by many names. Some of these are given below (see Note 7.12).

1. *Management Teams.* These are major decision-making groups including senior leaders such as CEOs, senior executives, and planners. Examples include corporate planning groups, steering committees, and boards of directors.

2. *Permanent Work Teams.* Groups of people who work daily with each other and have organizational names like branch, section, and so on, are called permanent work groups. They often have a common boss.

3. *Temporary or Special-Purpose Teams.* These are groups of people who have to perform a specific task jointly together—for example, the groups of people representing quality circles, performance improvement teams, tiger teams, horizontal teams, vertical teams, cross-functional teams, and standing committees.

4. *Interface Teams.* These are two or more units from different organizations that have a permanent work relationship and that must cooperate together to get the job done—for example, groups of users and suppliers, new product development teams of designers, manufacturers, and marketers; and so on.

5. *Networks.* Groups of people who may not have face-to-face contact but who depend on each other for information, expert advice, alerts, and so on, are often referred as *networks*. If their primary purpose is compiling, sharing, and distributing knowledge, they are called *knowledge networks.*

Thus we notice that different teams and groups are referred to by different names. Can we make a clearer distinction between them and classify them? Given below is a more systematic nomenclature for teams and groups of individuals.

Defining Different Types of Teams by Their Relationships. Let us try to classify different types of groups and teams and develop a nomenclature for the same. A systematic typology of teams will also help us understand the dynamics of the ways teams function.

Different groups of people can be classified based on two dimensions:

A. *Degree of Permanence of the Group Relationships.* This is measured by the duration of time for which a group of individuals continue to relate with one another. Some groups are formed for a temporary duration. Others continue relationships for a long time.

B. *Degree of Formalization of Relationship Between Members.* This is a qualitative measure of the formal or informal relationships between different individuals in a group or a team. As individuals we participate in groups that are informally formed. We also work in teams that are put together formally to carry out a specific mission.

By combining these two dimensions in a two-by-two matrix, we can classify teams into four different types. They vary in terms of the permanence and the formalization of their relationships (see Exhibit 7.2).

A. Informal Groups. There are two types of informal groups. The informal groups are formed by the members themselves, independent of any instructions from the hierarchy of their organizations. When the informal relationship among members of a team is temporary, they are called *interest groups.* And when the informal groups have relatively permanent relationships, they are referred to as *friendships.* The interest groups last only as long as their members share their common interests. Friendships, however, emerge as a result of long-lasting informal relationships between different individuals.

An example of the interest group can be the women's feminist groups that sprung up in the United States and worldwide from their informal gatherings during the 1980s. They were initially interested in coping with the

EXHIBIT 7.2. Relationships in Different Types of Teams

	Temporary	Permanent
Formal	Task teams	Command teams
Informal	Interest groups	Friendship groups

Degree of Formalization (vertical axis)

Degree of Permanence (horizontal axis)

male dominance in the workplace. Gradually they took the objective of increasing the representation of women in the workplace. These groups help female newcomers join the work force and network with their senior counterparts working in the organization (see Note 7.13).

B. Formal Groups or Teams. Formal teams are formed through legitimate organizational mechanisms. These formal teams are increasingly appearing on the organization charts and in the formal memos of a growing number of firms.

When the formal teams are relatively temporary and are put together to solve a specific problem, they are called *task teams.* Typically, a task team may be dissolved once it has solved the prescribed problems and made its recommendations.

When the formal groups are relatively permanent, they are called *command teams.* For example, quality assurance may be carried out by a command team that is assigned the objective of ensuring a desired level of quality of goods and services on an ongoing basis. The Japanese manufacturers organized command teams for the work groups in a Just-in-Time manufacturing department. In the automotive industry, General Motors organized its automated assembly lines into work teams of 5 to 20 members. However, sometimes such reorganization into teams was not accepted favorably by the organized unions (see Note 7.14).

WHY ARE INFORMAL GROUPS FORMED? Let us first reflect on why we join certain informal groups in our workplace, church, neighborhood, or community. Members join different informal groups for a variety of reasons. A clear understanding of the reasons for forming informal groups can help us understand how strong and high-performance formal teams can be developed.

1. Interpersonal Attraction. Members join informal groups because of their interpersonal attraction to the group's members. The members have common beliefs, behaviors, race, values, or personality.

2. Proximity. Physical proxity plays a very important role in interpersonal relationships. People who work near each other tend to go out together for lunch. Japanese employees, for example, rarely have much interaction with people outside their own organization. On the other hand, inside the Japanese organization the workers and staff members do not have separate office cubicles. Furthermore, years after the Japanese graduate from a particular high school, they remain in touch with their fellow graduates.

3. Common Activities. People also join an informal group because of their interest in its activities. For example, people interested in football are likely to join a football club. Similarly, people who enjoy baseball are likely to join a baseball club or go to the ball park together. In suburban areas, parents with children of comparable age are likely to get together and form an informal group because of their common interest in parenting. For example, the informal groups of women who took child care lessons together are likely to stay in touch with each other even after they have delivered their babies. They may even form an investors group. Or they could join same aerobic classes, shop for toys together, and share notes on baby sitters.

4. Prestige and Honor of Membership. Informal groups are also formed because of the prestige or honor associated with the membership of certain groups. Volunteering for the local United Way helps members network with some important people in that community. Membership to prestigious golf clubs or country clubs is sometimes less for the facilities they offer and more for the membership they boast of.

5. Need for Social Affiliation. Finally, some members join informal groups to satisfy their needs for social affiliations. By joining these groups they avoid their loneliness. The priority assigned to this, however, may vary from one individual to another. Sometimes it also depends on the chosen profession of the individual. A salesperson or a small business owner is more likely to join the local Rotary Club, not only to socialize and contribute towards helping social issues, but also to meet other influential people in the community. A university professor is unlikely to seek such encounters, because he or she spends most of his or her available time doing research.

How Are Formal Teams Formed? Stages of Team Development. Let us
now consider how formal teams are formed. Such teams are put together
formally by the managers with the objective of facilitating a smooth implementation of organizational changes and strategies. By involving all members into command or task teams, managers expect a quicker buy-in by the
employees.

Many teams generally go through five distinct stages of development as
they evolve over time. These stages are described below (see Exhibit 7.3).

A. Forming Stage. In this stage, team members try to define goals and try
to understand the tasks they are assigned to perform. Teams in this stage focus on getting to understand each other. This stage is driven by *cognition.*
During this stage, different members may react to each other in a number of
different ways. They may (a) hold back their feelings and opinions to themselves, (b) feel confused about what is expected of them, (c) be very polite
so that they do not offend anybody else's feelings, or (d) try to assess their
personal costs relative to the potential benefits they gain from associating
with the team efforts.

B. Storming Stage. In this stage, teams get into some form of serious interpersonal trouble. Generally, as members are trying to figure out their respective roles, some conflicts appear in terms of either (a) the relative priorities of the alternate goals, (b) who is responsible for what task, or (c) the
chosen direction of the team leadership. Competition for leadership and

EXHIBIT 7.3. Five Stages of Team Development

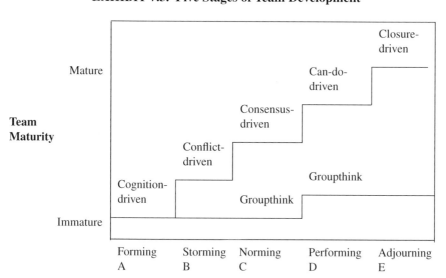

Stages of Team Development

conflict over goals to be pursued by the team are common occurrences. This stage of team development is driven by *conflict*.

The emotional tension thus generated may induce some members to withdraw from the conflict. Others may try to suppress conflict. The confrontation leads some other members to move towards formulating acceptable rules for working together. The teams cannot evolve into the next stage of development if the team leaders make team members suppress conflict or withdraw from it. Withdrawal from conflict can make the team fail prematurely, whereas suppression of conflict can create bitterness and resentment in members that will last for a long time during the life of the team.

C. Norming Stage. Teams that have evolved into this stage start focusing their efforts on tasks and ways that will help them share information and accommodate different opinions and approaches to arrive at the needed decisions. Norms are the acceptable behaviors for the team members. This stage is driven by *cooperation and consensus*. A sense of shared responsibility evolves to a collective sense of concern and empathy for each other.

D. Performing Stage. In this stage the evolving team is ready to perform and take actions. Different team members have been assigned different tasks, and these are accepted and understood by other members. Team members understand when they are to work independently and when they are to interact with others. This stage is driven by a *can-do* attitude by team members. Some members and teams start learning from sharing others' experiences. Other teams or members who did not fully recognize and accept their own and other team members' assigned roles tend to operate at a marginal level, barely sufficient for their survival. A low and barely adequate level of performance can result from either (a) self-centered roles of leaders or team members or (b) poor norms for the team's performance.

E. Adjourning Stage. Teams may eventually arrive at a stage of maturity when they have either achieved their assigned objectives or are in a stage where no further progress is likely. This stage is driven by *conclusion and closure*. In certain situations, teams take on a life of their own and keep going even after their mission is accomplished. They may take up new and bigger challenges or keep on assuming new tasks and problems for fear of losing their powers on adjournment.

It must be recognized that all teams do not evolve through these five stages in a fixed sequential manner. A team may fail at any specific stage, without maturing to the next stage.

Groupthinking. Not all teams work efficiently or effectively. Some teams seem to work very efficiently but they are not effective. They arrive at decisions too fast, without considering all the potential possibilities or points of views. Groupthink is an ineffective team decision-making mode. This is prevalent in teams driven by members with the mentality of making decisions and arriving at agreements at any cost. This is quite common in business and organizational settings (see Note 7.15).

The teams affected with groupthink have certain common characteristics of their team membership. These are listed below:

1. Teams have high cohesiveness among members, and they are insulated from others.
2. They lack ability to use systematic procedures for search and appraisal.
3. They have very strong directive leaders.
4. Team members are under a heavy stress to decide but they have a low degree of hope for finding a better solution than the one favored by the leader or other influential persons in this group.
5. They are facing a complex and fast changing turbulent environment.

Conformity-Seeking Tendency of Groupthinking Team. There are some common characteristics of the groupthink process of decision-making.

1. The team has an illusion of either vulnerability or excessive optimism.
2. Team members have a tendency to collectively rationalize while disregarding their assumptions.
3. Team members have a strong belief in the inherent morality of the team. Therefore the team finds it not necessary to check for any external threats.
4. The team holds stereotypical views of outgroups and considers them as stupid rivals or evil.
5. There is an illusion of unanimity. Silence is assumed to imply consensus.
6. The team exercises direct pressure on dissenters and brands them as disloyal. Inside the team there is a censorship of any differences or counterarguments. "Mind-guards" are appointed to maintain complacency and conformance of all members.

Defective Decisions. Even though groupthinking produces some fast decisions, the process produces mostly defective decisions. The decisions are defective for the following reasons:

1. Decisions are based on incomplete survey of alternatives, and goals.
2. Decisions fail to examine all the potential risks of the preferred final choice.
3. Decisions are based on poor search for information. Or there is a selective bias in the processing information readily available at hand.
4. The team has failed to work out contingency plans.

Finally, how can we avoid or decrease the groupthinking in teams? Given below are some steps that team members can take in ensuring that the groupthink behavior does not set in.

1. The leader should try to remain neutral and encourage criticism, counterpoints, and new ideas.
2. Encourage people who hold (or are sympathetic to) alternate views to present them.
3. Bring small subgroups or outside consultants to introduce different viewpoints.

Let us next consider the value of active participation by the members of a team.

POWER OF PEOPLE'S ACTIVE PARTICIPATION

Taiichi Ohno, who innovated Toyota's *kanban* (or Just-In-Time) production system, believed that the real secret to the success of Toyota production rested in the fine-tuning suggestions implemented by the Toyota employees working on the shop floor. Ohno felt that the impact of their 20 million suggestions in 40 years since their inception of Toyota Creative Idea System in 1951 was far more significant than that of all the decisions and policies by Toyota's upper management and engineers (see Note 7.16). How did such a powerful suggestion system originate?

Roots of Suggestion System

The first known suggestion system introduced in the American industry was at Eastman Kodak Company in 1898. The first suggestion was proposed by a worker, who suggested that the windows of the plant and office buildings should be washed periodically to improve lighting in the workplace. The Kodak plant was located in upstate New York, which had a winter season that lasted for 6 months with gray skies. Washing the windows brightened the work place. The suggesting worker received an award of $2, which was a significant amount in 1898.

In the United States in the early 20th century, as the outlook towards human relations in industrial organizations evolved, a growing number of organizations adopted a suggestion system.

Henry Ford II: The Father of Modern Suggestion System. Around 1951 the Ford Suggestion System was established at the giant automobile company. While launching the new program, Henry Ford II sought employees' buy-in, by saying the following (see Note 7-16):

In any organization that becomes as large as Ford, a mutual exchange of ideas gets to be very difficult. On this occasion, therefore, we have established a suggestion system.

In this system, you ladies and gentlemen submit your original ideas one after another on how to improve work and upgrade working conditions, and we will try them out at a special location. Those suggestions that obtain good results will be implemented on a company-wide basis. I have no doubts that the great majority of you have excellent suggestions on how to do things better, simpler, and more safely in order to increase production. However, these suggestions have yet to be proposed and implemented. This company of ours needs your original ideas. As to just how to make your work the best, there is no one with better knowledge than you.

To those of you whose suggestions are adopted and implemented from among those submitted, we will present a sizable cash prize, and give ample recognition for such cooperation. This suggestion system gives you the opportunity to make known your creative ideas. The suggestions will be considered fairly and without delay by a committee of experts. And we are committed to awarding appropriate prizes to those of you whose suggestions are adopted.

Japanese Suggestion Systems

The American-style suggestion system was first adopted by the Japanese electrical manufacturing companies. Hitachi adopted a suggestion system in 1930, Yasukawa Denki in 1932, and Origin Denki in 1938 (see Note 7.16). The Japanese suggestion systems then evolved differently. Given below are some significant differences.

1. Participation Rather than Savings as Primary Objective. In the American factories, the role of suggestion boxes was to reduce cost and save money for the investors. The person who gave the money-saving suggestion was also rewarded with money. On the other hand, the primary objective of a suggestion system in Japanese organizations was employee participation in decision-making.

2. Spirit of Continuous Improvement. Ohno, who developed the Toyota Production system and the suggestion system, believed that the suggestions sprang from continually asking "why" to the conventional wisdom. One must always ask how can one make one's organization or its processes more efficient and effective. However, continuous improvement may be a complex issue that is hard to measure. Ohno therefore suggested two ways to make the stream of suggestions flow smoothly.

A. Improvement Begins with "I". At Ford, suggestions were defined as constructive concepts; that is, the suggestions involved better and safer ways to work. They improved the process as well as the welfare of the employees. In general, the most useful ideas were the ones applicable to one's own work. The suggestion system therefore invited suggestions from people who can suggest best how their own work can be improved further.

B. Focused Suggestions. To make the suggestion system more manageable and targeted, periodically, say every few months, suggestions can be sought for different focused missions. These invitations to suggest could be focused on cycle time, employee participation, workers' health, and so on.

Implementation of Suggestions

The most critical component of a successful suggestion system is whether or not the suggestions are implemented. Given below are some ways suggestions can be implemented.

*1. **Active Top Management Involvement.*** The life of a suggestion system rests in the active involvement of top management. More than rewards, most workers would like to see their suggestions implemented. Even if a suggestion is not suitable for implementation, the suggester will like to know the opinion or the comments of top management for the suggestion (see Exhibit 7.4).

*2. **Empowerment to Implement Suggestion.*** The support of management can be nurtured by letting employees try out their suggestions before they

---

EXHIBIT 7.4. Basic Features of Ford Employee Suggestion System

1. **Examples of Good Suggestions**
   - Improving process
   - Improving physical operations
   - Improving handling of materials
   - Improving appearance
   - Improving quality
   - Reducing/eliminating waste
   - Improving safety/access
   - Improving service
2. **Examples of Suggestions Not Eligible**
   - Ideas submitted by other employees
   - Ideas already under consideration
   - Suggestions for the obvious/routine
3. **Occupational Categories Disqualified for Making Suggestions**
   (These are persons paid to improve work.)
   - Engineers
   - Operation standards personnel
   - Process technology personnel
   - Cost accounting personnel
   - Machine designers
   - Foreman

are submitted for evaluation. The empowerment of employees to implement their suggestions can open the flood gates for new suggestions and employee participation. Some clear and objective guidelines can be set up on how a suggestion should be implemented, the cost incurred, and so on.

3. Recognition and Rewards. One way to recognize the value of a suggestion is to reward or recognize it publicly. Recognition can be given by letting the suggesters present the most creative ideas to the senior-most management of the organization. Creative suggestions may be eligible to enter a "You Have a Point" Club or some other prestigious club.

The preceding discussion should indicate that the suggestion systems are very easy to start, but hard to sustain. Therefore, Exhibit 7.5 shows some of the questions that must be addressed to start an effective suggestion system.

EXHIBIT 7.5. Suggestion System: Some Prompt Questions

1. Why do we want a suggestion system?
   (a) For savings?
   (b) For continuous improvement?
   (c) For participation?
2. How do we distinguish suggestions from complaints and comments?
   (a) Is suggestion equal to constructive improvements?
   (b) Seek organization-wide or focused suggestions?
3. Who do we ask for suggestions?
   (a) From employees—faculty and staff?
   (b) From customers, suppliers?
   (c) Anyone not eligible to give suggestions; such as Directors, Vice-Presidents?
4. How do we collect suggestions
   (a) In a sealed box?
   (b) By letters? Should they be signed or unsigned?
5. Who will get the suggestions?
   (a) A committee of experts/representatives?
   (b) Boss?
6. What will be the role of top management?
   (a) To reward and recognize?
   (b) To assess?
7. How will the suggestions be evaluated?
   (a) For impact, innovativeness, or implementability?
   (b) Evaluate after implementation or before?
8. How will we implement selected suggestions?
   (a) Implement on pilot-scale first, and then organization-wide?
   (b) Implement directly organization-wide?
9. How will we reward or recognize the selected suggestions?
   (a) Monetary reward, small or a lot?
   (b) Ceremonial?
10. What do we call our suggestion system?

These questions should help lay a strong foundation for a long-lasting and effective suggestion system. The ultimate success of a strong suggestion system is measured by how creative and innovative the suggestions are. We will discuss creativity in the next section.

INNOVATIVE AND CREATIVE RISK TAKERS

Technology-driven enterprises thrive on the creativity and innovativeness of their employees. As we have stated a number of times in the previous chapters, such Schumpeterian "creative destruction" helps redefine the boundaries as well as the rules of competition. And therefore creativity and innovativeness of an organization act as the major sources of competitive advantage for a business enterprise over its rivals.

Defining Creativity and Innovation

Unfortunately, creativity is often misunderstood in many organizations. The folklore of creativity differs significantly from the facts about creativity. To nurture and grow more innovators, the technology-driven organizations must distinguish the folklore from the facts.

Creativity can be defined as the art and science of developing new meaningful associations and connections between different subelements.

For example, the field of chemistry gained a major leap when an innovative scientist likened a long chemical molecule to a snake biting its own tail. Thus the ring structure of a benzene compound was discovered. With this, for the first time many observed properties of a large group of aromatic chemical compounds could be successfully explained. Similarly, a charred piece of rubber dropped accidentally on an oven gave Charles Goodyear the idea for the vulcanization process of rubber with heat. Goodyear could make that association because of the experience he already had with the rubber goods imported from England.

Folklore and Facts About Creativity and Innovation

1. Creativity Is Everywhere, Not Just in Arts. Traditionally, the term "creativity" is often associated only with the arts, such as painting, music, dance and drama. The term "innovation," on the other hand, is generally linked with technological inventions. With the new technological developments in the 1990s, the digital and image technologies play an active role in the music, visual, and other multimedia areas. Thus the scopes of the two terms have overlapped and must be widened beyond their traditional boundaries. We will therefore use the two terms interchangeably and create new ones such as "creative innovators." Creativity can be found in a painted ceiling as much as in a canvas painting, as demonstrated by Michelangelo's Sistine

Chapel. Creativity is as much a part of a dance and drama performance as it is in a bowl of delicious chili. Any area of human endeavor can be converted into a fertile ground where creativity can bloom.

2. A Creative Innovation Must Be Different and Original. The innovative creative processes and products have two basic original ingredients:

A. They must "be different" in the means they use.
B. They must "be different" in producing better end results.

The means and ends are "different" in creative solutions relative to the established ways of doing things. A brief look at some of the TV infomercials shows how "different" means and "different" ends produce creative products.

For example, an innovative "power mop" may use a highly moisture-absorbing polymer that wipes and dries at the same time. It is the application of an industrial material to a new innovative domestic use. Or consider the compact burger grill promoted by the former heavyweight boxing champion George Foreman. The compact burger grill used two-sided heating surfaces with serrated stick-free coating. The base is inclined so that the fat melting from the meat is drained out by gravity into a draining tray. The innovative product uses a somewhat different method than a regular oven and produces remarkably different results, namely, the fat-free burgers.

3. Creative Innovation and Productivity Are Different. Being productive implies moving down a learning curve (discussed in Chapter 2) with high cumulative production and lower unit cost of production. Being innovative means shifting from one learning curve to another faster learning curve. Experts generally have more accumulated experience, and they have high productivity. They follow an established best way based on their practice and experience. They can often do a job more efficiently than others. Innovators, on the other hand, go beyond the established ways of doing a task and search for a new and improved way to do things differently. The innovators are generally looking for better effectiveness, rather than for a mere improvement in the efficiency.

4. Creativity Does Not Require Rarely Found Genius. Some believe in the folklore that to be creative and innovative one needs to be a genius—that is, creative people are born creative. The fact is that everyone can be taught and trained to be creative and innovative. Creativity is a skill that can be learned like many other skills. Creativity is within reach of everyone. All of us can learn to be more imaginative with a little training and practice.

5. Creativity Can be Learned Routinely. According to the corporate folklore, creativity cannot be learned. Others believe that creativity can be routinized.

The truth lies somewhere in between. Motivation and inspiration are required to generate original ideas and creative solutions for new or old problems. Success with creative innovation requires that the inspiration is coupled with sustained "perspiration." Creativity, however, is not likely to be observed by just doing business as usual.

6. *Creativity Is Spontaneous and Not Hard Work.* Some assume that the creative work is all fun and not much work. Creative work is not a mere diversion from serious work. Coming up with original and improved solutions requires intense hard work. This requires a careful collection of past information and a disciplined evaluation of potential choices using rigorous techniques.

Vowels of Innovative and Creative Problem-Solving. As a result of an extensive literature search and many class room discussions, this author has identified a concise way to learn about the innovative and creative process to make decisions facing innovators (see Note 7.18). The different steps in the creative process are described using the "vowels" of the English alphabet. These are described below.

A. For Awareness and Problem Recognition. The first step to creative and innovative thinking and problem solving is to be aware of a tough problem that is facing the technology-driven enterprise. Finding a challenging problem and clearly defining an unstable situation are the key first steps to seeking innovative and creative solutions. Most of the people are likely to discourage the potential innovator by saying that the problem is impossible to solve or that they have already tried everything. But the creative innovator likes the challenge posed by such past failures.

E. For Exploration, "Naive" Incubation, and Gestation. The creative innovators are generally engrossed by the tough problem identified by them. They personally get immersed by the challenge of seeking the elusive solution and spend considerable time reflecting on the problem and its much-needed solution. Quiet contemplation allows them to do subconscious manipulation, restructuring, and new pattern-seeking. The creative innovators are generally looking for the untried original solution for the problem.

I. For Immersion, Information/Knowledge Search and Detailed Preparation. With the gestation and reflective exploration mentioned earlier, the creative innovators get immersed in finding out all that they can about a problem. They want to learn everything they can about the problem. To do so they seek factual information as well as expert opinions. They do so from diverse viewpoints. The creative innovators prefer to collect all that is known without getting biased by the past failures. Sometimes the failures help innovators learn more about a problem than the successes do.

O. For Organization of Learning, and "Knowledgeable" Incubation/Gestation. The "jungle" of information and knowledge collected by the creative innovators from diverse sources must be organized. Otherwise, it just produces chaos. The innovators give personal consideration to the unusual approaches and ideas. They sometimes blend and incubate naive notions with factual information and expert opinions. In this "organization" stage, all the pieces of the puzzle are brought together and flattened out, right side up.

U. For Understanding, Formulating, and Evaluating Alternative Solutions. The creative innovators then start "piecing" different bits of information together. This may be done singly, or collectively in a group. The "understanding" stage is divided into two phases: (1) the intuitive phase and (2) the analytical phase. In the intuitive phase, numerous alternate possible solutions are generated using creativity-enhancing techniques such as brainstorming, analogies, and so on. In the second analytical phase, the alternate possibilities are rigorously tested using known scientific techniques and methodologies for evaluation.

Y. For Yielding Results by Recommending Short-Term and Long-Term Action Plans for Implementations. Finally, a "Eureka" moment appears when one or more creative solutions are conceptualized and clarified. The ideas and analysis of the creative solution are put into action to solve the problem facing the creative innovator.

The creative innovator must not stop at discovering and implementing a creative solution for the problem on hand. She or he must continuously learn by sharpening her or his skills for future challenges. The journey must continue to the next stage.

Continuous Growth by Reassessment and Learning

The creative innovators must carefully judge and evaluate the results obtained as well as the journey taken to arrive at the solutions and the results. Some time must be devoted to reflect on the ways the creative process could have been improved further, such as (a) how a smaller amount of resources could have been consumed to arrive at the same results or (b) how more could have been achieved using the same amounts of resources. Before rushing to take the next challenge of fighting a new fire, it is critical for the innovators to pause and reflect on the lessons learned from the creative process just completed. Finally, one must combine all of these skills to bring about growth and change.

ENTREPRENEURS, CHANGE AGENTS, AND INNOVATORS

Technology grows and develops with time because of entrepreneurs, innovators, and change agents. They create and commercialize new sources of corporate wealth. Often they are forced to do this by leaving their corporate employers. In the Silicon Valley on the West Coast of the United States, entrepreneurs leaving their jobs to start a company have become an essential part of the Valley's culture. For example, according to Taiwan-born Pehong Chen, who founded Gain Technology in 1989 and founded Siebel Systems and BroadVision in 1993, an entrepreneur "need(s) to create stuff: new technology, new solutions, new companies" (see Note 7.19). The entrepreneurs leave their employers when they are not allowed to create and innovate within their organizational setting. Technology-driven organizations could benefit enormously from nurturing such entrepreneurs and retaining them instead of letting them go.

Entrepreneurs can be defined as the dynamic individuals who proactively take risks, innovate, and invent opportunities by generating new actions and awareness of unfulfilled needs.

In some organizations, entrepreneurs are promoted and encouraged to exert their leadership power over others. In other organizations, the entrepreneurship is dispersed between many different individuals and departments.

Change Agent Innovators

According to Professor Rosabeth Moss Kanter, author of *Change Masters* and Professor of Business Administration at the Harvard Business School, the corporate entrepreneurs, whom she calls "change masters" go through three stages (see Note 7.20).

Stage 1: *Formulate and Sell a Vision.* In this stage the entrepreneur sees a vision that others cannot. He or she must then sell this vision to others to ensure that the vision is not just a day dream.

Stage 2: *Find Power to Promote the Idea.* Creative and entrepreneurial visions need sponsors—that is, sponsors with the authority to finance and provide other support needed for the vision to evolve to its next exploratory stage.

Stage 3: *Maintain the Mandatory Momentum.* Entrepreneurs and change agents must recognize that they will face obstacles. People will disagree with their vision, and others will not want the entrepreneur to succeed. The entrepreneurs must believe in their visions and their own commitments to their visions. Commitment is important. It involves one giving the best one has.

Six Essential Skills of Change Masters

Professor Kanter considered the essential skills needed for a person to become an entrepreneur or a change agent in an organization. She discussed the six most important skills. These skills for an innovating change agent are given below.

1. Kaleidescopic Thinking. The innovators use existing data and information, process it, and come up with new shapes of conclusions. They are able to use the information around them to develop new and interesting patterns—that is, patterns which produce wealth for their organizations.

2. Communicating the Vision. They champion the need for innovation, and then hard-sell it to others even when they face stiff resistance. Communicating in clear and easy to understand ways is very important. The change agent must communicate the vision from the listener's point of view, not from the presenter's point of view.

3. Persistence. The innovating change agents persist and put their full power behind their innovations and visions. At times the situation and progress seem hopeless. But the innovators cannot give up.

4. Coalition Building. To build momentum, innovators enlist others to rally behind their vision and "buy into" their innovative ideas. Most innovative ideas require support of many different persons and resources. Coalitions are therefore critical for new innovative developments in technology and global markets.

5. Working Through Teams. The innovating change agents operate through team-based joint ownership of innovations. They participate in others' teams, and they invite others to participate in their teams. In the globalized economy, more than one head is better than a lone head.

6. Sharing the Credit. Finally, the innovating change agents make heroes out of their subordinates and supporters. They share the credit of success with them. This keeps their contributions coming for future innovations. The innovator who corners all the credit is not likely to see much support in the future.

Environment for Nurturing Innovators

Finally, what can the technology-driven organization do to create an environment that nurtures the innovators? Professor Rosabeth Kanter suggested four conditions that would facilitate an environment for innovations and changes. These are discussed below.

1. Broadly Defined Jobs. Narrowly defined jobs restrict the sense of ownership of employees. They therefore are not encouraged to bring about a change. Broad definition of jobs gives a holistic feeling about their roles in the organization. The innovators can see their contributions clearly.

2. Small, Flexible But Complete Organizational Structures. For teams of people to be effective innovators, they require a variety of needed skills. But these groups should not be burdened with large unnecessary bulky organizational size. In large organizational units, the weakest or the slowest link defines the speed of the entire team.

3. Inculcate a Culture of Pride in People. Employees are the most critical components of bringing about innovative changes. People should have a sense of growth and achievement rather than a false notion of security. People who do not have pride, or lack a sense of their role in the organization, are not likely to innovate and bring about creative changes.

4. Empower People with Power Tools for Innovation. Words are not enough. People need three types of power tools to convert their innovative ideas into actions. These are given below.

A. Easy Access to Relevant Information. Current and accurate information is the backbone of innovation. People must have open lines of communication with their peers as well as their senior leaders. Innovators must be provided free access to the information they need to innovate. Obsessive hierarchical concern for secrecy may hinder innovation.

B. Open Channels to Support Networks. With the evolution of technologies, newer opportunities to innovate lie at the interface between two or more fields of knowledge. Therefore cross-discipline career paths and mentoring by senior employees help nurture the skills of innovators.

C. Receive Flexible Resources. Many senior leaders control their resources excessively. They closely monitor the resources used by their employees, including their time. Innovators require some resource slack to tinker with their innovative ideas and to see if they can be brought to fruition. Innovative companies such as Minnesota Mining and Manufacturing Company (3M) allow their employees free time to tinker with their ideas.

Technology-driven enterprises serious about innovation must provide an atmosphere that supports innovators.

SOME LESSONS LEARNED IN THIS CHAPTER

In this chapter we discussed the different roles of people in technology-driven enterprises. We noted how the 1990s saw the departure of job security under a continuous assault of downsizing. Yet progressive companies like 3M, Federal Express, Johnson & Johnson, Microsoft, Intel, and others continue to put people before their profits. These companies have great hope regarding the humanware of their organizations. We referred to people who are the assets of technology-driven organizations as the creative innovators.

We then discussed the seven roles innovators play in technology-driven enterprises. They learn and relearn on a continuous basis, and they make fast decisions in systematic ways. We discussed a seven-step problem-solving and a four-step decision-making process. The innovators are sensitive to their social and ethical responsibilities and support their company's values. The innovators participate in self-directed and high-performing teams. We discussed the different types of informal and formal teams, along with the reasons behind formation of the informal groups. We also learned about the different stages of evolution of formal high-performing teams. We discussed how dysfunctional teams produce groupthink. These are fast but poor solutions. We then discussed the suggestion system to encourage innovators to proactively participate by making their suggestions, and we discussed the ways by which creativity and innovation can be nurtured in technology-driven enterprises. Finally, we noted how innovators can be change agents and entrepreneurs, and we described what organizations can do to create an atmosphere that supports them.

SELF-REVIEW QUESTIONS

1. What is the impact of downsizings on the human capital of technology-driven enterprises?

2. Discuss the learning organization and the role learning innovators play in such organizations.

3. What are the different types of decisions agile decision makers make? Describe a multistep decision-making process.

4. How do human values influence decision makers? Should a corporation exclusively maximize its profits, or be socially responsible?

5. What are different types of teams? Define them by their relationships. How do formal teams evolve over time?

6. Discuss suggestion systems, and how companies should implement their suggestion systems.

7. Define creativity and innovation. What are the folklores and facts about creativity and innovation? What are the vowels of innovative problem-solving?

8. What are the essential skills of innovative change agents? How can an organization nurture an environment that supports the innovators?

NOTES, REFERENCES, AND ADDITIONAL READINGS

7.1. O'Neill, Hugh M, and Lenn, D. Jeffrey. 1995. Voices of survivors: words that downsizing CEOs should hear. *Academy of Management Executive,* 9(4): 23–34.

7.2. Gephart, Martha, Marsick, Victoria, Van Buren, Mark, and Spiro, Mechelle. 1996. Learning organizations come alive. *Training & Development,* December: 35–45.

7.3. For Stephen Covey's ideas in general see *The 7 Habits of Highly Effective People.* New York: Simon and Schuster. For ways to build trust, Covey can be contacted at The Covey Leadership Center in Provo, Utah. The news about Covey's comments appeared in the December 1996 issue of *Training & Development,* under "News You Can Use."

7.4. Haggerty, Patrick E. 1981. The corporation and innovation. *Strategic Management Journal,* 2(2), April–June:97–118.

7.5. See Xerox Home page, 1997. Also see Xerox documents describing strategic learning.

7.6. Senge, Peter M. 1990. *The Fifth Discipline: The Art and Practice of the Learning Organization.* New York: Doubleday Currency. The author discusses these subjects in more behavioral and cognitive details.

7.7. Gardner, Howard. 1985. *The Mind's New Science.* New York: The Basic Books.

7.8. Argyris, Chris. 1982. *Reasoning, Learning and Action: Individual and Organizational.* San Francisco: Jossey-Bass. For more details see Argyris, C., and Schon, D. 1978. *Organizational Learning: A Theory of Action Perspective.* Reading, MA: Addison-Wesley; and Argyris, C. 1985. *Strategy, Change, and Defensive Routines,* Boston: Pitman.

7.9. See Johnson & Johnson Annual Report, 1996, and home page on the Internet.

7.10. Weber, Max. 1958. *The Protestant Ethic and the Spirit of Capitalism.* New York: Charles Scribner's Sons. This is a translated edition of the original in German.

7.11. Thurow, Lester. 1982. *Head to Head.* New York: William Morrow and Company.

7.12. Kinlaw, Dennis. 1991. *Developing Superior Work Teams: Building Quality and the Competitive Edge.* Lexington, MA: Lexington Books.

7.13. *Business Week,* 1985. Women at work, January 28:80–85.

7.14. *Business Week,* 1987. Detroit vs. the UAW: at odds over teamwork, August 24:54–55.

7.15. De Cenzo, David A., and Robbins, Stephen P. 1994. *Human Resource Management,* 4th edition. New York: John Wiley.

7.16. See Ohno, Taiichi. 1988. *Toyota Production System.* Cambridge, MA: Productivity Press. Also see Yasuda, Yuzo. 1990. *40 Years 20 Million Ideas.* Cambridge, MA: Productivity Press.

7.17. Excerpted from *Ford Plant Handbook* by Shoichi Saito in his 1952 book *America, The Automobile Nation.* For details see 7.16.

7.18. Gehani, R. Ray. 1995. Vowels of creativity. Teaching notes. For related discussion see Kuhn, Robert L. 1989. *Creativity and Strategy in Mid-Sized Firms.* Englewood-Cliffs, NJ: Prentice-Hall. 244–245; and Kuhn, Thomas S. 1962, 1970. *The Structure of Scientific Revolutions.* Chicago: The University of Chicago Press.

7.19. Port, Otis. 1997. Starting up again—and again and again. *Business Week,* August 25:99–100.

7.20. Kanter, Rosabeth Moss. 1983. *The Change Masters.* New York: Simon and Schuster. Also see Kanter, Rosabeth M. 1989. Becoming PALs: pooling, allying, and linking across companies. *Academy of Management Executive,* 3:183–193. Also see Kanter, Rosabeth M. 1990. How to compete. *Harvard Business Review,* 14(6):7–8. The global aspects are covered in Kanter, Rosabeth, 1995. *World Class.* New York: Simon and Schuster.

CASE 7A: TECHNOLOGY MANAGEMENT PRACTICE

Nominal Group Technique for Team-Based Decision-Making

The Nominal Group Technique (NGT) is a structured team-based decision-making process. It facilitates generation of creative solutions for problems for which team members either lack agreement or possess incomplete information. The NGT method values individual inputs in arriving at the team decisions, and it gives an even representation to all the team members. The team members must contribute their judgments to solve the team task. NGT is also helpful in arriving at identifying critical variables for a problem situation.

NGT is not very effective for routine problem solving which requires extensive coordination and information exchange. NGT is also not good for negotiation-based decision-making, such as between unions and workers.

NGT contains four steps. These steps are described below.

Step 1. Generating Ideas for Defining a Problem

In this first step, the team members are involved and invited to select the problem on which the team wants to work. If the problem has been already identified and assigned to the team, then this step can be used to generate creative solutions.

In this step a problem statement or a stimulus question is announced and written down so that all team members can see it. This could be as simple as one of the examples given below.

"What do you wish to learn in a technology management course?"
"What should your university do to improve its enrollment?"
"What problems should we consider over the next year?"

Once a problem statement has been clearly defined, each member is given a sheet of paper. They are asked to take 5 to 7 minutes to individually write their ideas (or the problems they want to work on) on the paper provided to them.

The private generation of ideas by individual team members avoids (a) the direct pressure caused by differences in their status or (b) competition between team members to dominate others or be heard. It also avoids the "higher weight" grabbed by team members who speak the loudest. Such distortions cause a feeling among some team members that their problems will not be worked on, and therefore they withhold their commitment to work on the problem. By avoiding such "public discourse" a number of unwanted effects that could compromise the generation of most creative solutions are eliminated. Yet, one must also understand that sometimes the presence of other team members provides a creative pressure to produce innovative ideas.

The first step, along with the structured approach of the following three steps, provides adequate time to each member for thinking and reflecting, and it avoids the premature selection of ideas for the problem to be solved.

Step 2. Recording Team Members' Ideas

In the next step, ideas (for problems to be solved) are recorded in a round-robin manner on a board or a flip-chart so that all members can see them. Each member is asked to contribute one idea at a time. That idea is carefully written down. (Or else each group member is asked to submit the ideas anonymously on a 3×5 index card). The facilitator must clear each idea with the different team members so that the same idea is not written twice, though an idea may be expressed in slightly different words. If an idea or a problem is repeated, these ideas are combined and consolidated. Each idea or problem statement thus recorded is assigned a letter from A to Z.

This process is repeated until the displayed list satisfactorily records all the ideas generated by the team members individually. The round-robin process provides equal opportunities to different team members to present their ideas which they consider significant. The public listing of ideas depersonalizes the ideas and reduces conflicts among their originators.

Team members are sometime impressed by the ideas presented by other team members, which contributes to their enthusiasm for further participation in the process. If there are too many ideas or problems to be considered, then they are prioritized to a smaller number of most important ideas. These ideas are usually kept under ten for their effective comparison with each other.

Step 3. Clarifying Ideas

In this step, each listed idea is discussed, one after another. The meaning of each idea is clarified by the team members so that the logic and thinking behind each idea is clear to all the team members. Team members are encouraged to express their agreement and disagreement openly. This helps reduce any hidden misunderstandings.

Step 4. Voting on Ideas

Step 3 is not intended to evaluate the relative merits of each idea. The differences in opinions are resolved in step 4. A voting procedure decides the issue of importance of each idea. This will be illustrated with an example given below.

Example: Topics to Be Covered in a Quality Management Course.
Let us say that there are six ideas generated in step 1. These were
recorded and clarified in steps 2 and 3. These are assigned letters A
through F. Thus the ideas for topics to be learned in a course on total
quality management may include the ideas listed below:

    A: Leadership
    B: Process control
    C: Empowerment
    D: Customer satisfaction
    E: Supplier relationship
    F: Performance

The team members are then asked to rank each of these ideas from
the most important idea ranked as 6, to the least important idea
ranked as 1. Thus if a team member #N considers that "customer satisfaction" is the most important idea in a course on quality management, then she or he marks a 6 in front of the D idea. Other ranks are
similarly given to each idea. These ranks are indicated below:

Ranking of Team Member #N

A: Leadership	____5____
B: Process control	____4____
C: Empowerment	____1____
D: Customer satisfaction	____6____
E: Supplier relationship	____2____
F: Performance measurement	____3____

Each team member thus submits a list of rankings for the identified
ideas A to F, ranked in the order of their importance.

A nominal group then compiles a consolidated list of all the team
members' rankings for each of the identified ideas. The rankings are
added across each row, and the item with the highest score gets
attention first.

	Rankings						
Member #	#1	#2	#3	#4 ...	#20	Total	Rank
A: Leadership	3	5	4	6	4	22	3rd
B: Process Control	2	3	3	3	2	13	4th
C: Empowerment	5	6	5	4	5	25	2nd
D: Customer satisfaction	6	4	6	5	6	27	1st
E: Supplier relationship	1	2	1	1	1	6	6th
F: Performance	4	1	2	2	3	12	5th

The relative advantages of NGT for decision-making over the usual interacting team methods arise from (a) the higher significance assigned to idea generation and (b) more attention paid to each idea in NGT. If the team members know each other very well and are willing to communicate freely with each other, then the use of NGT may not be particularly advantageous. NGT is particularly effective where teams are dominated by a few loud team members.

CHAPTER 8

Project Management and Integration

PREVIEW: WHAT IS IN THIS CHAPTER FOR YOU?

We have already covered a number of management aspects related to projects. In the previous six chapters, we learned about the six pistons of the turbo-engine of technology-driven growth—for example, how production operations have evolved, or what are the ways to develop or use proprietary intellectual know-how needed to do the same. Or how to achieve high quality promise and reliability in the products produced. We also learned how customers are satisfied and how people are motivated to be creative.

In this chapter we will learn what makes a technology project and how managing a project is different from managing a process. Projects are defined by the project goals. These goals define what is it that a project is expected to accomplish at the end of its completion. These goals often include constraints of time of completion and budgeted cost of a project. These have to be monitored effectively to avoid overruns. We will also learn about some of the tools used to monitor and control the projects effectively.

The six technology-related competencies, like the six pistons of the turbo-engine of technology, can be used with the management of projects. Therefore, instead of repeating them here again, we will focus on only those aspects which are unique to the management of projects. This includes a project life cycle, parametric metrics for monitoring projects, types of projects, and so on.

PROJECTS ARE PARAMOUNT

As illustrated in the case of the Manhattan Project for the development of atomic technology during World War II (see Case 8A), projects play a critical role in the management of technology and manufacturing operations. In technology management, there are many events and operations that are conducted for the first time. These projects, which are a one-of-a-kind set of activities, need to be managed in a different manner than the routine ongoing process activities of a technology-driven organizations (see Exhibit 8.1).

EXHIBIT 8.1. Process Versus Project Management

Management Aspect	Process Management	Project Management
1. Activities	Repetitive	Unique
2. Event scheduling	Ongoing	Long-duration
3. Functional responsibility	Functional	Cross-functional
4. Interdependence	Low	High
5. Risk	Limited	High
6. Returns	Marginal	Wide-ranging

The projects, at least some of them, make history. Consider the following examples (Note 8.1).

1. The Empire State Building, New York

This building construction project defined the word "skyscraper" once for all times to come. In the 1920s, architects were successively conceiving building taller and taller building projects, dwarfed only by their own egos—until the Great Depression and the crash of banks halted the new construction projects. For 40 years the 1248-foot-tall Empire State Building in Manhattan ruled the skyline of New York City.

More than its height, the speed of construction was astounding. The construction of this landmark on Fifth Avenue in mid-Manhattan began the day after the stock market crashed in 1929. The project was completed 410 days later, without spending a nickel for the overtime. The construction project ran like clockwork. The 58,000-ton structural frame was riveted together in 23 weeks. In 8 months, the platoon of masons finished the exterior surface. The building consumed 17 million feet of telephone wire, and plumbers put in 51 miles of pipes. The Empire State building project redefined the management of building projects.

2. The Golden Gate Project, San Francisco

This bridge is often admired as one of the most elegant bridges of the world. But it is also the pinnacle of the civil engineering projects. It stands proud and unpurturbed in the San Francisco fog and winds of more than 70 mph at the top, the conditions under which the construction crew built the project. The base is constantly lashed by the Pacific Ocean tides. Its 764-foot-high towers are the highest ever constructed. The bridge is supported by yard-thick cables hanging from the towers. For the first time in history, a bridge was constructed across the mouth of an ocean.

The Bridge survived the 1989 Loma Prieta earthquake, which destroyed the San Francisco Candlestick Stadium. The span is being reinforced so that it will withstand a 90-second earthquake that is up to 8.3 on the Richter scale. Weather has shut down the bridge only three times in 60 years.

3. The Panama Canal, Panama

President Woodrow Wilson blasted 40 tons of dynamite to shatter the Gamboa Dike to finally connect the Pacific Ocean to the Atlantic Ocean. An 8-mile-long ditch was already carved out between the continents of North America and South America. The French had tried to do the same first, but failed and gave up in 1898. The American engineers picked the challenge in 1902 with the signing of the Spooner Act. The project was completed within budget and on time.

The engineers dug enough dirt to pour 12 feet high over Manhattan island in New York City. There was so much malaria and yellow fever among the workers that the mosquito was fought with warlike ferocity. Sanitary levels better than those of most American cities were maintained, while swamps were completely drained. Some say that the Panama Canal was really built by the doctors. Making the Panama Canal first required building a dam that formed world's largest man-made lake of its time.

The project was commissioned in 1914, and it has operated nonstop for the past eight decades. A ship from the Caribbean Sea travels through the 7-mile-long ditch. Then it is towed into the 1000-feet-long Gatun Locks, where the ship is lifted 85 feet, or more than nine stories high. The ship then travels through Gatun Lake and the Gaillard Cut. Two locks drop the ship by 31 feet and 54 feet height. Machinery is not used for filling. The project is designed in such a way that gravity does that job, saving significant energy.

These are some world famous projects. Let us next learn how projects are managed.

WHAT IS A PROJECT?

Defining a Project

A project is defined as a set of interrelated and nonrepetitive activities and operations required to accomplish a required set of objectives at desired levels of goals. Some project managers have explained that projects are work activities that produce. Projects usually have a well-defined starting point and often have a target finish point. Therefore projects have a well-defined project life, though sometimes projects seem to drag on and on.

As mentioned earlier, projects usually involve a new set of activities. Bechtel Corporation, a world-famous construction company, does many projects. For example, they may receive an order to do a petroleum refinery project. Bechtel has constructed other petroleum refineries before. But we will still call Bechtel's new order a project because it could have a set of unique new requirements and corresponding constraints. They may have new deadlines or new environmental and political requirements.

A project has a unique combination and relationship between the required activities and the desired objectives. Thus, building a custom home is considered a project, but making similar prefabricated homes is not. Assembling a customized motorcycle is a project, but manufacturing a Harley-Davidson in its assembly plant is not.

PROJECT PARAMETERS

By now, you may have guessed that most projects are defined by a set of project parameters. These parameters drive a project. These parameters are also used to monitor and plan a project's progress. The nonrepetitive nature of activities causes uncertainty and unpredictability. But this also provides the variety that most workers love to work with. These variations result in variability in the use of resources. At one time in the project, there is a lot to do. At another time, there may not be enough to stay busy. The most important parameters are discussed next.

1. Cost of the Projects—By Type

Each project comes with an explicit or implicit contract. There is an understanding between the sponsor of the project and the executioner of the project about the estimated use of financial resources for the project.

Payment terms also play an important role in the project management. From the project manager's point of view, the expenses start incurring from the word go. But from the project sponsor's point of view no revenue is generated until the project is completed. Furthermore, if all the expenses are paid by the project sponsor before its completion and commissioning (such as in the case of a petroleum refinery project), the project may not perform at the level it was expected to. Therefore a part of the payment is withheld by the client until the promised performance is guaranteed.

Costing of projects is done in one of many different ways. Some of the popularly used methods are briefly discussed below.

A. Fixed-Cost Projects. These projects are commissioned for a fixed price, for delivering certain prespecified performance levels. If the project manager can deliver these performance levels by incurring less cost, there is profit from the project. On the other hand, if the project manager is not able to contain its costs within the budgeted levels, the project results in a loss. Elimination of waste adds to the bottom line and therefore acts as an incentive.

A modified version of fixed-cost projects is to add inflation or foreign-exchange adjustments (especially for the international projects). These projects receive additional compensation for the inflation in costs of materials, labor required, and other supplies over the life of the project.

A limitation of a fixed-cost project is that the project managers may resist any changes because they add to projects costs. For the long-duration projects, this is not a good thing.

B. Cost-Plus Projects. Federal and State projects funded by the U.S. government often pay for the cost incurred plus certain margin for the project contractor. Most of the new fighter aircraft development projects and other defense contracts are projects commissioned on the cost-plus basis.

For the cost-plus projects, there is no incentive for a project manager to cut the waste. Such savings do not add to the project's bottom line.

One advantage with the cost-plus projects, which may be one of the reasons why the U.S. Federal government uses it so often, is that the project contractor does not complain about making certain changes, as long as the client is willing to pay for the additional cost.

In cost-plus projects, the project contractor's fees may also be computed in a variety of different ways. The fees may be held constant, irrespective of the cost of the project. Or the fees may be increased (like a commission percentage) in proportion to the rise in the cost of the project.

2. Completion Time of the Project

On-time completion of a project is very important for the sponsor of the project—whether it be a nuclear reactor project, a petroleum refinery project, or an aircraft construction project. Only after the project is completed can the sponsor of the project expect to generate revenue. Until then, its capital investment and its site is locked into the project.

During this period, the environmental or market conditions may change. Thus, sponsors of long-duration projects fear the time overruns. Most time-sensitive projects will offer a bonus for completing early and include a penalty for delays.

3. Performance of the Completed Project

At the completion of the project, it should perform the way it was expected to. Let us say that the project was for constructing a continuous production plant for a polyester packaging film of 10-ton-per-day capacity. Then at the completion of the project, the continuous film production plant should produce 10 tons per day of film of uniform quality, over a continuous run of a week or so.

For long-duration projects, it is also customary to specify intermediate checkpoints. For example, a project contract for a hotel construction project may specify that the land will be developed by such and such a date, the foundation will be poured by such and such a date, the building structure will be erected by such and such a date, and the interior fittings will be installed by such and such a date. This way the sponsor of the project can estimate that the project is progressing at the right pace, so that other related activities may be coordinated with the projected completion date of the hotel—for example, hiring of staff, advertising, room reservations, and so on.

THE PROJECT PLANNING PROCESS

Projects need careful planning. Otherwise they have a tendency to overrun their costs, take too long to complete, or perform below their expectations. The project planning begins with a project proposal, and it ends with a successful handing over of the completed project to its client. Planning involves carefully identifying what a project is expected to accomplish and what kind of resources are needed to do so. The purpose of planning a project is to achieve the maximum results with the limited resources available (see Note 8.2).

1. Search Stage: Proactively Searching Project Opportunities

Most projects begin with a project proposal, but the project life cycle begins before that. Project planners must anticipate their customers' requirements.

Savvy project managers keep their opportunity-seeking antennas up in the air all the time. They monitor macroenvironmental trends, to preempt where the demand for new projects is likely to crop up—before their competitors do the same. For research projects, researchers routinely check the U.S. Federal funding to anticipate where the new projects are likely to emerge. Knowing that these projects are awarded based on certain skills, researchers build these skill levels all the time—long before they have received a research project to do.

In many fields, such as aircraft or hotel construction, it is a cutthroat competition. Half the competitive battle is won by getting the contract for a project. Most of the marketing effort for project management is done upfront.

2. Conception Stage: Preparing a Project Proposal

The next step is to either (a) write a project proposal or (b) get a Request for Proposal, routinely referred as RFP.

Developing a good project proposal takes time, money, and a lot of effort. Even though the client does not pay (most of the time) for writing and submitting a proposal, this step cannot be overlooked and taken lightly. A project proposal locks in the resources that a project will get. It also specifies the expected performance levels for the successful completion of the proposed project. Time limits also may be locked in. These project parameters must therefore be estimated very carefully. Planners must provide for some buffer for unavoidable but likely-to-occur events.

Developing a project proposal or a bid for project will sometime involve carrying out 10–40% of the project work—without getting paid for it. Yet, this crucial upfront work must be invested in order to get the desired project.

For large projects it is not uncommon to develop a steering committee for the preparation of a project proposal. This steering committee may call on different functional managers to feed-in the data and information necessary to build a project proposal.

Teams assigned to prepare project proposals must pick and choose their project battles carefully. They should pick only those battles which have enough incentives, are crucial to the continuous running of their project team, and have a good chance of coming their way.

Some project proposals must be prepared in certain standard formats. They may demand attaching an executive summary sheet, or limiting the proposal to less than a certain number of pages of documents. Since an attractive request for proposal will bring in a large number of project proposals, it is not uncommon to specify the maximum page limit in the proposal.

Developing project proposals that win competitive bid contest and beat the competitors to it is an executive art. Since award of projects is sometimes quite arbitrary among some equally competent project contractors, the person awarding the project contracts has a lot of discretion. Some project managers, recognizing this "weak" human component of the otherwise tight technology and resource-driven project, will go an extra mile to meet the needs and requirements of the person awarding the project contracts. Sometimes they will even meet the personal needs not directly related to the performance of the project under contention. More than one politician and Congressmen have been known to appreciate contributions to their "reelection funds" when they are about to give out defense or government contracts, considered by some as "Congressional pork" projects.

Smart preparers of project proposals will have informal discussions with the person to whom they are submitting their project proposal, to get a sense of different priorities as well as likes or dislikes. The project proposal is then developed meeting these requirements and providing all the required information. Most project proposals include a brief description of the organization and its capabilities or past achievements, the technical details of how the client's requirements will be met and assured, and a cost estimate component that describes how different numbers were arrived.

Because of the factors discussed above, the selection of people assigned to write a project proposal must be done very carefully. They should have the required experience as well as maturity to get the job done well within the set time limit. They must be good negotiators, and be able to work on their own, with limited supervision (see Note 8.3).

In many U.S. corporations, it is quite customary for lawyers to review the project proposals before they are submitted. They verify that all legal issues are in order and that no laws of the nation(s) involved are violated. They also add certain standard clauses, about what is being promised and what is not. Often, there are sections which cover the mechanisms for resolving disputes, their place of jurisdiction, and so on.

3. Birth Stage: Project Presentation

In many corporate settings, each proposal for a project must be presented to a board of decision makers. Many senior executives prefer oral presentations, besides a written submission. During the project presentation of a project, the decision-makers can ask questions on aspects not clear in a project proposal. They also like to see a sense of enthusiasm in the people commissioned to do the project.

Presentation of a project proposal is critical. The presenters must develop a rapport with the decision-makers to whom the project proposal is being presented. Presenters must not just ramble out the technical details. They must explain that the numbers in the project proposal were estimated carefully and that the client is not being overcharged. Eye contact and a cordial relationship during a presentation are prerequisites to winning a project contract.

4. Growth Stage: Implementing and Monitoring Project

If the project proposal and its presentation are hits with the decision-makers, then the project contract is awarded. After thanking the sponsoring clients, a careful plan to implement the project must begin immediately.

Most projects are awarded accompanied with a clear definition of the scope of work. It often mentions the work to be done, in as much detail as possible at that early stage. It also specifies the guaranteed performance expected by the client.

The main challenge in project planning is breaking down the entire complex project into more manageable discrete pieces or activities. This division of work may not be always so easy, considering the fact that a project, by definition, is a one-of-a-kind activity. These project pieces are then assigned to different agencies, inside and outside the corporation, to make use of a parallel division of labor (if it is possible to do so). The parallel arrangement cuts down the total time required to complete a project. It also helps to draw on the best expertise available for each individual activity and its component task. The different pieces of the overall "puzzle-like" project must be integrated, so that the different pieces mesh together seamlessly.

To maintain an order in the complex distribution of work, a command-and-control hierarchy must be carefully developed. This must also clearly assign the authority for incurring expenses and the accountability of responsibilities.

In a later section we will learn certain tools used to maintain the progress of projects, the on-time schedule, and the budget.

5. Maturity Stage: Project Handover, Test-Runs, and Guarantees

As the project winds to its completion date, the project manager must schedule a suitable appointment with the client when the completed project may be handed over. Once such suitable dates are obtained, the project manager must demonstrate a satisfactory completion of the project. In the case of a housing project, this is done by listing all the fixtures fixed by the contractors. But in the case of a project involving a production plant or a performing equipment, the project contractor must demonstrate the performance at a prespecified satisfactory level.

One purpose of this stage is to get fully paid for the work done. But sometimes a more important function of such hand-over is to mark a clear transfer of responsibilities. After the project is handed over to the client in a satisfactory condition, responsibility for damages or safety is that of the new owner of the project.

Besides these the manager of a project must use the hand-over exercise to review the lessons learned from this project. This includes a list of what practices worked well, along with those that did not work too well. Procedures for the latter type of practices must be improved for the next projects. If there were any major obstacles or problems encountered in a particular activity, these should be highlighted, so that future managers can be alerted in advance. Any gaps in skills or information encountered during the project must be also recorded. Suggestions can be made on how other parts of the corporation can fill these gaps for the future projects. The project manager must give recommendations for the next project manager, even though the next project is most likely to have new unforeseen challenges of its own. New and the latest project tools are procured.

PROJECT PLANNING TOOLS

Wars have resulted in discontinuous amounts of growth in the U.S. economy. As the young American soldiers returned home to a highly industrialized economic machinery, they needed goods and services. During the post-war economic boom of the 1950s, U.S. organizations were faced with growth. Many new projects emerged for construction of plants, roadways, and housings. Given below are some tools developed to meet this challenge.

1. Gantt (Bar) Chart

This is a tool commonly used to monitor the progress of a project. A bar chart was first proposed by Henry L. Gantt. In 1915, during World War I, Gantt developed and used bar charts to monitor the production and delivery of ammunition to the military forces.

The Gantt chart lists the various major activities to be performed in a particular project, in the required chronological sequence. The corresponding starting and finishing times are indicated in front of each activity.

Let us consider developing a Gantt (bar) chart for a project to build a fictitious Acme Pilot Plant and R&D Laboratory for a technology-driven corporation. Before a Gantt chart can be constructed, the different activities required to complete the project must be listed. The time required to satisfactorily complete each one of those activities must be estimated too (see Exhibit 8.2).

Besides these the project manager has to carefully think through the *precedence relationships* of these activities. The precedence relationship of an activity specifies its prerequisites—that is, those activities that must be completed before a particular activity, say "p," can be started. For example, in the case of a pilot-plant project, the equipment cannot be installed until

EXHIBIT 8.2. Pilot-Plant & R&D Laboratory Project Planning

Item Activity Description	Prerequisite Relationship	Duration (Weeks)	Man Power Level
A. Select location and site	Start	3	4
B. Choose subcontractors	Start	3	3
C. Site preparation	A, B	4	6
D. Procure materials	B	4	2
E. Place equipment orders	A	4	2
F. Lay the foundation	B	3	4
G. Build building structure	E, F	8	6
H. Build roof structure	C, D	4	4
I. Install large equipments	G, H	2	6
J. Raise main prefabricated walls	I	4	4
K. Connect utilities	J	4	4
L. Finish interior fittings	K	4	4
M. Exterior decoration	L	2	3
N. Clean up for commissioning	M	1	2
O. Test-run pilot plant	N, K, L	2	4
Q. Inspection for hand-over	O	1	2

activity involving fabrication of the building structure is complete. Similarly, the doors and windows cannot be fixed until the pilot-plant equipment is installed, because the large equipment will not go through the standard doors and windows of a building.

The activities, their brief descriptions, precedence relationships, estimated time duration required to complete each activity, and the corresponding manpower requirement are all tabulated. This is shown in Exhibit 8.2.

2. Resource Allocation Balancing

In the Gantt (bar) charts we noted that the estimated duration of an activity varies depending on the allocation of manpower to that activity. When an activity is required to be completed fast, managers allocate more people to that high-priority activity. When an activity is not a critical activity, meaning that it does not delay the overall completion of the project on hand, then less manpower may be allocated to this activity. In other words, there is an implicit relationship between (a) the time required to complete an activity and (b) the manpower or other resources allocated to it (see Exhibit 8.3).

A second planning concern is about hiring and firing workers. Ideally, a project manager should be able to hire employees when he or she needs them, and fire them when they are not needed. But each time a manager needs to hire or fire an employee there is a hiring cost or a firing cost involved. To minimize such hiring and firing, either to reduce the cost involved with it or to minimize the agony that causes to others, a resource balancing may be done.

EXHIBIT 8.3. Gantt (Bar) Chart

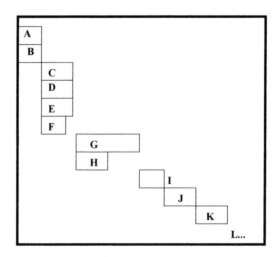

TIME, WEEKS

By carefully reviewing the manning requirements for the different activities, along with their precedence relationships, certain activities may be delayed in order to carry out more critical activities when some employees are transferred from other activities waiting to be completed (see Exhibits 8.1, 8.2, and 8.3).

3. Network Path Methods

In the case of Gantt (bar) charts, the precedence constraints were visually incorporated into the project planning. This is not always possible to do, particularly if the estimated duration of completion of an activity cannot be determined precisely, or if many activities rely on a large number of controllable and uncontrollable variables. Network path methods were developed to incorporate these added complexities.

There are two alternate, though similar, network approaches. The two approaches were developed at about the same time in different contexts. Both these approaches help project managers plan their projects in the following four ways (see Exhibit 8.4):

1. Provide Clarity. The network path methods must clearly identify the various activities and events required to complete a project.

2. Network Link. The different activities and events must be logically interrelated into a network.

3. Estimate Duration Time. Each activity and event must be defined by its estimates. Activities are assigned duration time, whereas events are assigned estimated time to start and finish.

EXHIBIT 8.4. Critical Path Method and Program Evaluation Review

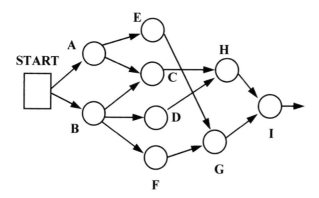

4. Identify Critical Activities and Slacks for Noncritical Activities. Different activities are classified for their criticality in completing the project at the minimum possible time. Activities which define the total completion time for the project are considered and are called critical activities. For other activities, called noncritical activities, slack times are estimated. These are the time durations for which the project completion time is not affected.

Some of the advantages of using the network path methods are as follows:

1. They reduce project completion time and reduce idle times.
2. They save project costs.
3. They facilitate smoother planning.
4. They provide indicators to coordinate with other suppliers and vendors.
5. They reduce frequent troubleshooting and crisis management.
6. The provide more time to make crucial technological decisions.

The network path methods help highlight crucial project-related information that is not obvious by simpler planning tools. These methods allow the statistical sophistication to work with probabilistic and estimated duration time.

To plan simple projects, manual coordination can be used. For complex projects, such as big construction projects, assembly of large equipment, and so on, more systematic analysis is required.

Department of Defense prefers its weapons contractors to use network path techniques and project management software which specify intermediate milestones.

Project managers and project clients wish to stay in control of the following project variables:

1. What is the most likely time of completion of the project?
2. Which of the activities must be completed on time or they will delay the entire project? These are called critical activities.
3. Activity planning schedule. When should each activity be started so that the project is completed on time?
4. What is the slack? How can the non-critical activities be delayed without delaying the completion of the entire project?

3A. Critical Path Method. The critical path method (CPM) was developed in 1958 by E. I. Du Pont, to monitor their chemical plant construction projects. This technique, though modified over time, continues to be used in the construction industry.

This method is based on a distinction between any path, defined as a series of sequential activities, and the critical path, which is a continuous path of the most optimal sequence of the project's critical activities.

In CPM there is only one deterministic estimate of the duration of completion of each activity. This method offers the advantage of not only controlling the optimum time of completion of a project, but also optimizing its total cost. For more details see Note 8.2.

3B. Program Evaluation Review Technique. About the same time when the critical path method was being developed by the Du Pont Company to monitor their construction projects, elsewhere Booz, Allen, and Hamilton consultants were retained by the Lockheed Aircraft Corporation (later Lockheed–Marietta Corporation) to implement a contract to build the Polaris ballistic missile project. In 1958 the two developed the Project Evaluation Review Technique (PERT). PERT continues to be the method of choice in the aircraft, space, and defense related industries.

PERT uses three estimates of the duration of each activity. These estimates are for the most optimistic time, the most pessimistic time, and the most likely time of duration for each activity. One definition of the most optimistic times and the most pessimistic times was determined by (a) the 1% chance of improving the most optimistic time and (b) the 1% chance of taking longer than the most pessimistic time.

PERT allows for skewed stochastic distribution of such time estimates. Project managers know that for different activities the likelihood of improving time over mean is not the same as the likelihood of doing worse than the mean. In these skewed distributions (called beta distribution), the equivalent expected times and their corresponding standard deviations can be mathematically estimated. See Note 8.2 for details.

TECHNOLOGY PROJECTS ARE LIVING, LEARNING THINGS

We have discussed some of the scientific sides of project management. But technology-related products are really more creative—like art. There is a dialectic trade-off between (a) the pursuit of control and order for better efficiency in completing a project and (b) a more chaotic and creative tolerance to create a more effective outcome of a project. While a project is being implemented, many new ways are discovered. New enabling technologies emerge. Cheaper materials become available. New learning takes place. Project managers must carefully weigh how such discoveries and their implementation in the project would improve the outcome and the desired goal for a project with new intellectual content. More control-driven managers would compromise such opportunities for improvements for a more predictable and on-time completion of projects. Enlightened sponsors of such projects would favor pushing the frontier of knowledge further—for a little additional price.

The dialectic struggle is illustrated in Case 8A on the early stages of the Manhattan Project. This project to develop the atom bomb during World

War II is perhaps one of the world's most strategic projects ever undertaken. It clearly illustrates the dynamics of managing a project.

SOME LESSONS LEARNED IN THIS CHAPTER

We defined a project by leveraging what we have already learned in the previous six chapters about production process, proprietary intellectual property, promised quality, and more. Projects such as the Empire State Building in New York City, the Golden Gate Project in San Francisco, and the Panama Canal in Panama were briefly described to illustrate the decisions involved in managing a project. A project was defined by its cost, completion time, and performance parameters. Multiple stages in a project planning process were explained along with the project management tools. These topics showed how technology projects take the life of living and learning species.

SELF-REVIEW QUESTIONS

1. Outline the progress and evolution of a project you undertook recently.

2. Discuss how you will plan your research paper for this course using the principles of project management.

3. Compare and contrast different project planning tools.

NOTES, REFERENCES, AND ADDITIONAL READINGS

8.1. Pope, Gregory T. 1995. The seven wonders of the modern world, *Popular Mechanics,* December: 48–56.

8.2. Kerzner, Harold. 1989. *Project Management: A Systems Approach to Planning, Scheduling, and Coordinating.* 3rd Edition. New York: Van Nostrand.

8.3. Role of human resource in projects is well illustrated in House, Ruth Sizemore. 1988. *The Human Side of Project Management.* Reading, MA: Addison-Wesley.

CASE 8A: TECHNOLOGY MANAGEMENT PRACTICE

Manhattan Atom Bomb Project: How Is a Project Born?

The project for developing atomic technology is an example of how a systematic scientific project, given the right kind of support and allocation of resources, can produce and provide amazing results, which, in the case of the Manhattan Project, was enormous force and control over other people. Yet, in the early stages of the Manhattan Project there was a great deal of resistance to sponsor what later proved to be a major scientific breakthrough. The support came only after the external political circumstances made it very attractive to pursue this project.

The Manhattan Project is also the story of how scientists and engineers, like mountaineers trying to conquer Mount Everest, journey forward based on the experiences of those who have gone before them.

The Final Result. On August 6, 1945, the city of Hiroshima saw a man-made sun. It was a bomb that used the fundamental principles of the sun, along with its primary source of energy. This man-made sun killed about 100,000 people, most of whom died instantly. Three days later, on August 9, 1945, the city of Nagasaki was also completely destroyed by the fury of a second atomic bomb. On August 14–15, 1945 the Emperor of Japan agreed to an unconditional surrender to the United States and Allied forces, thus marking the end of World War II and human genocide in different parts of the world. The Rising Sun of the Far East succumbed to the fury of the sun-like man-made force developed in the West.

In December 1945 at the Nobel Anniversary Dinner in New York, within 5 months of the Hiroshima and Nagasaki atomic bombings, Albert Einstein, who gave the starting signal to construct the atom bomb, spoke of his remorse like that of Alfred Nobel, the father of dynamite explosives. Einstein had sent the signal to President Franklin Roosevelt to start the atomic bomb project. He and other scientists had helped create the technology for the most formidable weapon ever known to mankind, to prevent the enemies of mankind from achieving it first. He felt that the atom bomb helped win the war, but not the peace (see sources at the end of this case).

Whereas Einstein and his fellow scientists felt a sense of responsibility and even some guilt, he said that they had chosen the development of atomic technology over letting German scientists do the same under a dictator who would not hesitate using it. Their choice was as clear as their anguish over its use.

Einstein's Request for Proposal to FDR. The two unprecedented nuclear explosions over Japan crowned America as the undisputed new leader of the world. But the efforts to develop the atomic bomb technology did not exist until 1939 when Albert Einstein agreed with a small group of scientists to draw President Franklin D. Roosevelt's attention to the potential military uses of atomic energy. Einstein and other Jewish immigrant scientists, who escaped Hitler's extermination, warned FDR about the progress that the German Nazis were making in developing this technology. President Roosevelt acted on the proposed opportunity quite swiftly. In October 1939 President Roosevelt commissioned the Manhattan Project to develop atomic bomb technology for military applications. It was a pioneering joint project between the U.S. military, industry, and university scientists.

The Prelude to Project Completion. In the spring of 1945, at the tail end of World War II, many significant events were unfolding in rapid succession. In mid-April 1945, America and the Allied Forces were shocked by the death of America's four-term president FDR before he could share the secrets of the Manhattan Project with his successor Vice-President Harry Truman. Nobody could have foreseen this in 1944, when Senator Truman was picked as FDR's running mate as a compromise candidate. Either as Senator or as Vice-President, Truman was not informed of the massive Manhattan Project conceived primarily to develop an atomic bomb. In fact, even after becoming the President, for 12 days Truman did not know about the Manhattan Project. Secretary of War Henry L. Stimson finally told President Truman of this highly secret project that could have a decisive effect on America's military strategy. Even though the primary target Germany surrendered on May 7, 1945, the members of the Manhattan Project continued to work towards the development of atom bomb technology. Truman, according to his biographer David McCullough, was in favor of using the bomb that had already cost the American people $1.8 billion to develop. This was equivalent to more than $100 billion in 1997 dollars.

The Ground-Zero Explosions. The full potential of atomic bomb technology did not become clear until July 16, 1945. At 5:30 a.m., about half an hour before the usual sunrise, Alamagordo desert air base experienced the first man-made nuclear explosion of an atom bomb. Its impact was very clear. President Truman was informed about this successfully controlled atomic explosion in New Mexico while he was attending Potsdam Conference with other Allied leaders Atlee and Stalin. With this knowledge of a superior weapon technology, the three Allied leaders issued a Potsdam Declaration to Japan, to surrender or suffer severe destruction immediately.

Nuclear Uncertainties. This was even though Secretary of War Stimson told President Truman that there was much uncertainty associated with the atomic bomb as a military weapon of any significance. Secretary of War Stimson also recommended President Truman to form a high-level policy-making committee to make the decisions about the use of an atom bomb for military applications. President Truman conceded. This Committee recommended to him that "the (atom) bomb should be used against Japan as soon as possible . . . against a military target surrounded by other buildings . . . without any prior warning of the nature of the weapon."

Humanitarian Voices. Many young scientists associated with the development of atom bomb technology opposed this recommendation on humanitarian grounds. They wanted an international organization, such as the United Nations, to oversee the test blast. They wanted the United Nations to issue a warning to Japan about the severity of a nuclear bomb. The policy committee put together by Secretary of War Stimson rejected their recommendations. The military planners believed that the surprise and severity of an atom bomb, like any other military strategy, were critical to defeat the highly stubborn and resilient Japanese military rulers.

War Drums Calling. According to some historians, many policy makers, in the face of Japanese military persistence, preferred the dropping of atomic bombs over other alternate war efforts. They feared loss of more American lives with alternatives such as a naval blockade, continued air bombardment, or even Russia's invasion of Manchuria and Japan. They believed that the decision to drop atom bombs over Hiroshima and Nagasaki were governed by these sound military considerations.

Project Deliverables: Threat Versus the Real Thing. Some historians have noted that both President Roosevelt and Prime Minister Churchill in their meeting in Hyde Park in September 1944 discussed and deliberated the possibility of actually using the atom bomb against the Japanese or merely using it as a threat. But even making a threat required the capability to carry out the threat.

In a fireside chat on the eve of his last U.S. election in November 1944, FDR might have deliberately referred to the development of an atom bomb when he proposed the possibility of using terrible new weapons to bring a lasting "end to the agony of war." He said that "Another war would be bound to bring even more devilish and powerful instruments of destruction to wipe out civilian populations. No coastal defenses, however strong, could prevent these silent missiles of death, fired perhaps from planes or ships at sea, from crashing deep within the United States."

In 1944 President Roosevelt told his office secretary Grace Tully, who was not aware of the Manhattan Project, that if the atom bomb works, "and pray God it does, it will save many American lives." In January 1945 Roosevelt told his son James that a new weapon could make the invasion of Japan unnecessary.

Let us pause and review how the Manhattan Project was planned.

Project Start: Approach Royal Belgian Queen via Friend Einstein.

As Hitler rose to power, Leo Szilard, a Hungarian physicist, crossed Atlantic and moved to the United States. He joined the nuclear scientists who were doing nuclear research at Columbia University in upper Manhattan. They had just replicated new nuclear experiments discovered by French nuclear scientists at Madame Curie Center in Paris. In 1939 Szilard and his colleague Wigner decided to contact Albert Einstein on summer vacation in Long Island, New York. They were concerned that Germany might obtain large quantities of uranium from the Belgian Congo, then the primary source of uranium. They knew that Einstein was on friendly terms with Queen Elizabeth of Belgium.

On July 15, 1939 the two scientists visited Albert Einstein. They lost their way and almost returned back to Manhattan. That would have been the end of the Manhattan Project. As a joke they asked a 7-year-old boy about Einstein. To their great surprise the boy knew where Einstein lived, and he took them to Einstein's cottage. The two researchers told Einstein about the recent developments in nuclear fission reaction. Until they mentioned it, Einstein was not aware of the possibility of a nuclear chain reaction and its potential military use. Einstein immediately understood the impact of the recent scientific developments and agreed to his fellow scientists' suggestions for blocking supplies of uranium to Germany. The three scientists discussed a number of alternate ways to implement the proposed action. Their two alternatives were to either (a) contact Queen Elizabeth of Belgium directly or (b) contact her through the Belgian Ambassador in Washington D.C., in consultation with the U.S. State Department.

Prior to visiting Einstein, Szilard had already approached a financial expert in the Wall Street district of New York City for additional funding for his experiments with Fermi, for which the Physics Department of Columbia University did not have sufficient funds. While the three scientists were meeting in Long Island, New York, this financial expert contacted Dr. Alexander Sachs, a famous economist with Lehman Brothers and an unofficial advisor to President Roosevelt. These experts from New York's financial world proposed that Einstein's letter should not be addressed to the Belgian Queen or the Belgian Ambassador but instead to President Roosevelt. Sachs even agreed to ensure that Einstein's letter would reach FDR. In the meantime, Szilard mailed a draft letter to Einstein at his summer cottage.

Szilard and Edward Teller, a visiting professor at Columbia University, called on Einstein once more in July 1939 at his Long Island summer home. Einstein agreed to their suggestions and dictated the draft of a letter in German to Teller. Based on this German draft, Szilard prepared two English versions of Einstein's letter to President Roosevelt. Einstein picked the shorter version of the letter given below:

> Albert Einstein
> Old Grove Road
> Nassau Point
> Peconic, Long Island

F. D. Roosevelt
President of the United States
White House
Washington, D.C.

Sir:

Some recent work by E. Fermi and L. Szilard, which has been communicated to me in manuscript, leads me to expect that the element uranium may be turned into a new and important source of energy in the immediate future. Certain aspects of the situation seem to call for watchfulness and, if necessary, quick action on the part of the Administration. I believe, therefore, that it is my duty to bring to your attention the following facts and recommendations.

In the course of the last four months it has been made probable—through the work of Joliot-Curie in France as well as Fermi and Szilard in America—that it may become possible to set up nuclear chain reactions in a large mass of uranium, by which vast amounts of power and large quantities of new radium-like elements could be generated. Now it appears almost certain that this could be achieved in the immediate future.

This new phenomenon would also lead to the construction of bombs, and it is conceivable—though much less certain—that extremely powerful bombs of a new type may thus be constructed. A single bomb of this type, carried by boat or exploded in a port, might very well destroy the whole port together with some of the surrounding territory. However, such bombs might very well prove to be too heavy for transportation by air.

The United States has only very poor ores of uranium in moderate quantities. There is some good ore in Canada and the former Czechoslovakia, while the most important source of uranium is the Belgian Congo.

In view of this situation you may think it desirable to have some permanent contact maintained between the Administration and the group of physicists working on chain reactions in America. One possible way of achieving this might be for you to entrust with this task a person who has your confidence and who could perhaps serve in an unofficial capacity. His task might comprise the following:

(a) To approach Government Departments, keep them informed of the further developments, and put forward recommendations for Government action, giving particular attention to the problem of securing a supply of uranium ore for the United States.

(b) To speed up the experimental work which is at present being carried out within the limits of the budgets of University laboratories, by providing funds, if such funds be required, through his contacts with private persons who are willing to make contributions for this cause, and perhaps also by obtaining the cooperation of industrial laboratories which have the necessary equipment.

I understand that Germany has actually stopped the sale of uranium from the Czechoslovakian mines which she has taken over. That she should have taken such early action might perhaps be understood on the ground that the son of the German Under-Secretary of State, von Weizsacker, is attached to the Kaiser Wilhelm Institute in Berlin, where some of the American work on uranium is now being repeated.

Yours very truly,
A. Einstein

For the role of the key liaison person, Szilard recommended K. T. Compton, Bernard Baruch, and Charles Lindbergh, with Lindbergh as the most favorite choice.

Amazingly, the letter by Einstein, the world's greatest living scientist at that time, to President Roosevelt, at that time the most important political leader of the world, on the developments of one of the most crucial technology at that time, was probably not delivered for the next 2 months. Szilard was not sure until October 3, 1939 if the letter was delivered by Sachs to President Roosevelt. In the meantime, war broke out in Europe on September 1, 1939, almost 4 weeks after Einstein had signed his letter. Sziland and Wigner got concerned that they may need a replacement for Sachs to deliver the letter.

World War II Breaks Out. As World War II broke out, in the September 1939 issue of *Discovery,* its editor C. P. Snow (later Sir Charles Snow) discussed the possibilities of a weapon based on the nuclear chain reaction. He pointed out that many leading physicists around the world, in the laboratories in the United States, Germany, France, and England, were working on a belief that within a few months a military bomb may be produced with the explosive power more than a million times more violent than dynamite. But such journalistic speculations were not new. Reporters had been making similar speculations since the 1920s. However, the belief and disbelief in nuclear energy went up and down like the waves in the Atlantic Ocean. Yet, the outbreak of World War II caused a fear among all concerned that if Hitler's Germany was the first to produce a nuclear bomb, it would not hesitate in using it against others under some pretext or another. This alone was the main reason in 1939 for the United States to consider investing large amounts of money in developing the technology for atom bomb.

Yet during this crucial stage in world history, Dr. Sachs held onto Albert Einstein's letter to President Roosevelt, and waited for an opportunity to meet the U.S. President. Finally, on October 11, 1939, Sachs met President Roosevelt and submitted Einstein's letter. He gave a more technical memo by Szilard, where he elaborated that fast neutrons could make it easier to construct extremely dangerous bombs . . . with destructive power far beyond all military conceptions. Sachs also gave FDR his own written statement, elaborating his thoughts on the U.S. exploration of the nuclear fission.

President Roosevelt acted fast. He immediately appointed an "Advisory Committee on Uranium," reporting directly to him and as soon as possible. Barely 8 days after his receipt of Einstein's letter, he wrote the following reply.

The White House
Washington

October 19, 1939

My Dear Professor,

I want to thank you for your recent letter and the most interesting and important enclosure.

I found this data of such import that I have convened a board consisting of the head of the Bureau of Standards and a chosen representative of the Army and Navy to thoroughly investigate the possibilities of your suggestion regarding the element of uranium.

I am glad to say that Dr. Sachs will co-operate and work with this committee and I feel this is the most practical and effective method of dealing with the subject.

Very sincerely yours,
Franklin D. Roosevelt

President Roosevelt set up his Advisory Committee on Uranium with Dr. Lynam J. Briggs, Director of the National Bureau of Standards, as its chairman. Most of the other members of the Briggs Committee were military officers in the U.S. Army and Navy. In 10 days they studied the potential of nuclear fission reaction and concluded that any military and naval applications were pure speculation at that time. However, they allocated a grant budget of U.S. $6000 to study the subject during 1940.

The Briggs Committee was drastically restructured in the summer of 1940. It was governed by the National Defense Research Committee, an agency under the Office of Research and Development. It also gave out many major contracts to a number of U.S. universities to carry out fundamental research on nuclear fission. Besides this, a special committee of the National Academy of Sciences was charged to inform new developments in the field of nuclear fission affecting military defense to the U.S. Government. The declassified reports of these committees indicated that from May 1940 to July 1941 the U.S. government was made aware of the possibility of atomic bomb and radioactive poisoning, but the uranium-based atom bomb was not likely to be of a strategic significance in World War II. The possibility was considered more likely in the area of using nuclear fission to generate energy. This became the U.S. government's view. On the other hand, nuclear physicists working in the U.S. universities, and funded by other means, held very different views.

The Atlantic Alliance. By the summer of 1941, the opinions of Americans started changing drastically. This was primarily a result of informal and word-of-mouth communication between the different scientists. For example, in the September 1941 issue of the *Bulletin of the Atomic Scientists of Chicago,* Arthur Compton wrote that he met Ernest Lawrence of University of California and President Conant of Harvard, where Lawrence showed them calculations proving that a powerful atomic bomb could be made using much smaller amounts of uranium-235 isotope than what was earlier considered necessary. These calculations had convinced the British scientists to proceed with building the atom bomb.

In October 1941, Dr. Vannevar Bush had a long conversation with the President in which he reported the British understanding that an atom bomb could be constructed by a diffusion plant using uranium-235. However, these conclusions were based on laboratory-scale experiments only. President Roosevelt supported such exchange of information with the British. He proposed that all such information should be kept within a small group consisting of the U.S. President, the Vice-President, Secretary of War Stimson, General Marshall, Dr. Vannevar Bush and Dr. Conant.

In November 1941, the National Academy of Sciences prepared its third report, with a focus on the potential of an explosive fission reaction with uranium-235. Dr. Vannevar Bush attached a cover letter. There he pointed out that the Academy report was more conservative than the British report because the Committee had some hard-headed engineers besides some eminent scientists. Therefore compared to the British estimates, the new report estimated that the atom bomb was likely to be less effective, cost much more, and take longer to start production.

The British observers believed that what happened later in America did not involve many major technological leaps (see Clark, 1961) at the end. Compared to the British counterparts working in the middle of a raging war, the American physicists and engineers, away from the war, were able to contribute enormous energy and resources to pursue different lines of progress.

In late October 1941, two powerful and highly respected American physicists, Professors Pegram and Urey, were sent to see the British nuclear fission program. The British scientists and engineers were told to show them everything. They visited many scientific sites in London, Liverpool, Bradford, Birmingham, Cambridge, and more. They learned many new things—for example, use of neutrons without moderators. American research was primarily focused on slow neutrons. They also saw the isotope separation process, and the allocation of resources to the entire technology development effort was explained to them. American physicists were very impressed by the progress made by their British counterparts, working in the middle of the battlefield.

By the time the visiting American physicists returned and filed their report, the opinion of the Academy Committee had also turned around. They set aside their doubts expressed in their earlier reports. In their November 6, 1941 report they proposed that the possibility to build an atom bomb must be seriously considered by the U.S. government. They also believed that in a few years, atom bombs such as those based on uranium fission may provide a nation significant military superiority over other nations.

Based on the different reports, Bush, then the Director of the U.S. Office of Scientific Research and Development, decided to demand more resources for a program to expand research in nuclear fission. The decision on this request was made on December 6, 1941, barely hours before the Japanese attack on Pearl Harbor.

Pearl Harbor. The Pearl Harbor attack instantly forced America to become an active player in World War II. In a fortnight a much larger and reorganized Academy Committee was listening very intently to visiting physicists Pegram and Urey's report of the British nuclear program. In the early months of 1942, the emphasis of nuclear technology shifted from scientific interest to an industrial-scale military execution. The rest is history. Einstein, like everybody else, was later shocked by the power of the atom bombs produced by the Manhattan Project.

Remorse of a Researcher. Since the dropping of atomic bombs over Hiroshima and Nagasaki, for a decade until his death, Albert Einstein worked very hard for world peace. He wished to create a supranational world government organization of the United States, Great Britain, and the Soviet Union. He proposed that this supra-national government should be charged to monitor and oversee control of the bomb so that it would be never used again. But he refused to give that charge to the United Nations Organization, at that time a young organization. He did not think much of it because of its structure.

Albert Einstein did not consider himself the father of the release of atomic energy. He claimed that his part was quite indirect. He only contributed in that he believed that it was theoretically possible to unleash atomic energy. Atomic energy became practical through the accidental discovery of a chain reaction observed by German physicist Otto Hahn. But he could not interpret the experimental findings correctly. The chain reaction in nuclear fission was explained by Lise Meitner, a German refugee physicist who shared this with Niels Bohr in Stockholm.

Environment for Discoveries. Based on his lifelong experience with discoveries, Albert Einstein believed that organizations can help in the application of a discovery. But he believed that the organizations cannot organize the scientific discovery itself. He usually gave the example of Charles Darwin's discoveries. He postulated that organizations could not make such discoveries. Einstein believed that discoveries can be made only by free individuals. Organizations only help to the extent that they can ensure the freedom and working conditions conducive to discovery of new ideas.

Einstein also believed that because atomic energy was developed by the U.S. government, only the American government must control the atomic technology. He feared that atomic energy could lead to wars even more easily than capitalism did, because it represented even greater concentration of power for some people over others.

He cautioned that he did not foresee atomic energy to be a boon but a menace to mankind in the near future. However, he believed that its fear could act as a deterrence and lead to some sort of peace. Albert Einstein died on April 18, 1955. For 10 years he had lived with the devastating effects of the nuclear energy unleashed by his theoretical equations on the people of Hiroshima and Nagasaki in August 1945.

Source: Adapted from Nathan, Otto, and Norden, Heinz, eds. 1968. *Einstein on Peace.* New York: Schocken Books, p. 355. Einstein gave an interview to Raymond Swing that was published in the *Atlantic Monthly* in November 1945. This appeared in the Nathan and Norden book. Also see Clark, Ronald W. 1961. *The Birth of the Bomb.* London: The Scientific Book Club.

CHAPTER 9

Pioneering Vision and Leadership

A PREVIEW: WHAT IS IN THIS CHAPTER FOR YOU?

During the early 1990s, many technology-driven enterprises saw their top-level leaders change or leave. In the Japanese *Kaisha* (meaning company), many strategic initiatives bubble up from a bottom-up *Ringi-sho* process. On the other hand, in Corporate America, because of the relatively high compensation of the top leaders, a large number of significant strategic initiatives are formulated or sponsored by the top-management leaders. These are cascaded down for implementation. In the technology-driven enterprises, the top-level leaders play a critical role in the strategic choices and direction-setting for their enterprises. They make crucial choices defining their organizations' future. Yet many leaders focus their efforts on the short-term and the day-to-day transactions. They are therefore referred to as the *transactional leaders or managers.* On the other hand, other top-line driven leaders focus their efforts on the long-term opportunities for their enterprises. They develop new vision and are called *transformational leaders.*

This chapter discusses the different leadership styles of top-management leaders under the static and the dynamic competitive conditions. In the 1990s, the technology-driven organizations and their leaders faced immense pressure from the outside as well as the inside stakeholders. The transactional bottom-line-driven managers operated as agents of stockholders, meeting their principals' expectations of drastic reductions in expenses. The transformational leaders, on the other hand, were driven by potential new contributions to the top lines of their financial statements. They developed their original visions and sought out new opportunities for growth.

Traditionally, under relatively static market conditions, the corporate leaders used strategic planning process to formulate plans and strategies. They then developed their organizational designs to implement the formulated plans. In the large organizations or organizations facing dynamic technological changes, the leaders must craft their strategies in an interactive and negotiated manner. For the fast-shifting market conditions, leaders must develop core-unique competencies of their organizations to gain sustainable competitive advantage. Whereas the transactional bottom-line-driven managers perform well in the short run, focusing on cutting expenses, the sustained competitiveness and growth of an organization over longer periods requires proactive entrepreneurial risk-taking by transformational leaders.

The transactional and the transformational leaders have different leadership styles. They differ in their missions and their methods for implementing their missions. They also deploy different sources of power. The pull sources of power derived by transactional leaders from their organizational position do not last as long as the push sources of personal power deployed by the transformational leaders. These and the competencies of corporate leaders competing for the future of their enterprises will be discussed in this chapter.

CEOS: THE DISAPPEARING DECISION-MAKERS

The early 1990s were pretty rough on the corporate leaders in America. An epidemic seemed to be spreading fast among the upper-echelon leaders of the technology-driven firms. Let us recall some changes. On October 26, 1992 Robert Stempel, the top leader of General Motors, was leaving that position. Three months later, on January 26, 1993, the Chief Executive John Akers of IBM left that company. This was a day after James Robinson III of American Express decided to leave; and it was a day before Paul Lego, leader of Westinghouse Electric, left his employer. These senior leaders were later joined by their other counterparts: CEO Kay Whitmore left Eastman Kodak on August 6, 1993; John Sculley stepped down as CEO of Apple Computer on June 18, 1993; and CEO Anthony D'Amato left Borden on December 9, 1993. Similarly, Vaughen Bryson of Eli Lilly and Philip Lippincott of Scott Paper also changed their jobs in June and December of 1993, respectively. Many firms and their boards of directors chose to go outside their home-grown rank-and-file talent, and even outside their industries, to fill such coveted high chairs for top leaders. What was the reason behind this turmoil in corporate leadership? (See Note 9.1.)

Multiple Revolutions in Markets

One of the reasons for this chaos in leaders' chambers was that the markets in the early 1990s were in a state of intensely dynamic competitive conditions. In a historical convergence, a number of revolutions were taking place at the same time in customers' expectations, in competitors' interference, and in the development of critical technologies and regulatory conditions (see Note 9.2). According to another *Fortune* magazine article, "the competitive rigors of the new economy—the demand for speed and global reach, the struggle to respond to the increasing oomph and persnicketiness of customers, the need to exploit information technology—are too big and too urgent to be addressed by anything less than changes on grand scale" (see Note 9.3). Such megashifts in expectations from senior leaders were not just limited to the computers and communications industries. These were also observed in many other industrial sectors as well, such as automobiles, airlines, financial services, and retailing segments of the economy.

These profound changes in market competition made enormous demands on the roles and responsibilities of the top leaders of the U.S. organizations. In the past, the commanding Chief Executive Officer (CEO) of a corporation was above board and beyond any questioning. The CEOs frequently fired their vice-presidents, or they explained away their organization's poor performance on recession, economic cycles, unfair foreign trade, and so on. They themselves were rarely accountable for their firms' poor performances. Somewhere during the 1990s, things changed dramatically. Under chaos-like competitive market conditions, many of the top executives were expected to make enormous changes quickly. Incremental and risk-free slow changes were no longer effective and had to be ruled out quickly. According to CEO Jerre Stead of AT&T Global Information Solutions (GIS), "if you do incremental changes, you'll never get there."

Unfortunately, some top executives adapted to such megashifts in competitive conditions by readily resorting to cheerleading-like behavior. They wanted to please their outside constituencies. They resorted to myopic short-term decisions to cut costs and boost bottom lines—as demanded by their short-term oriented stockholders. Too many top executives holding leadership positions at the top of their organizational hierarchy either completely neglected or paid little attention to the "top lines" or the sources of generating revenues for their balance sheets.

Purpose of Corporate Leader. What is the primary reason for hiring the top executives in big technology-driven organizations, with big compensation packages, huge bonuses, and golden parachutes? Should the leaders take major risks to lead their organizations in new directions, or should they take a safe incremental approach? To improve the bottom lines of their firms, where should their efforts be focused? Should they focus on incremental cost-cutting, or should they seek to pioneer new directions by transforming their businesses into new forms and adding new revenue streams to the top line? Are senior executives hired to identify and implement radical and strategic actions, adapting more appropriately to shifts in their competitive environments? Or are they to manage incrementally by doing the business as advised to them by outsider stakeholders? These are critical questions for fast-changing technology-based corporations.

Historical Antecedents

In 1932, economists Adolf Berle and Gardiner Means wrote in their ground-breaking book, *The Modern Corporation and Private Property,* that the American capitalism had evolved from its entrepreneurial phase in earlier years, to a new corporate phase. Over two-thirds of the nation's wealth had shifted from the individual private ownership to a public-financed ownership under a corporation structure. This also shifted the power from multitude of invisible individual stockholders to the professional manager executives. Managers had thus evolved from being the mere hirelings of rich and powerful entrepreneurs. These professional managers, such as Alfred Sloan of General Motors, acquired significant power of their own.

In *The Wealth of Nations,* Adam Smith pointed out that the "invisible hand" operated the organizations managed and owned by the same person. If not, the professional manager was not likely to manage other people's money with the same vigor as he or she would have if it were his or her own money. The professional management of corporations has worked and survived for many decades. During this period the compensation of the chief executives also steadily skyrocketed. Over time, some leaders became increasingly insulated and less accountable for the fluctuating performance of the organizations they managed. Golden parachutes were invented side by side with multiple and cross-membership of different boards of directors. Corporate leadership became an "old boy's network," self-perpetuating and sustaining the members' well-being. Economist John Kenneth Galbraith of Harvard University noted that the big public corporations had evolved into self-sustaining, powerful bureaucracies—for example, General Motors. Many executives at General Motors often claimed that "If it was good for GM, it must be good for the U.S. of A."

To remedy these, some companies' board of directors tied their top-level leaders' compensations, particularly stock options offered to them, to their firms' performance in the marketplace. Gradually the corporate leaders became the agents of the investing principals. Unfortunately, the leaders' positions were not linked to other measures, such as the organizations' top-line growth or the employees' well-being.

Stockholding Concentration

Peter Drucker, in his 1976 work *The Unseen Revolution,* noted that in the early 1970s the shareholding of public corporations had become less diffused and more concentrated in the hands of a few big investors. In 1970 the institutional ownership accounted for 19% of the equity raised. In 1994, this had increased to about 50% overall. For example, the institutional owners owned about 75% of equity of the American Express. In the mid-1970s, Drucker postulated that the institutional investors will remain passive and will let the professional managers and leaders manage their businesses autonomously. This changed drastically by the 1990s.

The giant pension funds were the first ones to get concerned with the social responsibilities of leaders and managers in upper echelons of business corporations. For example, they significantly exercised their financial muscles by barring many U.S. companies from doing business with the apartheid South Africa. They forced executives to be more vigilant and to avoid polluting oil spills such as that of the Exxon Valdez in the William Sound in Alaska.

Gradually, however, the interest of institutional investors shifted from social responsibility of a corporation, to its very day-to-day economic governance. And these institutional investors twisted the arms of managing top leaders through their boards of directors.

Corporate historians believe that the turning point came in October 1992 at General Motors (GM). In July 1990 GM's Chief Executive Roger Smith retired. The institutional investors wanted GM to pick an outsider, but the heir successor Robert Stempel was appointed as the next Chief Executive. CEO Stempel and GM's board of directors bestowed on Roger Smith additional bonuses, even though GM had lost significant market share during Smith's tenure. The institutional investors considered the company's actions as squandering away billions of dollars on "misguided automation." This was the GM tradition. As GM persisted with record-breaking losses in the billion-dollar range, the pension fund holders tried to oust Chief Executive Robert Stempel. The California Public Employees Retirement System, with an $80 billion investment kitty, aggressively went after GM and its lack of independent directors on its board.

The institutional investors also got scared with IBM's steep fall in stock price. They pressured Chief Executive John Akers with regard to his compensation, and they pointed out the lack of outsider directors on IBM's key committees. Akers left under persistent external pressure. Similarly, at American Express, between 1989 and 1992 the profits fell by 62%, and between 1989 and 1992 the stock price declined by 50%.

During these few months, the institutional investors made deep inroads into corporate governance and its leadership. Is the separation of management from ownership of an enterprise good or bad for business? Should the source of leadership of a corporation reside outside its hierarchy? The shareholders in America are known to be impatient and short-term-oriented. They may not care much for the management's long-term action plans for corporate vitalization and sustained competitive advantage (see Note 9.4). Who will control their greed?

According to Edward Regan, who until 1993 was the head of the New York State and Local Retirement Systems, and America's second-largest public pension fund,

> The last thing in the world that you (as an investor) want to do . . . is to discourage chief executives from taking risks. So while there is a nice, healthy tension now (between the institutional owners and the corporate managers), there's a danger of going too far (towards short-term orientation of owners).

TRANSACTIONAL LEADERSHIP

Wall Street's Concerns

To some investment analysts on the Wall Street, fixing the bottom line of firms through radical and rapid expense cutting seems to be the paramount managerial responsibility for top-level leaders. However, to arrive at a quick and healthy "bottom line," these Wall Streeters can only look to squeezing the firms' expense items. They overlook altogether the starting point—that is, the top line of the business enterprise. The top-level leaders are hired in the first place to identify new entrepreneurial opportunities. This was particularly true for the technology-driven organizations facing dynamic changes.

1. Preference for Outsider Leaders. Because of the earlier mentioned historical developments in Corporate America, the stock analysts called 1993 the year of the Outsider CEOs. Many firms reached outside their organizations and even their industries to make their big leadership appointments. Corporate directors believed that outsider CEOs, without any allegiances to any of the entrenched factions in an organization, can "breathe fresh air into organizations" and "shake a company out of its rut" (see Note 9.5). This belief led to the appointment of outsider CEOs at corporate giants, such as Louis V. Gerstner of Nabisco at IBM, George M. C. Fisher of Motorola at Eastman Kodak, Randall L. Tobias of AT&T at Eli Lilly, Charles W. Harper of ConAgra at RJR Nabisco, and many more. The year 1993 marked over 35% of the new CEOs appointments from outside their organizations, the highest-ever level since 1949.

2. Popularity of Cost Choppers. Why was there such a strong desire to bet the barn on an unknown rather than a known leader in upper echelons? Wall Street often appreciated the appointment of outsiders, because they coldly chopped their expenses and boosted additional profits into their bottom lines. For example, Lawrence A. Bossidy at Allied-Signal and Stanley C. Gault at Goodyear Tire & Rubber Co., started wielding their cost-cutting axes as soon as they took their new jobs in 1991. To restive and short-term-oriented stockholders the appointment of outsider CEOs by board of directors became a surrogate for a significant strategic change within an organization.

3. Loss of Leadership Capital. Typically, in the technology-driven firms where outsider CEOs were appointed as leaders, the key executives who were bypassed for the outsider CEO's appointment would often leave the company. If they were good at their trade, particularly in managing technology under dynamic changes, they themselves were often hired by other organizations as outsider CEOs. Their previous employers lost significant resource competence, which was much needed for its competitiveness in the dynamic marketplace. Often this came at a time of crisis when the organization could use these key leaders' familiarity with the organization's internal and external environments.

4. Blunders in a Hurry. Sometimes such quick returns to stockholders were masked by drastic blunders by outsider CEOs. These leadership mistakes had adverse long-term implications on their firms' competitiveness. Shareholders were promised double-digit growth by slashing R&D or marketing expenses or raising the prices of products. These resulted in sharp erosion of their market shares. Sometimes the board of directors realized such gimmicks and then booted out the outsider CEO, who then became available for appointment at another short-term-oriented myopic organization. There never seemed to be a shortage of U.S. firms seeking a quick-fix troubleshooter from outside.

5. Some Sensible Shifts. In 1994 there was some change in the stereotypical short-term thinking about top-level leadership in Corporate America. Some outsider CEOs, during and after their appointments, pointed out that they were not going to be able to deliver instant improvements in their firms' bottom-line results. CEO Michael H. Jordan informed Westinghouse shareholders that because of the outdated mature businesses (such as the nuclear power plant construction business), their quarterly earnings would fall before they recovered over a longer time horizon. Some of the new CEOs arriving from outside were also resolved to building (rather than chopping down) their new home bases. The outsider CEO George Fisher vowed to take Eastman Kodak for a new look at silver halide photography, and he attempted to build a better future in digital imaging (see Note 9.6).

Myths About a Transactional Manager's Job in a Dynamic Economy: Planning, Organizing, Coordinating, and Control

In 1916, Henri Fayol proposed that in business organizations the managers perform four specific functions. These four functions were planning, organizing, coordination, and control. Until recently, there were widespread beliefs that a manager's work still involved these four functions. Managers were reflective and systematic planners. Managers primarily organized others' efforts. An effective manager had no regular day-to-day duties to perform and had all the time needed to do their reflective planning and organizing. Another important role they played was controlling. Managers in senior positions controlled others by using aggregated information, best provided by management information system. And finally, there was a general belief that managers did coordinating. They coordinated the different contributions of their subordinates in a systematic manner. Based on this understanding, management was believed to be a systematic science requiring precise analysis.

In 1975, Professor Henry Mintzberg of McGill University tested this understanding for an empirical validation. (see Note 9.7). He and his research team questioned manager respondents. They shadowed managers in the United States, Canada, the United Kingdom, and Sweden, and they included five U.S. CEOs. This study included foremen, factory supervisors, field sales managers, hospital administrators, presidents of companies, and so on. The researchers studied the managers' work by studying the regular work flow, or how long the managers worked on different tasks. They also studied when and how the managers were interrupted, who interrupted them, and their mode and medium of communication. The researchers spent about an hour interviewing each of their respondents. The researchers also studied the work content of the respondents, the activities managers did, and why they did these activities. Based on these investigations, the researchers discovered that the facts were very different from the prevailing beliefs (see Exhibit 9.1).

1. Planning. The first belief was that the manager worked in a reflective, systematic manner. Empirical investigation indicated that the managers worked at an unrelenting pace. Their activities were best characterized by brevity, variety, and discontinuity. The managers were strongly action-oriented, and many of them disliked reflective activities. More than 50% of the activities of a typical manager lasted less than 9 minutes. Only 10% of the activities of the investigated managers lasted more than 60 minutes. The research sample of 56 foremen did 583 activities in 8 hours, with each activity taking on average 48 seconds.

2. Organizing. There was a belief that the leaders and managers do not have regular duties to perform. They do planning and organizing all the time. Researchers found that this was a myth. Besides handling exceptions, managers and leaders also performed a number of duties routinely, including attending formal rituals, ceremonies, and negotiations. They processed the soft information from within the organization and from the outside environment. This bound the organization together.

3. Controlling. It has been generally believed that the managers and leaders in senior positions thrive on aggregated and formal information. They control their organizations or departments by management information system (MIS). Many reports come to them, and they write reports to others. Controlling was considered to consume a large part of a manager or leader's work day. Focused investigations indicated that the managers strongly favored person-to-person verbal communication over written reports. They processed large bits of information collected informally. This helped them form their overall image about the situation facing them from time to time. Such integration was hard to delegate to others.

4. Coordinating. It was generally understood that the managers and leaders systematically analyzed the challenges around them and took carefully considered actions. Observations of the leaders' and managers' work indicated that many managerial tasks and leadership initiatives for decisions were locked deep inside their brains. They did not coordinate but used their judgment and intuition to develop a Gestalt, or an integrated view of the whole. They reflected on their challenges and often worked alone in implementing key elements of their missions.

Facts About a Leader's Job

According to the above-mentioned team of researchers, a person in charge of an organization or its subunits played many different roles. They identified about 10 different roles, which could be classified in three separate groups. These three clusters were (A) the interpersonal roles, (B) the informational roles, and (C) the decisional roles. (see Exhibit 9.1).

A. Interpersonal Roles. These roles involve leaders and managers interacting directly with other people.

1. Figurehead Role. Managers and leaders must carry out many ceremonial duties. This, researchers noted, took about 12% of their time and effort. Of their mail, about 17% was for acknowledgments and requests related to their status. In this figurehead role, managers and leaders exchanged little serious communication and made no important decisions. They acted as the symbols of their organizations.

EXHIBIT 9.1. Roles Executives Play

Traditional Myths	Modern Facts
1. Planning	**A. Interpersonal Roles**
2. Organizing	1. Figurehead role
3. Controlling	2. Leadership role
4. Coordinating	3. Liaison role
	B. Informational Roles
	4. Knowledge-base-role
	5. Monitor/disseminator role
	6. Spokesperson role
	C. Decisional Roles
	7. Entrepreneurial role
	8. Troubleshooter role
	9. Resource allocator role
	10. Negotiator role

Source: Adapted from Mintzberg (1975). See Note 9.7.

2. Leadership Role. Managers and leaders lead others directly by getting involved in hiring and training of new and old employees. They also provide indirect leadership by motivating and encouraging employees to work towards their organizations' goals and vision for the future.

3. Liaison Role. Managers and officers often acted as a liaison with a wide variety of people, many outside the vertical chain of their commands. They spent about 10% of their time with their superiors and spent 45% each with subordinates and peers or outside units.

B. Informational Roles. These roles involved use and flow of information. Leaders and managers collected information and disseminated information.

4. Knowledge Center Role. The leaders and managers often knew a lot about their organizations. They often got information from many sources inside and outside their organization. They also had long years of experience in the organization or in their profession.

5. Monitor/Disseminator Roles. Leaders and managers monitored a lot of information regarding the actual performance of the various subunits under their control and supervision. They compared the monitored information with the corresponding targets. Leaders and managers also play the role of a disseminator. They disseminated either the compiled information or new directives of their own to their subordinates.

6. Spokesperson Role. Leaders and managers acted as spokespersons of their organizations to those outside their organizations. For example, the Chief Executive Officer is a spokesperson of the organization to the company's board of directors. They also addressed the analysts and the representatives of the investment community. In annual meetings the CEO and his or her team of top-level managers and leaders addressed the shareholders and other community members.

C. Decisional Roles. The third set of roles involved decision-making by the managers and leaders. They were hired and compensated heavily to make the tough decisions.

7. Entrepreneurial Role. Leaders and managers must interface with shifts in external environments and adapt their units to changes in environmental conditions. They looked for emerging opportunities in the environment. They urged and initiated innovations, so that the organization was able to exploit those opportunities.

8. Troubleshooter Role. Leaders and managers regularly handled disturbances within the organization, such as the workers' strikes and bankrupt customers. When there were hurdles in the way of achieving targets, the leaders and managers helped and intervened to remove those hurdles.

9. Resource Allocator Role. Managers and leaders allocated resources of their organizations. They were caretakers and agents of these resources on behalf of the investors as principals. The corporate resources included the use of their own time. Often resource allocations involved finding funds for expenses beyond planned budgets.

10. Negotiator Role. Leaders and managers spent a lot of their precious time in negotiating contracts and relationships. They negotiated with different constituencies/stakeholders of their organizations. Thus, they managed the expectations of others for their organizations. Often these expectations were mutually inconsistent. For example, the society expects contributions to charities, while customers expect low-cost products, and the government expects the organizations to abide by their laws. The Environmental Protection Agency (EPA) wanted auto makers to produce automobiles with low or no emissions in auto exhausts. But producing an automobile with no emissions would cause the large auto makers to charge additional amounts to the customers, which the customers may not want to pay for. The CEOs, managers, and leaders of the auto makers therefore must negotiate for the realistic emission levels which customers are likely to be willing to pay for.

Role Integration

These 10 roles of managers and leaders are not separate, but form a Gestalt, or an integrated whole. Different managers and leaders have different preferences and may pay different attention to different roles. Production managers tend to place more importance on the decisional roles, and they spend more time smoothing the work flow. Staff managers devote more attention to their informational role, and sales managers focus more on their interpersonal roles.

TRANSACTIONAL MANAGERS VERSUS TRANSFORMATIONAL LEADERS

Transformational leaders do things differently from the transactional managers. A transactional manager is obsessed with doing the right thing. A transformational leader, on the other hand, is concerned about doing the right thing. Therefore, the managers measure and are driven by improving efficiency, whereas the leaders strive to improve the effectiveness of their actions. A transactional manager may continuously demand extensive and detailed plans, along with a strict adherence to them—to the last dime, nickel, and penny. Thus a manager will devise devious control mechanisms to monitor every single action of every manager. The transformational leaders are more interested in setting the overall direction for their organization's endeavors. They genuinely urge employees to give their best, or suggest changes for further improvement.

Transactional managers relate with others based on their official titles. The transformational leaders respect and admire the skills and the competencies others contribute. The transactional managers tend to be risk-aversive, whereas the transformational leaders see opportunities where risks exist. The transactional managers tend to penalize people for their failures. The transformational leaders are concerned when people are not making enough mistakes. They encourage others to experiment and take risks. Every failure is converted into a significant step toward learning and toward future success.

This distinction between transactional managers and transformational leaders raises important questions. Are transformational leaders always preferred over the transactional leaders? Or are managers and leaders preferred under different circumstances? When will the manager-style supervisor work better than the leader-style supervisor?

Based on the above discussion, the transformational leadership can be defined as the art and craft of creating a vision for the organization, sharing and discussing it with other members of the organization, and having the influence to translate the shared vision into reality over a sustained period of time (see Note 9.1).

Future-Focused Renaissance Leader

Much has been written in strategic management and total quality management literature about the critical need and characteristics of leadership for organizations facing intense competition. Leaders are expected to play different roles under two different approaches to gain competitiveness. The role of leadership in the transactional management approach tends to stress inward-looking management control. Such executives are driven by the efficiency and cost-cutting objectives of their organizations. He or she may also be risk-aversive. On the other hand, the role of leadership under the strategic management approach to gain competitiveness and change is, by definition, outward-looking and more integrative. Such a transformational leader is driven by effectiveness criteria and is often willing to take some calculated risks. Under the panicky investment climate of the 1990s, the latter was far less commonly found and tolerated than the former.

New Leaders at IBM and Kodak. During the mid-1990s, two major American corporations saw change of leadership in their upper echelons. At Eastman Kodak, leadership changed from Kay Whitmore to George M. C. Fisher, while the I.B.M. board of directors brought in Lou Gerstner to replace John Akers. The two new leaders offered a great contrast in leadership. IBM's CEO Gerstner persisted that IBM did not need a new vision.

Instead he preferred a more systematic control of its internal transactional processes. On the other hand, CEO George Fisher hit the road running, selling his new vision for Kodak in digital imaging to his potential employers, even while they were still interviewing him for a potential appointment at Eastman Kodak (see Note 9.6).

According to the historian James MacGregor Burns, who has studied many political leaders, there is a clear distinction between the two different types of leaders: the transactional leaders and the transformational leaders. The "transactional leaders" set modest goals, enlisting widespread populist cooperation, in exchange for immediate gains meeting basic needs. In organizational settings, these "transactional leaders" may be called "scientific" managers. On the other hand, the "transformational leaders" go for "stretch" goals. They often achieve these by appealing to people's potential for satisfying higher needs, such as the individual leaders' need for high achievement, prestige, and so on. Transformational leaders give higher priority to effectiveness and market fit (for details see Note 9.1).

Next we will discuss the different styles of leadership.

TOP-EXECUTIVE LEADERSHIP STYLES

Michael Maccoby, author of *The Gamesman,* suggested that there are four characteristic types of top managers in large corporations. These four types of top executives are: the craftsman, the company man, the jungle fighter, and the gamesman (see Note 9.7). Maccoby's model of different leadership styles can be extended by considering its basic underlying dimensions.

A Leadership Typology

A careful analysis of Maccoby's typology of four modes of leadership styles of top executives and leaders, discussed earlier, shows that these four different top-management leadership modes are anchored in two basic characteristics of leadership. These are discussed below.

A. Leader's Mission. Different leaders define their missions differently. The primary mission of a leader may be toward sustaining past successes for meeting present challenges. Or the mission may be to provide a vision for a better future (see Exhibit 9.2).

B. Leader's Methods. Different leaders use different methods for implementation of their missions. The leaders tend to either (a) focus primarily toward internal strengths and weaknesses to meet their purpose or (b) stress the external opportunities and threats to build their vision.

Based on these two leadership dimensions, different leaders can be classified into six types. Remember that real-life leaders are hybrids.

EXHIBIT 9.2. Leadership Styles

	Preserve Past	Promote Present (Short-Term)	Progress for Future (Long-Term)
External-Oriented	Fighter	Gamesman	Architect
Internal-Oriented	Archival	Craftsman	Engineer

Leader's Methods (vertical axis, External-Oriented top, Internal-Oriented bottom)

Leader's Mission (horizontal axis: Preserve Past, Promote Present (Short-Term), Progress for Future (Long-Term))

Type 1. The Archival-Type Leader. These leaders thrive on building their "traditional" institutions. They live and thrive on the historical archives and past achievements of their organization. This company-person type of leader may promote status quo and a career for himself or herself, while compromising or suboptimizing the overall organizational goals. Some planners and quality inspectors of yesteryear behaved in this manner when they became leaders.

   John Akers as CEO of IBM assumed that the computer pioneer enterprise would grow to be a $100 billion company in a year with mainframe computer technology, when actually IBM reached only about two-thirds of that target. For many years, after Apple's introduction of the desktop computers, the leaders and top managers at IBM continued to believe that the customers would return to the mainframe computers, IBM's historical strength.

Type 2. The Craftsman-Type Leader. The craftsman-type leaders are motivated to build masterpieces of high quality, and they focus on severe control and inspection of the current processes needed to deliver the same. They pursue perfection no matter how much time and effort it takes. The common weakness of this type of executive is his/her inflexibility and insistence on using "one best technical solution." Sounds familiar? Many former quality managers and technical development experts seem to follow such a pattern of leadership style.

Paul Allaire and his team of cautious top-level managers and leaders at Xerox outsourced a number of corporate functions (such as payroll data processing) to continue to run a highly profitable photocopier manufacturing and servicing company during the 1990s.

Type 3. The Fighter-Type Leader. The fighter-type leaders thrive on power, strength, and a ruthless need to dominate over others—inside and outside their organizations. With their mission to preserve their glorious past heritage, they fight to regain what is rightfully theirs. To them, winners should win and losers should be annihilated. They think they are brave, but such domineering and ruthless intimidating authority can stunt independent and creative thinking of others. This tendency may also arise out of their own fear for the unknown. These leaders therefore lack a sense of adventure and are therefore incapable of leading organizations facing highly competitive and innovation-driven market conditions.

The predecessors of George Fisher of Eastman Kodak strongly felt that Kodak was not getting the market share that it deserved and that it had in the past. For many years, the leaders and top managers at Eastman Kodak assumed that the demand for photographic film markets grew at an annual rate much faster than it actually did, and they also assumed that global markets (especially in Japan) were not accessible to them to the extent that they should be.

Type 4. The Gamesman-Type Leader. Gamesman-type leaders thrive on taking calculated risks. They are excited by new ways of attacking current tough challenges. They thrive on change, and they enjoy winning under tough and adverse odds facing them in the market. Their weakness lies in sometimes fantasizing about what may seem to others as too rash. They may risk everything in the pursuit of a rash goal or a rival.

Michael Dell of Dell Computers took a big gamble on competing with IBM by marketing his computers by 1-800 direct-mail telephone lines. He was therefore able to cut significant distribution and other intermediary costs.

Type 5. The Process Engineer-Type Leader. The engineer-type top-level leader is oriented toward controlling internal processes to minimize losses due to deviations in finished goods and services. They offer high organizational and process stability, but are less sensitive to external shifts in environmental changes and special causes. Like the engineers of the Brooklyn Bridge or the Empire State Building, the engineer-type leaders are very detail-oriented.

Bob Galvin and his team of leaders and top-level managers adopted six-sigma-process reliability to drive their quality as well as strategic efforts at Motorola. Jack Welch, the CEO of General Electric (GE), also adopted the six-sigma goal and journey as the great challenge and opportunity for GE. GE chose this engineering-type rigorous approach for process quality for a number of reasons. Their customers demanded it, their competitors had already adopted the same, and their future gains in productivity and inventory turns depended on it. (see Note 9.9).

Type 6. The Architect-Type Leader. On the other hand, the architect-type leaders are oriented and driven by their new vision. They try to foresee the future changes in environment, and they act proactively. They use their perception of these shifts to satisfy and delight their demanding customers' changing expectations. While doing so, they use the fundamental principles of success. They adhere to regulatory codes and to the natural laws of nature, engineering, and business. Yet they are keen on creating something new and original. They are more likely to succeed in exploiting untapped opportunities in the changing marketplace. They value agility, and they rapidly adapt their responses to shifts in external environment and regulatory requirements. However, while doing all this, the architect-type top executives may turn too dreamy. They can become extremely spendthrift, building their utopian visions. The conservative investors may feel uncomfortable with radical architects. Yet different times require different solutions. The architects invariably tide over their hurdles with their persistence.

Fred Smith, the founder of Federal Express, developed a dream based on the hub-and-spoke system for a highly dependable service for overnight parcel delivery. To do so, he created a new vision based on a research paper he wrote while studying at Yale University. A low grade on his college paper, along with many financial hurdles on the way to starting his first-of-its-kind company, did not deter Fred Smith from giving up his dream. This is described in detail in a case (Case 4B) in Chapter 4 on new product development and customer trust in technology-driven enterprises.

STRATEGIC LEADERSHIP: PROPERTIES AND PROCESSES

Based on the discussion above, strategic transformational leaders can proactively monitor external environmental shifts as well as competitors' actions and thereby bring about cross-functional changes that add value for their customers and investors. One of the most significant changes is to achieve positive changes in the human intellectual capital of an enterprise.

Strategic leaders can be defined by the properties or the traits they possess and by the processes they use to bring about changes.

A. Properties of Leaders

According to the leadership traits model, people with certain traits are likely to be more effective leaders than the people who do not possess those traits. Many researchers have been observing a large number of leaders, both effective and not so effective, and have compiled a list of traits that may explain whether a new leader will be effective or not. After identifying hundreds of traits of successful leaders, researchers were still unable to use them in successfully predicting the success of new leaders based on these traits.

Despite this overall failure, the behavioral scientists have recognized four traits that are common in most, if not all, successful leaders. These are:

1. Intuitive Intelligence. Effective leaders have a high level of intuitive intelligence. They have a superior ability to see connections across apparently dissimilar subjects and events that many of their counterparts are not able to see. Because of this X-ray 20/20 vision, the leaders are able to see opportunities which many others around them miss. Some of this intuitive intelligence is acquired by disciplined effort and persistence.

2. Emotional Maturity and Breadth. Transformational leaders are generally emotionally mature. They can tide over ups and downs and stay focused on the distant goals. Any significant transformational vision and change is likely to attract criticism from some quarters. The leader must have a strong faith and conviction in his/her vision to remain on track. Emotional maturity helps the leader in this regard.

The leaders also tend to have a broad breadth of interests. This helps them relate to a wider group of people and their concerns. Narrowly focused leaders cannot do that. Because of their breadth of interests, more people can relate to the leader.

3. Inner Motivation and High Need for Achievement. In general, all leaders are ambitious in their own ways. They want to challenge, achieve, and accomplish difficult and demanding tasks, and they want to leave a mark behind or to challenge what has not been done before. For example, the U.S. president John F. Kennedy predicted a space trip to the moon in less than a decade to win leadership for America in the space technology race. The leaders have a high need for achievement. To achieve, they do not wait for others to motivate them into seeking these challenges. They have an inner source for motivation. This inner motivation keeps them going and crossing hurdles.

4. People-Centered. Successful leaders are able to work successfully with other people in a wide range of situations. They recognize that respecting others and considering their opinions or expectations is critical to the success of most people-driven activities.

What about the physical characteristics of leaders? Do height, weight, color, and beauty make a difference with regard to the effectiveness of a leader? Whereas taller and bigger people seem to dominate among the leaders, there are leaders who more than compensate for their lack of these physical attributes. The above-mentioned four attributes have a higher significance on leadership than do physical attributes such as height.

Next we will discuss the processes involved in leading others.

B. Processes of Leadership

Influencing Others and Sources of Power. All supervisors, transactional managers, and transformational leaders must influence other people to do desirable things. This is done by using different sources of power. Power is the ability of a person to motivate and influence others. In an organizational setting, there are many different sources of power. And a supervisor, manager, or leader must clearly understand his/her own power base. In the absence of such understanding about self, the organizations are run by rules. The individuals are then no longer needed except for policing the enforcement of rules.

What are the different ways transactional and transformational leaders exert influence over others? Influence and power are the keys to leaders' work. To influence others the leaders must have access and ability to meet some needs of those whom they intend to influence. There are a number of alternate sources of power in an organization, and there are alternate ways to have influence over others. These are briefly described below (see Exhibit 9.3).

1. Legitimate Power. This source of power comes from the official hierarchical or legal position of a person within an organization. A leader has the official right to ask the subordinates to do certain things, and the employees are obligated to comply with it. This understanding is the basis for the subordinate employee's contract of appointment with the organization.

For example, two individuals, Neal and Roy, may be equally smart and knowledgeable. But if Neal has a higher position in the hierarchy of the organization, and Roy reports to him, then Neal holds a legitimate power over Roy. Neal can ask Roy to do a number of job-related tasks. But Roy cannot ask Neal to do the same.

2. Reward Power. A leader generally has the ability to reward actions of other subordinates with promotion, bonus, incentives, or recognition. The key to success with this source of power is that the leader must offer something that the subordinate genuinely desires. If the reward is insignificant in the eyes of the receiver, then it will not bring about the desired change in the recipient's behavior. If the receiver does not value the reward, then the reward will not be able to motivate a change in behavior.

EXHIBIT 9.3. Leader's Sources of Power

Push Powers	Pull Powers
1. Legitimate power	4. Referent power
2. Reward power	5. Expert power
3. Coercive power	

For example, in many organizations monetary rewards are provided to individuals who are economically very well off. These individuals may instead appreciate a recognition in public in front of their peers and family. That recognition is likely to be of more value to them than a silent monetary reward. At other times the economic rewards may be perceived as insignificant by the person receiving them. For example, in many technology-driven organizations, the submission and approval of a patent is rewarded by a small monetary reward. Very often it is barely equal to the price of a dinner for two at a good restaurant. Employees who pursue a patent do so not for the monetary reward they receive, but for the prestige associated with receiving a patent. However, if a technology-driven organization attaches a high importance to the employees submitting applications for patents, then the reward or the recognition for doing so must adequately reflect that expectation. Employees use the rewards and recognitions as measures of the true intent of their top-level managers and leaders.

3. Coercive Power. A leader can very often use coercive force and authority to punish others if they do not do certain things or if they do not meet their work-related expectations. For example, a supervisor can assign a subordinate an unimportant task or can move the subordinate to a risky and hard-to-live location. Of course, each employee has the ultimate freedom to refuse to take such orders by quitting their job. Some leaders are good at assessing how badly a subordinate needs that job. The subordinates are sometimes protected by contracts or union negotiations. The leaders cannot easily force the union member to perform certain actions which are beyond the scope of their collective bargaining.

The leaders must be sensitive to find out what they can do or cannot do under the law of the land. Under the new laws in the United States governing sexual harassment, racial discrimination, and the employees' disability, the employers and therefore their leaders cannot force their employees to do certain practices inconsistent with these laws. The employers, with deeper pockets than their employees, are liable to be sued by affected employees for such violation of laws. The investors are liable to promptly fire such employees, even if they are leaders.

4. Referent Power. This source of power comes from a leader's ability to act as a role model to his or her followers. A follower is influenced by the leader because of his or her association with a task or a program. We dress the way our heroes do, because we admire them and wish to be like them. We join certain membership clubs because the people we admire are members of those clubs. This influence arises from the personal charisma, integrity, courage, or other admirable traits in the leaders we admire.

Millions of people followed Nelson Mandela in South Africa, Mahatma Gandhi in India, Martin Luther King, Jr., in the United States, and Mother

Teresa worldwide because they admired the courage and integrity of these leaders. Many young computer professionals who do not work in Microsoft admire the way Bill Gates, the CEO of Microsoft, created a multibillion-dollar company with his entrepreneurial ability to see new opportunities. Many employees of Mary Kay cosmetics company admire the ground-breaking courage of their founder/leader.

5. Expert Power. This source of power comes from the expert knowledge or skill base of a leader. If others believe that a leader has the expert knowledge and knows what is needed to carry out a task, they are more likely to follow such a person. Microsoft employees willingly follow their CEO Bill Gates, because they know that he has the expertise to program and develop innovative software. Andy Grove commanded a similar following at Intel, because of his doctorate in quantum physics from the University of California at Berkeley. The expert power of a leader influences the behavior of others, the way medical doctors can influence our behavior when we are sick or suffering from an ailment. Sometimes the expert power is based in the educational qualifications of a leader, and sometimes it is based on the many years of experience that a leader has in tackling tough situations.

Effective Uses of Multiple Sources of Power. A leader may have more than one source of power to influence others' behaviors. Different types of leaders tend to use these alternate sources of power differently.

The first three sources of power are anchored in the leader's relative position in the hierarchy of his or her organization. The last two sources of power are anchored in the leader's personal attributes. A transactional leader uses his or her organizational power to push others to behave in a certain manner. A transformational leader uses his or her personal sources of power, based on referent power and expertise power, to pull others into wanting to do what the leader wants them to do. The influence of the organization-based sources of power lasts only as long as the leader holds a particular position in the hierarchy of an organization. If a leader loses the hierarchical title, then the "push" sources of power cease to have an influence on the followers. On the other hand, the personal or "pull" sources of power continue to influence the behavior even long after the leader is gone. Long after Mahatma Gandhi, Martin Luther King, Jr., and Mother Teresa were gone, they continued to inspire and influence people's behaviors in different parts of the world. The key to effective use of different sources of power is that a leader must use the organization-based sources of power as sparingly as possible. An exclusive reliance on organizational power does not produce a sustained high performance from others. The personal sources of power, on the other hand, though harder to acquire, can produce long-lasting influence on others.

LEADERSHIP NEEDS FOR THE NINETIES

In the past, it has been noted in leadership literature that different times require different types and styles of leaders. The Nifty Nineties, characterized by historical convergence of multiple revolutions in customers' expectations, competitive intensity, and critical contextual conditions, demanded a new type of leadership for sustained competitive advantage. The time was up for the transactional leadership, relying exclusively on expense-cutting and bottom-line dressing. Instead we needed innovative and risk-taking transformational leaders, ready to exploit external opportunities with agile core-competencies of their organization. In this chapter we have tried to develop a framework for desirable characteristics of such a transformational leader. Let us next look at the link between leadership and business strategy.

Concept of Strategy and Fast Markets

Leaders and top managers are actively involved in the development of strategies for their organizations. According to *The American Heritage Dictionary,* the word *strategy* is defined as (1) the overall planning and conduct of large-scale military operations, (2) the art or skill of using stratagems, as in politics or business, or (3) a plan of action. The English term *strategy* is based on the Greek word *strategos,* meaning "a General leading an army to destroy enemy through effective use of his resources." In turbulent times, the conventional knowledge about long-term plans and strategies fail to deliver the promised improvements in competitive performance. In highly globalized markets, with intense rivalry and fast-shrinking product life cycles, the formulation of plans and strategy must take place almost concurrently with its implementation (see Note 9.10).

Planning Paradigm. Traditionally, leaders and managers were intimately involved in strategic planning. The planning paradigm, presumably proposed and promoted by the Business Policy Group at Harvard Business School, considered the planning process as a two-phase process. The first phase involved formulation of plans and strategies by doing external environmental analysis and internal organizational analysis. The second phase involved implementation of the formulated plans and strategies by designing a suitable organization structure and control system to monitor the execution of plans and strategies.

Crafting Paradigm. Professor Henry Mintzberg (of McGill University) and his colleagues have proposed an alternate paradigm for plans and strategies, particularly in large organizations (see Note 9.11). Their strategy-crafting paradigm argued that a clear-cut division between (a) the formulation of plans and strategies and (b) the strategy implementation phase is forced and artificial. Professor Mintzberg suggested that the strategy and plans were formulated and implemented concurrently in an interactive manner. The leaders' strategic intentions were referred to as "intended strategies." These were different from what the organizations actually did, which were called "derived strategies." A part of the "intended strategies" became "unrealized strategy," and the dynamic interactions between the leaders' intentions and the rest of the organization resulted in "emergent strategies."

The strategy-crafting paradigm required a longitudinal study inside an organization. The emergent strategies followed in time after the intended strategies. This paradigm also required access to internal memos, meetings, and actions. This may be hard to use for many external observers, such as the investors interested in the future earnings performance of a publicly owned organization. But the key difference in the crafting paradigm from the previous planning paradigm was that the strategic process was not assumed to be a linear sequential process. This was critical for the leaders and top managers. Under the crafting paradigm, they must provide adequate time and resources for such additional delays due to the political negotiations within the organization.

The Competency Paradigm. In the 1990s, transformational leaders had to adapt themselves quickly to the external and internal changes if they wished to tide over multiple radical revolutions taking place in their competitive markets. The top-line-driven transformational leaders must, in dynamic markets, synergize external environmental changes, with the development of their organizations' core competencies. There are four basic competencies of a top-line-driven transformational leader interfacing external opportunities with organizational competencies. These leadership competencies are:

1. The transformational leaders have to respond to, respect, and exceed the expectations of their challenging customers.
2. The transformational leaders must proactively anticipate shifts and opportunities in their critical environmental conditions. These opportunities emerge from shifts in regulation, or new technological developments (in computer-communication technology, etc.).
3. The transformational leaders rise far beyond "benchmarking" their competitors,
4. The transformational leaders build core-competencies of their organization and its human capital.

Each of these competencies need to be explained further.

1. Responding and Exceeding the Expectations of Challenging Customers. In the "Nifty Nineties," the technology-driven organizations became increasingly dependent on satisfying and delighting their highly demanding customers. The impact of "market pull" forces exceeded the intensity of "technology push" force. The leaders in technology-driven organizations realized that "a better mousetrap" did not "automatically" attract millions of customers. As discussed earlier, the newer products and technologies demanded more persistent education of current and potential customers.

David Kearns, the CEO of Xerox, pointed out to all the Xerox employees that they had their jobs only because their customers bought Xerox goods and services.

Chairman Andrew S. Grove of Intel, a world-leading manufacturer of semiconductor chips, realized that the customers' criticism of the math coprocessor in the Pentium chip, along with a hammer blow from its major account customer IBM, could nullify millions of dollars invested by Intel in developing its radical new Pentium chip technology. CEO Grove's biggest challenge was to transform Intel's technically oriented culture into a customer-focused culture. Dr. Andrew Grove learned so much from the "Pentium problem" that he wrote in a book stressing that "only the paranoid (leaders) survive." He explained that many technology-driven organizations, from time to time, faced "strategic inflection points" (SIPs), which involved a "10×" order of magnitude of change. During this phase, the conventional age-old wisdom of that industry ceased to be effective. Grove suggested that the challenge for a leader facing a SIP was to avoid the temptation of taking complete control of the chaotic situation. Instead the leader should hold back and let the front-line employees experiment alternate solutions to resolve the chaos. After a reasonable solution has been identified, the leaders should sponsor and facilitate its careful and meticulous implementation.

Therefore, a transformational top-line-driven leader must not fall in the trap of overrelying on satisfying its current customers only. Professor Michael Porter of Harvard University, in his study of international competitiveness, indicated that the successful companies are those companies that proactively seek out customers that are most demanding and hardest to satisfy (see Note 9.12). The Japanese manufacturing firms, such as Toshiba, Hitachi, Matsushita, and so on perfect their products for global markets by first offering them to highly quality- and cost-sensitive Japanese customers. Transformational leaders take the challenge of meeting and exceeding the expectations of such demanding customers. For example, the photography pioneer Eastman Kodak always took pride in satisfying the demanding needs of professional photographers. Sony, a leader in consumer electronics, always paid a lot of attention to the requirements of professional sound and video recording studios and broadcasting stations. Meeting those requirements helped Sony and Kodak develop new products for the mass consumers as well.

2. External Environmental Opportunity Orientation. Transformational top-line-driven leaders must be extremely sensitive and proactive to external shifts, which take place rapidly beyond the boundaries and control of their organizations. According to Jerre Stead, CEO of AT&T Global Information

Solutions, who spent 3 days a week talking to external stakeholders, the transformational leaders must "Get back to the outside world." Furthermore, "all the indispensable tools of transformational leadership, such as vision, trust, rewards, compassion, and so on, cannot carve the masterpiece out of the marble, unless they're hammered home by facts about what's going on beyond the organization's boundaries" (see Note 9.3).

PROACTIVELY ANTICIPATING EXTERNAL MEGASHIFTS. Transformational leaders accept the fact that the changes in external environment are beyond the control of their organizations. They therefore proactively anticipate strong as well as weak shifts in their critical external variables. Noteworthy among these external changes are (a) shifts in regulatory practices required of firms in regulated industries and (b) technological developments for firms in high-tech industries. Many transactional leaders tend to wait too long or only react to drastic external changes. Transformational leaders act proactively and move fast.

3. Rising Far Beyond Benchmarking Competitors. Transformational leaders are also very sensitive to their competition. They continually benchmark their competition and then rise far beyond it. CEO Andy Grove of Intel could perhaps have made more profits from its highly successful 486 computer chips for a longer time. But to stay ahead of competition, Grove chose to develop Pentium chip technology and rush it to market, even though the 486 chips could be perfected further.

Tom Peters and Nancy Austin, in their 1985 best-selling book, *A Passion for Excellence: The Leadership Difference,* called this "the 1000% factor" (see Note 9.13). They noted that the competitive leaders, even in declining and regulated industries, perform hundreds of percent better than their competitors. They gave the example of Kelly Johnson of the Lockheed California Company, who introduced the term "skunk-work" to Corporate America. Johnson and his team pioneered the military jet aircrafts after the end of World War II, as well as the YP- or F-80 planes, the F-104, C-130, U-2, SR-71, and other major innovations in aircraft technology. Beside being an innovator, Johnson was also amazing with productivity. In one instance, Johnson successfully completed a defense project with 126 people, achieving what another defense contractor could not do with 3750 people. Similarly, Frank Purdue had a dominating 60% market share for chickens in many major urban markets of America. Purdue earned many times more profit margin on each pound of chicken, compared to that industry's average. Du Pont had a safety record that was 6800% higher than the average value for America's manufacturing sector.

4. Building Organizations' Core-Competencies. The transformational leaders, after soaking in opportunities as well as assessing the threatening shifts in their external environments, turn to developing their organization's strengths and eliminating effects of its weaknesses. The transformational leaders develop human resource capital, the key to survival and growth of firms facing hypercompetitive market conditions.

Building Core-Competencies. According to Jerre Stead, CEO of AT&T Global Information Solutions (GIS), many organizations are driven by rules and fat books of policies. These rules are aimed at controlling about 1% of employees, while they actually "handcuff the other 99%" (see Note 9.3). As the top-level leader took over his $7.3 billion a year AT&T division (acquired from former NCR), in one of the world's fastest changing economic sector, Stead "trashed two fat books of policies and replaced them with just 11 guidelines, such as the guidelines for capital spending. Stead relied on results and rewards, rather than on the rules. He linked all objectives and rewards to key results of GIS, such as the customer or shareholder satisfaction and profitable growth. Rewards were not always in the forms of money and promotions. They could come in other very different forms, such as attaboys, letters, notes, trips, cash, dinner, gift coupons, and so on.

Unlike the transactional top-executive relying on his/her authority, charisma, or specialized expertise, the transformational top-executive leader is invariably, intensely interpersonal. Stead advised top executives to avoid the "Moses-like trap," whereby a top executive believes in "going up the mountain and hearing God talk and getting a vision" for other people working under him/her. Transformational top-line-driven leaders, work intimately with their people, bringing along "each person, one by one, with compassion and patience."

The transformational leaders develop the core-competencies as the key sources of competitive advantage. Companies like Wal-Mart, Nucor Steel, and Southwest Airlines have consistently outperformed their competitors in their respective industries. Therefore, the industry structure, or the competitive forces in an industry do not completely explain the competitiveness of a firm. The competency paradigm unique to individual firms can explain some of this differential.

To make core-competencies contribute to competitiveness, a leader must make sure that they meet the following three conditions.

1. Value-Adding Competencies. A firm's competencies and resources must add value, as perceived by its customers. Microsoft's Bill Gates puts an enormous stress on hiring the right employees for his firm's human capital. Senior executives are involved actively in defining the hiring needs and actual recruitment of new employees. Honda's core-competency in designing engines, Sony's core-competency in audio electronics, and 3M's competency and passion for developing new products have been nurtured by their top leaders. These competencies have helped them gain significant competitive advantages over their competitors for a variety of products and businesses. Honda used its core-competency of engine design for developing competitive winners in the motorcycle, automobile, and outdoor power equipment industries. Sony used its core-competency of audio electronics to develop camcorders, Walkman compact disc players, and other electronic appliances. These firms broke ranks from their traditions and developed new value-added competencies to take the lead over their rivals. On the other hand, leaders in many "dinosaur" giant companies of the yesteryear either did not or could not adapt to the dynamic changes in their competitive environments. They would not or could not develop new core-competencies fast enough.

2. Rare Competencies. Leaders must be careful that if a firm's competency is quickly matched by its rivals, then that competency is no more likely to contribute to the firm's competitive advantage. Such a competency forms the baseline hurdle for competing in that industry or technology. For example, if Toyota and Honda achieve similar competencies in cycle time reduction or in jig changeovers, these competencies will not contribute to competitive advantage for either competitor in automobile market.

3. Hard-to-Imitate Competencies. To gain a sustainable competitiveness for a reasonably long period, the leaders must also ensure that their core-competencies are not only value-adding and rare, but are also hard to imitate by rival competitors. This is why policies for product pricing, cash-back guarantees, and low interest financing for automobiles are often quickly matched by rival competitors. Processes are not. However, even when competencies are hard to imitate, leaders should be aware of continual threats from new competitors or substitutes for the firm's competencies.

The premise of core-competencies also implies that the different competitors in an industry or a technology can compete on the basis of very different competencies. Leaders must take a hard look within their organizations to find out the centers of excellence.

COMPETING FOR THE FUTURE

How do leaders manage their strategies to compete favorably in future? Professors Gary Hamel of the London Business School and C. K. Prahalad of the University of Michigan Business School, authors of a 1994 book *Competing For the Future,* are also the originators of such popular strategic concepts as "Core-Competence," "Strategic Intent," and "Strategy as Stretch." According to them, very few senior executives in corporate organizations are strategically competing to dominate in their industry's future. This ignorance is suicidal for technology-driven companies. They urge that "the vital first step in competing for the future is the quest for industry foresight." This industry foresight includes deep understanding of trends and discontinuities relevant to the business. These insights could be used to carve out leadership position in the future, by forcing a redefinition of the boundaries of competition. This can provide a significant and sustained competitive advantage over rivals.

To find out about the degree of executive concern for strategic foresight, Professors Gary Hamel and C. K. Prahalad urge senior executives to answer three questions:

1. What percentage of the leaders' time is spent on *outward* (O%)-looking issues (such as impact of a new technology), rather than on inward-looking issues (such as debating corporate overheads)?
2. Of the time spent on outward-looking issues, what percentage of leaders' time is spent on *forward* (F%)-looking issues (such as 5 or 10 years in the future), rather than worrying about immediate concerns (such as reacting to a competitor's pricing policy)?
3. Of the time spent on outward- and forward-looking issues, what percentage of leaders' time is spent on building up a *shared* (S%) collective view with colleagues and subordinates inside an organization, rather than on pushing their own individual functional or idiosyncratic agendas and views?

According to Hamel and Prahalad, the senior executives' responses typically conform to the 40%–30%–20% rule. In other words, senior executives and leaders spend only 40% of their time on outward issues. Of this time, only 30% is spent on forward looking issues. And of the time spent on outward and forward issues, only 20% is spent on building up a shared future view of their business with their colleagues. Thus, they calculated that the senior executives spend only $0.40 \times 0.30 \times 0.20 = 0.024$ or less than one-fortieth of their time on competing for the future. Is this suficient?

Finally, we will consider the competencies needed in a corporate leader to compete for the future of a high-performance technology-driven enterprise.

The Six "C" Competencies for Corporate Leaders

Effective leaders share certain core competencies. These are as follows:

C1. Clear Self-Understanding. The effective leaders have a relatively clearer understanding of their own strengths and weaknesses. They strive hard to leverage their strengths and continually work on their own perceived weaknesses. They keep asking themselves about the areas they are good at and about the areas they have opportunities to improve further.

C2. Creating a Vision. Leaders use their ability to create new visions as magnets to attract other people to commit to their vision. Leaders must go beyond long-term planning. Planning is a management process, often for orderly results but not radical change. Setting a direction requires assimilating information from diverse sources, identifying new patterns and discontinuities, addressing uncertainties, and making bold decisions to define where a leader wishes to take his or her organization in the future. This is articulated by the leader as the strategic vision for the organization. Hamel and Prahalad have proposed that leaders set their organization's strategic intent to set guidelines and mobilize the entire organization. For example, Komatsu's strategic intent was its decision to "encircle Caterpillar." Canon and Fuji Photo intended to catch up with Xerox and Eastman Kodak, respectively.

C3. Communicate the Vision Clearly. A transformational vision by a leader is no good if it is not effectively communicated down to the rank and files of the organization. Leaders must also be good at clearly conveying their vision to others. Good leaders are able to clarify the desired state of effort. Such clear communication and illustration provides meaning to others in terms of what they are getting into.

C4. Create a Conducive Culture. Leaders must also recognize that transformational changes cannot be carried out by doing "business as usual." Different strategic visions demand different organizational designs. Strategy and structure of an organization go hand in hand. An organizational vision based on innovation requires new organizational structures and cultures that nurture openness and mutual trust. An excessive bureaucracy and elaborate reporting structure and reporting protocol can stunt such creativity demanding spontaneity and responsiveness.

C5. Catalyze Others with Empowerment. Leaders, having created a vision and communicated its meaning to others, must hold off and let others devise innovative ways to execute it. They must act only as the catalysts and not do everything themselves. Employees who are excited and stimulated by the vision that leaders create may get engrossed in doing what the vision demands, even though it involves some personal sacrifice. Leaders should allow them to do so.

C6. Clear Demonstration of Commitment. A strategic leader must also unequivocally demonstrate his or her commitment to excellence and ethical behavior. Employees tend to follow the leaders' footsteps and actions rather than follow what the leaders say and urge others to do.

A SUMMARY OF LEADERSHIP LESSONS LEARNED IN THIS CHAPTER

In this chapter we examined the critical role played by leaders in the technology-driven enterprises. We began by examining how so many corporate leaders and chief executive officers of technology-driven companies lost their jobs during the economic turmoil of the early 1990s. We learned that the role of corporate managers and leaders had evolved a lot since the 1930s. Initially, the professional managers had a lot of independence and freedom in making their corporate decisions. Gradually, by the 1990s, the institutional investors penetrated the leaders' domain of influence.

We examined the properties and characteristics of leaders, and we also identified the processes they used and the way they influenced and exercised their powers over others. We discovered the six leadership styles based on the leaders' mission and their methods. We learned from Professor Mintzberg and others that the conventional understanding about the job of top managers and leaders has changed significantly. The leaders no more developed their strategies and plans by using planning, crafting, and competency paradigms. For dynamic market conditions, when technological trajectories move discontinuously, the transformational leaders in corporations could develop six "C" core-competencies to produce a high-performance change.

SELF-REVIEW QUESTIONS

1. Why did so many CEOs and corporate leaders in different technology driven enterprises lose their jobs during the early 1990s?

2. What are the roles that a senior manager or leader plays? Compare and contrast this with our traditional understanding of planning, organizing, coordinating, and control model.

3. What are the common properties or traits of leaders?

4. How do leaders influence others? Describe push and pull sources of power.

5. What are the different leadership styles? Illustrate with examples for each style.

6. Compare and contrast the planning paradigm, crafting paradigm, and competency paradigm for developing plans and strategies.

7. Define the 6 "C" core-competencies of corporate leadership.

NOTES, REFERENCES, AND ADDITIONAL READINGS

9.1. Earlier versions of parts of this chapter were published in Gehani, R. Ray. 1995. Transformational leadership for quality: bottom-line micro-management versus top-line leadership, *Proceedings of the 51st Annual Conference of ASQC—Rochester,* 28 March 1995. Also see Gehani, R. Ray, 1994. The tortoise vs. the hare. *Quality Progress,* May:99–103.

9.2. Stewart, Thomas A. 1993. Welcome to the Revolution. *Fortune,* December 13:66–77. Also see Huey, John, 1993. Managing in the midst of chaos. *Fortune,* April 5:38–48.

9.3. Stewart, Thomas A. 1994. How to Lead a Revolution, *Fortune,* November 28:48-61. Also see Huey, John, 1994. The new post-heroic leadership, *Fortune,* February 21:42–50.

9.4. Linden, Dana Weschsler, and Rontenier, Nancy. 1994. Goodbye to Berle and Means: nineteen ninety-three was the year when shareholders and boards of directors showed the boss who was the boss. *Forbes,* January 3:100–103.

9.5. Lesley, Elizabeth, Schiller, Zachary, et. al. 1993. CEOs with the outside edge. *Business Week,* October 11:60–62.

9.6. Gehani, R. Ray, 1993. How would George do it? *Democrat & Chronicle,* October 12:9A

9.7. Mintzberg, Henry. 1975. Manager's job: folklore and fact. *Harvard Business Review,* July–August:48–61.

9.8. Maccoby, Michael, 1981. *The Leader,* New York: Simon & Schuster.

9.9. GE Document, 1996. The global journey. 6 Sigma quality. Fairfield, CT: General Electric.

9.10. Dertouzos, M. L., Lester, R. K., and Solow, R. M. 1989. Made in America: Regaining the Productive Edge. New York: Harper Perennial. Also see Berger, Suzzane, Dertouzos, Michael, Lester, Richard, Solow, Robert, M, and Thurow, Lester C. 1989. Toward a new industrial America. *Scientific American,* 260(6), June:39–47.

9.11. Mintzberg, Henry. 1987. Crafting strategy. *Harvard Business Review,* July–August.

9.12. Porter, Michael. 1990. *The Competitive Advantage of Nations.* New York: The Free Press.

9.13. Peters, Tom, and Austin, Nancy. 1985. *A Passion for Excellence: The Leadership Difference.* New York: Random House. Also see a number of publications by Warren Bennis. For example, Bennis, Warren, and Nanus, Burt, 1986. *Leader: The Strategies for Taking Charge.* New York: Harper and Row.

CHAPTER 10

Projected Future of Technology and Success in Society

PREVIEW: WHAT IS IN THIS CHAPTER FOR YOU?

So far we have looked at technology in an analytical way. We broke down the V-6 engine model of technology management into its component cylinders. In this final chapter, we will take a different approach. We will look at technology from a synergistic point of view. We will discuss how technology interfaces with society at large, and people in particular. How do people perceive technology? Is technology perceived by them as a friend or a foe? How do societal context influence the success and failure of a technology? And given the potential differences in perceptions, along with the influence of context and environment on technology, can we effectively forecast the projected future of technology?

TECHNOLOGY AND SUCCESS IN SOCIETY

In the previous chapters we reviewed how enterprises can effectively manage technology by managing production, products, proprietary know-how, people, processing of information, and promised quality—the "cylinders" of the V-6 engine model of management of technology. Then we saw two ways to integrate these—as projects and as pioneering leadership. In this final chapter we will learn how society and people interface with technology. Some people are fascinated by technology and what it can do to improve their quality of life. Many others are afraid that technology will turn into an uncontrollable monster. Should we, or can we, forecast the future of technology and how it influences us?

Peter F. Drucker, the best-selling author of *Technology, Management and Society,* suggested that in the 19th century the Industrial Revolution and its technological fruits were largely confined to a small minority of mankind in Europe and the North Atlantic region (see Note 10.1). Many Western observers have overlooked the centuries of "technological" developments in regions beyond the pyramids of Egypt—such as in China and India. The Industrial Revolution in England adversely affected India and China with their colonization. These age-old civilizations were colonized and ruled by others partly because they could not adapt to the assault of new weapon technologies developed elsewhere in Western Europe. Therefore an internal and insulated view of technology can prove to be detrimental to the growth of an organization or a society. Very few people in these large countries benefited from industrialization of textiles in England. In many cases, large populations of colonized people lost their livelihoods based on their preindustrial textile and farming technologies. They could not adapt to the rapidly emerging textile machines driven by steam and more productive sources of energy. Only in large cities did people benefit from the new technological developments of electric generators, street electric lights, newspapers, telegraph, street cars, and steam-driven railways.

In Europe, authors like the Frenchman Jules Verne (1828–1905) popularized the fictionalized view of technology to common people. In earlier chapters we discussed why these technological developments converged and took place in England. These past developments and new technological inventions in the past few decades clearly illustrate a close link between technology and its societal context. More recently, in the fast emerging global economy, the technological developments in one part of the world have significant impact in many distant parts of the world. It is therefore important to consider in this final chapter how people and society perceive and interface with technology.

Exploring the Technology Elephant

To many common people in society the technology is like an elephant. On one hand, technology seems complex and big, but on the other hand it seems friendly and fearsome at the same time. Technology is often either misunderstood or partially understood. For many people, their perceptions of technology are like those of the group of blind men exploring an elephant for the first time. Like a well-trained elephant, technology can help mankind move mountains and eradicate poverty. Using technology, the Europeans were able to build modern industries and gain world leadership. New Yorkers were able to build skyscrapers (like the Empire State Building) and the Brooklyn Bridge many decades ago. In the 1960s, with President John F. Kennedy's vision the American people were able to catch up with the Soviet space program within a decade and go to the moon and come back. On the other hand, like a wild elephant, technology sometimes runs wild, thereby threatening to devastate humanity. The nuclear reactor technology which produces much needed electricity for a high concentration of people in cities can go out of control and cause havoc on the people in a vast territory spanning across many countries.

Perceptions of Technology. Let us recall the world-famous story of the *Elephant and the Blind Men from Hindustan* by Rudyard Kipling about the seven blind men visiting and exploring an elephant. Every one of the blind explorers was confident that his perception of the elephant was the right one. Yet they all perceived only a part of the whole elephant. People often do the same with their perceptions of technology.

1. Technology Is a Perception Multiplier. Blind man #1 touched only the ears, and he thought the elephant was a fan-like thing. He believed that the elephant was a powerful wind machine. To some people, technology provides powerful "ears and eyes," which enhance our perceptions. The most visible role of technology is its ability to multiply and leverage man's capabilities. It can make us fly high above the ground, see microscopic particles and germs, and hear voices and sounds that are a long distance away.

Even though we are not birds, aircraft technology can fly us from one corner of the world to another corner. Even though our hearing is not as sensitive as that of some other species in the animal kingdom, telephone technology helps us hear messages from thousands of miles away. And television broadcasting technology can show us events from far-off places, such as an erupting volcano, real time, without being there. Sonar technology lets us hear the geological rumblings tens of miles under the surface of the earth. We know where the tectonic plates are located, and we know where to build and where not to build our homes, hospitals, and highways. Technology helps the deaf to "hear" with ear implants, and it helps the blind "see" with Braille and computer-aided technology. The computer amplifies memory and thought-processing capabilities for mankind.

2. Strong Pillars of Technology. Blind man #2 was impressed with the solid strength of the pillar-like feet of the elephant. The Manhattan Project during World War II clearly demonstrated and established the "supremacy" of America's superpower in the world. The industrialized nations rest firmly on the strengths of their technological advancements. They can reach into the vacuous space, get water during draughts, and transport men, materials, and machines over thousands of miles along the ribbons of coal tar and concrete we call roads. Technology adds strength to people's physical capabilities. Computers and new information-processing technologies help us estimate distances to distant galaxies and predict the forthcoming weather patterns in the next few days.

3. Technology's Wall of Obstacles. Blind man #3 felt the side of the elephant and concluded that the beast was nothing but a big wall. Many people consider technology to be a wall of obstacles. Not many decades ago, using a bathtub full of water was considered a sin. Mirrors were feared to capture people's souls. The computerization of the workplace was often considered not as a way to enhance our abilities, but as a way to block workers' growth and development. Computer-driven machines took away unskilled workers' jobs and sources of living. In more than one way, technology establishes the bar of challenge that companies must cross to compete effectively with global competitors with superior technology. For many years, Intel manufactured and marketed semiconductor microprocessors which established the frontier of technology that Intel's competitors must match or exceed.

4. Dual-Purpose Technology. Blind man #4 explored the teeth and was surprised to note the two sets of teeth in the elephant. One set of teeth seemed sword-like and appeared to be a symbol of elephant's power to other species. The other set of smaller teeth seemed actually more functionally useful, and the elephant used this set for chewing delicious stuff like sugar cane. Technology has dual purposes too. The U.S. military weapon technology was used as a deterrent to the Soviet assaults and expansion during the Cold War. But the same technological strengths were used internally to run the industrial machinery that produced goods and services for the world markets. The Jeep development program that provided the general-purpose transport to the GIs during World War II also provided the sports utility vehicle (SUV) for peacetime urban and cross-country transportation. Technologies developed from NASA's space exploration and Space Shuttle program produced a variety of technologies for civilian use. Velcro, polymer resins, and lightweight composites made of graphite fibers came in handy for lightweight high-strength golf clubs, tennis rackets, and skis.

5. The Platform of Technology. Blind man #5 sat on top of the elephant and found the elephant to be like a comfortable platform. Computer and information technology provide a platform to communicate, collect, save, and exchange lots of information. The Internet is the new platform of interconnected technology that will allow us to do a variety of things, from electronic commerce to virtual visits to distant museums and hotels. In many ways the new information infrastructure is almost as significant a platform as the network of interstate roadway and railway infrastructure laid down by earlier generations of Americans. The Genome project mapping human genes would generate the next generation of genetic engineering technologies, providing cures for our hard-to-understand medical aberrations.

6. The Technology Conduit. Blind man #6 felt the trunk of the elephant and concluded that the elephant should be used as a pipeline for fluids and gases. Technology is very often a conduit and a facilitator, thus adding value. Oil pipeline and natural gas-line technologies transport the natural resources from where they are found in abundance up in Alaska, to where they are needed most in the highly industrialized parts of California. The construction of the Erie Canal connected the industries of New England with the newly emerging markets of industrial goods in the Midwestern prairies. The canal also brought back the agricultural goods from the Midwest to the high urban concentration of population in the East coast cities.

The Space Shuttle is nothing but a conveyor belt of men and machines to space. Many technologies are perceived as the conveyors of ideas, goods, and services.

7. The Frail Tail of Technology. Finally, blind man #7 saw not much ado about nothing in the frail tail of the elephant. He argued against all the hoopla about the elephant being large and strong. Every now and then we hear similar muffled sounds of "so what" about the new developments of technology. With all the sophistication at our disposal, we still do not have a cure or a clear cause for cancer or the AIDS epidemic.

The Invisible Sides of Technology   The seven blind men could feel the tangible attributes of the elephant. However, they could not grasp some other intangible aspects of the elephant. In the same manner, we can also see and feel the tangible side of technology. But there are many more significant but intangible aspects and attributes of technology. These are discussed next.

1. Technology Is the Embodiment of Knowledge.   Like an elephant's long memory, there are certain aspects of technology which are hard to see, such as the accumulated knowledge embodied in a 1997 Ford Taurus. It carries with it all the successes and failures of auto-making since Henry Ford's Model T.

In this day and age, technology has transformed almost every facet of our life. Some of it comes in the form of hard tangible objects made of metals, glass, and plastics. On the other hand, other technologies provide us services and software, made of nothing but bits and bytes arranged in a particular way. These are hard to touch or see. It then becomes hard for us to pinpoint the essence of a technology. What is so high about high technology? Is an ordinary-looking Post-it note pad produced by the 3M Corporation a high-tech or low-tech product? The product is simple and is easy to use. But the process to produce Post-its involves very closely controlled surface-coating technology. That surface-coating process is invisible to the user of a Post-it. In the same manner, we can fly an airplane and go from one continent to another continent without having to learn how to fly an aircraft. We can collect and compile large amounts of information from widely distributed sources with the help of easy-to-use web browsers. We do not need to understand or have sophisticated programming skills to do so. Each technology embodies large amounts of accumulated knowledge.

2. Technology Learns and Is Cumulative. Like an elephant, technology learns and remembers for a long time. We often forget that what we see today represents the tip of the iceberg of accumulation of knowledge over many generations. The technological developments that we claim to be our own are based on the foundation of thousands of inventions and incremental improvements made by millions of inventors and innovators before us. A laptop computer is able to pack a large amount of information processing capability and memory because of what was learned in the development of a one-transistor microprocessor many years ago. Many of the modern life-saving pharmaceutical drugs are based on the wisdom accumulated over centuries. Yet it is often very tempting to claim it all and give little credit to what was learned from the past. Lawyers, as guardians of our rights, will repeatedly convince us about this.

3. Technology Represents Values. Another invisible aspect of a technology is the values that it embodies. The popularity of jeans in America has represented the high value for physical work and respect for a hard day's work. In many other societies, including the industrialized economies of Europe, physical work is considered menial. The rampant use of fast-food drive-in restaurants represents Americans' distaste for waiting in line. Easy access to guns, a source of thousands of deaths and crimes in America, to many is a symbol of freedom and liberty (at times from their own government). Some people argue that guns do not kill, but gun owners do. Yet, an easy access to these lethal weapons represents a value of the American society and its people.

4. Benefits of Technology Accompany Its Liabilities. One aspect of technology that is often overlooked but often comes back to haunt us is its liability. Most technologies are developed for some desired benefits. Yet these benefits often accompany certain side effects. The home furnaces produced 100 years ago are still liable for any accident that they may cause. A newspaper takes decades to decompose. Highly insulated polystyrene cups for coffee at fast-food restaurants will survive long after the last drop of the coffee is consumed. Landfills are full of technologies past their full use. Therefore, cars are being designed in Europe so that after they have been discarded, they can be disassembled so that some parts can be recycled for repeated future use. Eastman Kodak's Funsaver single-use "weekend" cameras are disassembled and reused with the same goal in mind.

Furthermore, these attributes of technology do not stay constant.

Changing Perceptions of Technology: A Friend or a Foe?

In the 1990s, the term "technology" conjures up a mixed bag of feelings. There are some who believe that technology is an enabler. Technology enables them to go to distant places without going there, and it enables them to see news from distant places as it happens. Technology provides access to the latest developments in lifesaving drugs and medical practices. There are others who are afraid of the way technology is encroaching on their lives. They have to learn to use computers if they want to hold onto their own jobs. Even the lowest-paying janitors and sanitary workers must learn about the safe ways to dispose of hazardous materials. How would this discussion of technology differ if we had discussed the same question with individuals born one or two generations earlier—that is, parents or grandparents of college students?

For a generation or two earlier, technology would have been recalled romantically as the ultimate savior of the society burdened with fast-growing population on a finite planet earth. Technology supplemented the fast-depleting reserves of resources in the earth's belly with man-made synthetic materials. Technology was that ultimate weapon of mankind unleashed to match against the fury of nature. New housing technology provided shelter when it rained and snowed, and medical technology saved our lives when the vicious viruses invaded our helpless bodies. Farming technology produced a bountiful amount of food, despite the plentiful pests. Technology also helped humanity make a giant step into the dark depths of star-studded space. Technology produced almost limitless power for our unsatiable need for electricity. For the earlier generations, technology seemed to ever-extend the power of humans over nature. Technology tilted the balance in favor of humans, when pitched against nature.

Roller Coaster of Technology: From Romance to Rejection

Somewhere down our memory lane the romance of mankind with technology ended. The honeymoon was over, and what seemed beautiful and bountiful earlier looked full of blemishes. The very success of what technology did for us resulted in the excesses that ruined our lives. Mankind became too greedy and started going for the golden goose. One golden egg at a time from technology was not good enough for our greed. The accelerated use of fertilizers and pesticides destroyed the soil so permanently that big stretches of land lay abandoned forever. The growing dependence of densely populated urban centers on nuclear power generators made us careless enough to cause the Three Mile Island and Chernobyl episodes. A schoolteacher died in a space exploration that turned into a space explosion—permanently etching her memory on the minds of millions of students and citizens watching the space launch. The augmented beauty of breast implants turned into unbearable pain and misery for the users of the implants.

Humanity seemed surprised at all this. To the average person, the technology was supposed to be failproof. They argued: How could technology fail them when they had paid for it? Even though scientists and technologists always estimated the unavoidable risk factor alongside the potential returns, others preferred to ignore these. The overenthusiastic marketers and investors oversold the merits of technology. They told the beneficiaries of customers that even though there was some risk, it was not likely to affect them. Technology was a miracle that could not go wrong.

We forgot that what was surprising about America's space exploration was not that a space flight failed on its journey toward the Final Frontier in space, but that a man-made space vehicle could be designed and developed with millions of parts and components to break through the earth's atmosphere and reenter the same, withstanding thousands of degrees of heat and abrasion. What was surprising was not that there was an occasional runaway nuclear reaction, but that mankind could break into the atomic code of matter and produce almost infinite usable energy by burning a small amount of mass.

Is there a way to get off this emotional roller coaster of euphoria for what technology can be enslaved to do, accompanied by the paranoia for what technology will do on its own to harm us? One way to grasp technology comprehensively is to develop a balanced understanding of pros and cons of a technology. We must accept that technologies can be the friends of society and can be sources of revenue for a businesses. We must also realize that if a technology turns into a foe, we will learn to protect ourselves. Then instead of becoming hysterical or hyperexcited about technology, we want to learn to manage technology successfully to its full growth potential.

In the earlier chapters we learned to manage the V-6 engine of technology by managing its component "cylinders." In this chapter let us introduce the impact and influence of the external environment on the effective management of technology. In the next section we will see that the birth and growth of a technology depends to a great extent on the fit between technology and the external contextual environment. This is the contingent view of the evolution of technology.

CONTINUAL CONTEXTUAL EVOLUTION OF TECHNOLOGY

If we examine the history and development of different technologies carefully, then we will notice that most technological progress and successful changes in technology can be shown to be continual and contextual. The technological developments are the fruits of creative and inventive thinking of risk-taking entrepreneurs. But the technologies that succeed in society do so because of their fit with the unfulfilled needs and wants of society.

Technologies Change Contextually

The technologies in general evolve continuously, with aggresive bursts punctuated by the intermittent lulls of seemingly limited action. Sometimes these lulls in the growth of a technology are periods waiting for new developments in the other enabling technologies to catch up.

From time to time some technologies seem to change discontinuously, either when they achieve a comparatively large amount of progress or when one technology borrows incremental changes from other dissimilar technologies. The mainframe computer cards for automated calculating machines were borrowed from textile weaving machines and gave the impression of a radical technological development in computing. The first printing press with moveable metal type in Europe was an adaptation of the Chinese, Korean and Japanese wood-block printing processes and the wine-crushing machine. The Eli Whitney cotton gin was a modification of the Indian gin, which borrowed from the sugar cane press that was used centuries before the start of the Christian era. These examples illustrate that different technologies are connected together.

The evolution of technology is almost always heavily influenced by the environmental and societal context. Thus the lull periods in technology may be caused by the contextual stunting of technology in society. The Dark Ages in Europe saw little technical progress, whereas the Renaissance period produced a burst of new scientific and technological ideas. America's faith in liberty to pursue happiness and prosperity has produced a steady stream of entrepreneurs risking their resources to develop and commercialize new technologies. Therefore, the success of a technology in the competitive markets or in the society depends on the fit between the development of technological competencies and its environmental or societal context.

It can be clearly shown that, like species in nature, technological artifacts also evolve in a manner quite similar to what Charles Darwin proposed. Technologies, like living beings and organic species, interact and adapt with the contextual environment. Let us consider some classical examples of technologies that have changed the American way of life significantly and that were cited by George Basalla in his ground-breaking book, *The Evolution of Technology* (see Note 10.2).

Case A. Eli Whitney's Cotton Gin: Catalyst to America's Industrial Revolution in the South.

America's early industrialization has been often attributed to Eli Whitney's invention of the cotton gin. Some called the ginny a radical and discontinuous discovery. But a careful study shows that Whitney's gin was the combination of a continuous stream of technological artifacts. The ginny drew heavily from the Indian gin or *charkha,* used in ancient Indian civilization for many centuries prior to the Christian era to produce cotton cloth (see Note 10.2).

The Indian charkha was developed from a sugar cane press. Like the sugar cane press, the Indian gin used a pair of long wooden cylinders fluted with fine lengthwise grooves. The cylinders mounted on a frame were pressed together. They were rotated on their longitudinal axis by a crank. As a cotton boll passed between the two rotating cylinders, it separated into the cotton fibers and the cotton seed. By the early 12th century, the Indian gin was known to Italian artisans as *Manganello* and it appeared in Diderot's *Encyclopedie.* In 1725 the Indian gin was introduced into the Louisiana region from the Levant. By 1793 it was a standard agricultural tool in America's Southern cotton belt.

Eli Whitney's Arrival in the South. Eli Whitney, a New Englander, visited a Georgia cotton plantation in 1793. He was hired as a tutor but he brought with him great mechanical skills. He quickly observed that a farmhand could spend up to 3 hours separating 1 pound of short staple Inland cotton from its sticky green seeds. The short staple cotton grew all over the South; but because of its stickiness, it had be cleaned manually. Whitney visualized that a machine could be designed to do the tedious job more efficiently.

However, some of the more scholarly studies of Whitney noted that some mechanical cotton gins were already available and were being used in the South to separate the long staple or Sea Island cotton with a less sticky black seed prior to Whitney's arrival (see Note 10.2). Unfortunately, the plants for this long staple variety of cotton grew in a very limited region of the South. But wherever available, this variety of cotton was used with the primitive Indian gin.

Eli Whitney took the challenge of developing a machine to clean more commonly available short staple Inland cotton with a sticky green seed in it. His invention also included a rotating wooden cylinder, just like in the wool-carding machines, with rows of bent wire teeth. These wire teeth passed through a slotted metal breastwork, with openings wide enough to allow the passage of wire teeth and cotton fibers, but not the cotton seeds. The cleaned cotton fibers were brushed from the teeth by a second rotating cylinder with rows of bristles. Whitney introduced a slotted plate (or breast-work) to strip fibers while the seeds were immobilized. This facilitated cleaning of more common short staple cotton. An efficient cotton gin was what the American South needed most at the turn of the 19th century. Whitney's gin served as a fountainhead for the development of many more types of modern cotton gins.

Eli Whitney's cotton gin illustrated that technological developments emerge and succeed in conjunction with their impact on society. For the success of a new technology, the societal expectations and the utility of the technological development are almost as important as the inventiveness of the inventor and sophistication of the technology. A technology is termed great only if the society or a culture puts a high value on it. Otherwise the inventor and his technology are considered a mere tinkering with the trivial laws of physical sciences.

Case B. Barbed Fence Wire: A Stimulus to America's Westward Expansion. Let us illustrate the significance of value assigned by a society to a technology by another interesting and pioneering example. A barbed wire will rarely come to mind in a discussion on modern technology. But the invention of barbed wire was surely one of the greatest technological stimuli when Americans were pioneering into the seemingly endless Wild West (see Note 10.2).

Barbed wire is simple technology. It consists of a long strand of wire with short strands of wire intertwined in it at regular intervals. The ends of shorter pieces are clipped at an angle so that they form sharp edges. Barbed wires are fixed to fence posts, acting as a deterrent to cattle wanting to roam away. The invention of barbed wire technology hardly needed advanced principles of scientific knowledge or intricate production processes. Yet barbed wire first appeared in a major way in America only in the late 19th century when three men invented it in 1873 in DeKalb, Illinois.

National Economic Significance of Fences. Early American settlers in the New World, as long as they lived closed to the Atlantic shores, did not worry much about elaborate fencing to protect their cattle. The sea provided the ultimate natural fence. Immigrants from England had brought with them skills to build fences with stone or wood around their agricultural and animal property. But as the settlers moved westward towards plains and prairies, fencing with wood turned out to be quite expensive. Historians have noted that between 1870 and 1880, fencing matters were far more important to America's farming population than the political, military, or even economic issues. In 1871, the U.S. Department of Agriculture estimated that the total cost of fencing the country exceeded the total national debt, and it also estimated that the annual cost of repairing the wooden fences far exceeded the total revenue generated by federal, state, and local taxes. Thus the development of alternate and cheaper ways to fence was of national significance to the fast-expanding America in the 19th Century.

Natural Hedges. One alternative to wood fence, used popularly in Europe, was to grow rows of tall thorn-bearing hedges around the boundary. Cactus, rose, and other plants were tried, but Osage orange proved to be the best plant for hedging. The latter grew in 3 to 4 years to a tall shrub, and it was native to parts of Texas, Arkansas, and Oklahoma. In 1860 its seeds were traded to the northern prairie states. However, these hedges posed some practical difficulties. They could not be moved. They sheltered weeds, took space, and cast a shadow on the valuable crop next to it.

Despite its disadvantages, Osage orange bush hedge served as an inspiration for design and development of man-made barbed wire. Osage branch, when observed closely, has strong and sturdy thorns, oriented perpendicular to the branch stem. These thorns protrude out at almost equal intervals, encircling around the stem in a spiral-like manner.

Around that time, in the second half of the 19th century, smooth wire fences were also in common use in regions with scarce supply of wood. Wire for fence was cheap, was easy to install, did not cast significant shadows, and did not encourage growth of weeds. Wired fence could also be easily dismantled to move the fence to new boundaries elsewhere. The major disadvantage of smooth wire fences was that the cattle frequently ran over it, even when multiple layers of wire were used for the fence. Smooth wire fence was good in all respects except for its deterring effect, which was the primary objective of fencing.

Imitating Nature. In 1868, Michael Kelly ventured to patent (No. 74,739) a fence wire that had the character of the thorn bush hedge. For thorns he took diamond shaped sheet metal pieces and attached them to a single strand of wire at a regular interval of 6 inches. In 1876 Kelly started The Thorn Wire Hedge Company to produce his newly invented thorny fence wire. But he was too late to make a big success out of it. There were other fences invented between 1840 and 1870. But none were produced in large commercial quantities.

Farmers and mechanics of DeKalb, Illinois, located at the edge of the Western prairie wasteland were aware of the need for an economical and effective fencing means. In 1873 at the DeKalb County fair, Henry M. Rose displayed a new fencing device (patent no. 138,763) that could be attached to the existing smooth wire to discourage cattle from breaking through fence. This was a 16-foot-long piece of wood, 1 square inch in cross section, through which long braids were driven so that sharp points protruded out of the wood.

Three visitors to the DeKalb County fair, Isaac Uloid, a hardware merchant, Joseph Glidden, a farmer, and Jacob Haish, a lumberman, saw the exhibit and decided to improve upon the fencing device. These three men knew about and had used osage orange thorn hedge. They thought they could embed the wire barbs as integral to the fencing wire—just like the Osage orange twig with its thorns. They succeeded in doing so and received patent no. 157,124. Haish founded one factory, while Uloid and Glidden started a competing firm. They took two strands of wire and twisted them with perpendicular pieces of short wire pieces attached at regular intervals. The new barbed wire was very effective in deterring the livestock and had all the advantages cited earlier for the smooth wire fences.

The barbed wire was well received by farmers around DeKalb County and beyond. By 1874, over 10,000 pounds of barbed wire fence were produced. This increased sixfold in 1875 to a total of 60,000 pounds. In 1877 the annual production reached over 12,000 pounds, and it exceeded 80,000 pounds a year in 1880. Barbed wire, though simple in technology, revolutionized America's fencing, facilitating its faster westward expansion. Barbed wire came to be used extensively ranging from farms to homes and from prison walls to the national borders. The success of barbed wire was based on its ability to meet an important unfulfilled need of the society.

Technology During the World War II Years

Let us move our time machine a few decades forward to World War II. The inception and development of the Manhattan Project to develop atom bomb technology during the World War II years is legendary. There is no doubt that the nuclear technology was developed at a fast pace because of the pressing wartime military requirements at that time. Adolf Hitler was bent on restoring the pride of German Fatherland lost during the previous war at any price. In the Far East, Japan seemed unstoppable. The American prestige was hurt badly by the sudden and shocking attack on Pearl Harbor. These were wartime circumstances. Everything seemed fair in the war. What America produced from its arsenal of smart educated immigrant scientists and engineers, who were mobilized nationwide in an unprecedented manner, was an original and unprecedented "bomb." Actually there were three bombs. One secretively used for trial, and the Fat Man and the Little Boy were bombs which the whole world noticed in Hiroshima and Nagasaki.

Technologically speaking, and not considering the ethical side to the nuclear technology, the Manhattan Project for the development of nuclear technology was a great scientific and technical success for its time. But one could counterargue that the Manhattan Project was an exceptional technological endeavor, not likely to be easily duplicated elsewhere.

What about more commonly used technology, like a car? Let us next consider the Jeep Project, which technologically perhaps played a very significant role during World War II, not far behind that of the Manhattan Project for atomic technology.

Case C. Jeep Technology: A Symbol of Ingenuity of U.S. Auto Technology. Before Pearl Harbor caught the attention of Americans and changed the course of America's involvement in World War II, the Jeep was described by General George C. Marshall, the highest ranking official in the U.S. Armed Forces, as America's main contribution to World War II. Jeep substituted the stage horse and became the reliable, versatile, and ever-present transportation workhorse of World War II. The Jeep was christened by its designation by the Army as a "general purpose," or G.P., vehicle. As a small, jumping, four-by-four quarter-ton vehicle, it was also called a jumping jellopy, an experimental railroad engine, and even a tractor. The Jeep was versatile, a very practical machine, and rightly called the military counterpart of Henry Ford's Model T car for the millions of American soldiers (see Note 10.3).

Outstanding Military Technovation. The Jeep was also the product of Yankee ingenuity and tinkering, just in time for the wartime requirements. The Jeep had a low silhouette and was very strong and hard to destroy in front-line usage. Its four-wheel drive helped it to get anywhere—from climbing up a 40-degree steep slope, to driving over soft mushy muddy ground. The Jeep could be driven from 3 to 60 miles per hour. In March 1941 photographers fell in love with a Jeep flying in mid-air at Fort Sam in Houston, Texas. In a precommissioning demonstration in September 1941, the Jeep helped position antitank guns in strategic and unforeseeable positions.

Jeep's In-Field Performance. Jeeps were first dropped into action in North Africa. It was the only vehicle that kept running in tough desert territory. A brigade of 18 Jeeps with 75 men drove in flying wedge formation, and it appeared from the desert onto an enemy airfield. They mowed down 25 aircrafts, damaged another 12 parked aircrafts, and then disappeared unscathed into the desert.

Jeep's Versatility. The Jeep was very versatile. With a makeshift stretcher attached on the outside, the Jeep became an ambulance moving injured soldiers from the front line. A radio fixed to a Jeep made it a mobile communication post. Jeeps towed planes. A 52-ton supply train in the Philippines with modified wheels was moved by a Jeep for a distance of 20 miles in an hour. According to a famous war time cartoon by Bill Mauldin, Jeeps also provided American soldiers hot water for shaving.

Because of this versatility, soldiers fell in love with their constant companion Jeeps. Many were very sad when their Jeep was shelled out of commission. Emotions ran high just like when a favorite horse or a pet is put away.

A Pain in the Back. The bone-jarring Jeep was a vehicle for the young. A drive of more than a few miles shook up most of the bones in a human body. A Jeep could not be locked because its ignition key was located in place of a lock and key. So to make it immobile, a Jeep either had to be chained or its distributor head had to be removed. A Jeep was not so easy to get in and out of, and the driver had to stand upright to get out. This made Jeep an entertaining vehicle for young male soldiers, especially as tight-skirted women maneuvered in and out of Jeeps.

Jeep's Core Engine. The Jeep was endowed with a great engine. It had the heart that kept on ticking no matter how tough the terrain. The engine for Jeep was designed by Delmar "Barney" Roos. Before designing Jeep's engine, Barney Roos had designed a 100-mile-per-hour Locomobile car in World War I. He later worked as chief engineer for Pierce-Arrow, Marmon. There he designed Studebaker and Marmon Straight 8.

In early 1938 Barney moved to Willys and was asked to improve a four-cylinder engine for a Willys light car. The light car could continuously run for barely 2–4 hours, and it developed 48 horsepower at 3400 rpm. It guzzled oil, shook off its starter violently, wore out bearings fast, and its cylinder head and water pump leaked. Barney modified this engine with a goal to (a) run it for 150 hours without failure and (b) race up to 65 horsepower at 4400 rpm. In 1939–1940 Ford had a competing engine that developed 60 horsepower at 3500 rpm. It took Barney 2 years and many improvements to match Ford. These included tighter machining tolerances, use of a tougher alloy and aluminum for cylinder pistons, and a lighter flywheel reduced from 57 pounds to 31 pounds.

Barney then heard of the Army's requirement for a vehicle light enough to be trucked around the world with low silhouette platform on 9-inch airplane wheels. It was expected to be concealed easily, though it had very low ground clearance. Barney shared his ideas with Major General Walter C. Short, who was later a commander at Pearl Harbor.

The Army's technical committee was testing compact cars made by American Bantam Company. They designed the specifications as a four-wheel-drive vehicle, with a maximum weight of 1600 pounds including 600 pounds of payload. The Army was also in a rush, and only 75 days were allowed for delivering the first 70 assembled vehicles. The Army requested bids from 135 manufacturers, but knew that only Bantam and Willys were able to meet the requirements.

Bantam responded with lightning speed with a bid and a plan within a week. They also promised to supply the first pilot model in 49 days. Willys requested more time, because Barney Roos's engine weighed more than the specified weight. The request for extension was refused, but Barney managed to demonstrate to the Army the performance of Willys. Willys promised higher-than-expected performance at the expense of 2425-pound-weight instead of the specified 1600 pounds. The Army was willing to compromise with a 2175-pound limit but no more, while another major competitor, Ford, was also invited for the bid.

Barney Roos still did not want to outsource its engine, as Bantam had done. Instead he systematically reduced whatever weight he could to shave off additional weight from his vehicle. Unnecessary body and chassis parts were eliminated. Long bolts were sawed off and made shorter. Heavy carbon steel frame was substituted with tougher and lighter metal alloys. Fenders were made thinner. Spray painting was invented to reduce 10 pounds of the paint required by the vehicle. The stripped-down Willys weighed 2164 pounds, well within the specified military limit. These Willys were tested side by side with Fords and Bantams.

The Infantry Board found the performance of Willys superior to that of Bantam and Ford in acceleration, maximum attainable speed, grade climbing, and cross-country driving. The main reasons were its greater horsepower and torque, engine, transmission, frame, and so on. The Infantry Board concluded that the standard Army vehicle should be based upon the Willys chassis, the Ford shift lever and hand brake arrangement, and the power and performance characteristics of the Willys. Soon Bantam withdrew, when it had produced only 2500 Jeeps. Willys' blueprints were shared with Edsel Ford, and between Willys and Ford more than 651,000 Jeeps were produced during World War II.

Jeep Mythology. Jeep's performance gave birth to many myths. World War II Willys bragged that "the sun never sets on the Jeep." During World War II, after the arrival of Japanese in South Eastern Asia, two American reporters drove 1500 miles from Burma to Imphal in India through thick jungle. They surprised the natives who had always pointed out that there were no roads to drive nearby. One of the reporters replied "Sh-h! Not so loud.

Our Jeep hasn't found out about roads yet, and we don't want to spoil it" (see Note 10.3). The Egyptian Army dug out Jeeps buried in the desert for 10 years and were able to commission them to use. In Germany, a 15-year-old Jeep engine from the North African Theater was still humming smoothly.

Postwar Potential for Jeep. Jeep's outstanding wartime performance raised the expectation that the Jeep would be popular even after World War II was won. The U.S. Department of Agriculture was interested in using Jeeps for farm labor. The Jeep was expected "to plow, harrow, or dig; plant corn or cotton," all on 1 gallon of gas to the acre. In November 1943, the first combat-retired Jeeps were put on the market for civilians.

By December 1960, about 100,000 Jeeps were produced each year in the United States. The United Nations Children's Fund (UNICEF) spread 2500 Jeeps throughout the world. The U.S. Navy took Jeeps to Antarctica, where the Jeep was the only four-wheel vehicle. In India, political candidates with more Jeeps than their rivals won their elections. In Mexico, bull-fighting rings used Jeeps instead of wasting horses to test the fighting strength of newly bred fighting bulls.

By December 1960 the Jeep also lost its charm with the U.S. Army, and orders trickled to none. The Army had been wooing Ford to develop an improved quarter-ton four-by-four vehicle. In June 1959, Ford complied and called its vehicle *Mutt*, for *Military Utility Tactical Truck*. For a few years the future fate of the Jeep seemed in doubt.

The Jeep's popularity was revived in the early 1990s. The original Jeep was the source of a new vehicle subspecies, called the sports utility vehicles (SUVs). The appeal of SUVs lay in their four-wheel driving capability and their versatility on the road as well as off the road. The biggest new ideas in the automobile industry in the 1990s were the $40,000 sport-utility vehicles and the two-seat sport cars (see Note 10.4). Chrysler Corporation retrofitted the World War II original Jeep into the 1997 Wrangler, which was marketed in 100 different countries. The Wrangler preserved the rugged features of the Jeep such as its hood hinges, round headlights mandated by customer clinics, and foldable windshield. The Wrangler used a truck-style ladder-type strengthened frame for four-wheel drive, and it used coil springs on all four wheels for rock-crawling and smoother ride. The company claimed that if the sun never set on the World War II Jeep, it rarely passed high noon for the Wrangler. According to the Jeep Platform General Manager Craig Winn, the new Wrangler design was "meshing the past with the future" (see Note 10.5).

The Postwar Role of Technology in the New Industrial State

The post–World War II victorious America saw a great potential for technology in the society. With rapid industrialization of America, economist John Kenneth Galbraith explored the imperatives of technology on the economics of society in his famous 1966 treatise entitled *The New Industrial State.* Galbraith defined technology as "the systematic application of scientific or other organized knowledge to practical tasks." He claimed that "nearly all the consequences of technology and much of the shape of the modern industry derive from the need to bring knowledge to bear on . . . the final need to combine the finished elements of the task into the finished product as a whole" (see Note 10.6).

Five Consequences of Technology on Society

Galbraith offered five consequences of technology on the society's economic activities in the 1960s.

1. Widening Gap Between Start and the End of a Task. Galbraith observed that with increasing sophistication in technology, "an increasing span of time separates the beginning from the completion of any task." For example, in the manufacturing of automobiles, there is a large time span between (a) the start of the process with the designers specifying steel and placing an order to the steel mills and (b) the completion of the finished automobiles. In the 1960s many of these tasks took place sequentially. Over the past three decades the cycle time has been shrunk significantly by doing some of the tasks simultaneously and running certain operations in parallel.

2. Increase in Capital Invested. With the increasing sophistication in technology, the capital investment required in production increased far more than that explained by the increase in production output alone. The increase in time span between start and finish of production meant more goods in process, more intermediate stages, and so on, which required more capital. Many myopic managers are unwilling to invest such capital from their quarterly bottom line to gain their future competitiveness.

3. Increase in Precision. Galbraith also noted that the tasks required for new technology must be performed more precisely than ever before. The American production system was revolutionized in the 19th century by the use of interchangeable parts. The rapid industrial growth was based on the precision machining of reproducible parts and components for guns, watches, and textile machines. The precision required for new automobiles, electric equipments, and more recently electronic appliances, semiconductors, and computers have increased significantly over the years.

4. Need for More Skilled Work Force. The increasing precision and sophistication of technology required more highly skilled workers. The work force was expected to be more specialized too. New specializations of engineers emerged. Chemical technology, for example, slowly specialized into many different fields such as polymer specialists for plastics processing, rubbers and fibers technologies, food technology, fertilizer technology, and so on. Similarly, medical technologies evolved into many specialized fields including workers specialized to deal with cancer and the AIDS epidemic.

5. More Organization and Planning. Finally, the increasing sophistication of technology also introduced many subsidiary functions to the production and operations functions directly related to the manufacture and assembly of products. These included secondary tasks such as planning, materials management, information processing, quality control, and so on. With larger organizations there was a greater need for controls and extensive planning. More recently the accountants and number crunchers have exerted significant influence on the management of technology and investment of resources for the same.

CHANGING LANDSCAPE OF MANAGEMENT OF TECHNOLOGY

Over the past three decades since Galbraith's ground-breaking book, new megashifts have redefined the economic landscape of the technology-driven industrialized societies.

The Tectonic Shifts in Economy

Professor Lester Thurow of the Massachusetts Institute of Technology, the world famous author of *Head to Head*, and *The Future of Capitalism,* looked at the five invisible economic forces changing the world's economic topology in a punctuated equilibrium. Thurow's five economic tectonic plates included (1) the end of communism, (2) a shifting demography, (3) globalizing economy, (4) lack of dominant economic, political, or military power, and (5) a technological shift to an era dominated by man-made brainpower industries (see Note 10.7).

With respect to management of technology, Thurow postulated that in the 19th and 20th centuries, most technologies and industries were anchored in the classical theory of comparative advantage and the endowment of natural resources and capital in different countries. Capital intensive technologies grew in the Western and Northern hemispheres where labor was in short supply. Because of natural endowments, steel technology grew in Pittsburgh, whereas the petroleum oil technology had a bigger base in Texas. The labor-intensive technologies, on the other hand, evolved in the highly populated poor countries. Compared to these technologies of the past, the newly emerging technologies are based on the knowledge, the brainpower, and creativity of people.

Thurow cautioned that after the late 1980s and in the 1990s some American industries have resorted to drastic downsizing, adding to a total of 300,000 to 550,000 a year in the name of restructuring their work force for changing technologies. However, their explanation does not fit the facts. Thurow pointed out that these technologies have evolved gradually over the past 30 years and did not emerge suddenly to warrant such drastic reductions in the work force.

Some of these shifts in the work force can be explained by the rapid growth of knowledge and human-intellect-based technologies. These new technologies do not rely on cheap access to the natural resources and materials. With the emergence of the global financial markets, many of these new technologies were not limited by their close access to financial capital either. All technology-based entrepreneurs borrowed their capital requirements from the same sources in the financial centers of New York, Tokyo, London, and Singapore. Thurow observed that knowledge and skills remained the main sources of comparative advantage, as illustrated by the Silicon Valley of California and Route 128 in Massachusetts. As a result of these tectonic shifts in the economic landscape of the world, the technologies need to be managed proactively and very differently than ever before.

The American Technology Renaissance

According to Michael Moynihan, author of *The Coming American Renaissance,* the United States continues to be the world's largest creator and exporter of technology (see Note 10.8). The United States runs the world's biggest and only technology trade surplus of $12.7 billion. The United States generally was way ahead of its global rivals in developing most of the new technologies, including biotechnology, software, artificial intelligence, virtual reality, and design of central processing units and semiconductor memory chips. However, Moynihan admitted that sometimes much of American capital was tied up in leveraged buyouts. Therefore in the 1980s, the United States had problems turning new technology into commercial products. Moynihan gave examples of the video cassette recorder (VCR) and liquid crystal displays which were invented by Americans who failed to commercialize them. Recently, many U.S. firms like Eastman Kodak, IBM, and Hewlett-Packard have re-strengthened their race for patents and achieved significant gains in this endeavor. The technological leadership for the United States was particularly noteworthy in information processing technology. In terms of the number of personal computers per capita, the United States had almost four times the number in Japan and two and a half times the number in Germany. In terms of the installed personal computers the United States has seven times as many computers in Japan and eight times as many in Germany.

If the U.S. businesses propose to retain their technological leadership in the future, they cannot afford to do business as usual. Some businesses have already geared up to meet the new challenges. Given below are three examples of how to manage technology in the years ahead.

1. The Future of Jet Aircrafts. What will the future jet aircrafts look like? Over the years the aircrafts have become so sophisticated that the flying of a sophisticated modern jet airline has been given to computers and digital networks rather than to the human pilots. Some observers in the aircraft industry jokingly stipulate that the future jet aircrafts will be flown by one pilot and a dog in the cockpit. The pilot's job will be to feed the dog, and the dog's role will be to keep the pilot from touching any switches (see Note 10.9). Even though the aircrafts produced over the years by Boeing, Mc-Donnell Douglas, Airbus, Lockheed, and others look similar on the outside, they have changed radically on the inside. Over the years the aircraft body parts have gradually used increasingly larger amounts of space-age lightweight and stronger metal alloys with high-performance polymer composite materials. The flying is controlled automatically by the fly-by-wire networks of on-board computers.

The first wide-bodied jet aircrafts (DC-8 and Boeing 707) and Lockheed Electra entered the airline markets in 1958. They all used similar technologies, replacing reciprocating pistons with rotating compressors and high-temperature gas turbine engines. The new turbine engines offered higher speed, particularly when they were used with supporting technological development of swept wings with higher stability. The 707 jet engine used 35-degree wings, and the DC-8 used 30-degree wings.

In his recent book, *Twenty-First-Century Jet*, British author Karl Sabbagh describes Boeing's 6-year project, from the conception to the commercialization of the Boeing 777 airliner full of high-tech bells and whistles (see Note 10.10). The senior executives at Boeing early on decided to use computer-aided designs instead of paper blueprints for each and every one of the four million components of the aircraft. Thousands of engineers, designers, and production workers were reorganized into 250 multidisciplinary design-to-build teams (DBT). These design-to-build teams were fully empowered with complete responsibility from conception to assembly of their part of the 21st-century jet. These teams often used chaotic group dynamics to make the four million parts move in close formation before successful completion of the future jetliner. Some aircraft industry observers believed that the success of the Boeing 777 aircraft was in not only the hardware used in the aircraft, but more so in the innovative management practices used to put the high-tech aircraft together. Customers participated actively.

2. Automotive Innovations. In 1996 the U.S. automobile industry celebrated its 100th anniversary and honored its pioneers such as Gottleib Daimler, Karl Benz, and Henry Ford. During this intervening period the automobile technology attracted a large number of technical talent all around the world. In 1912 many of the mechanical elements of the modern gasoline-based engine, overhead camshafts, multiple valves per cylinder, and so on, were well established. The electric starter in Cadillac made a major leap forward in automobile technology in 1912. By the 1950s other driver- and passenger-friendly refinements such as air conditioning, automatic transmissions, on-board audio, and power steering were common (see Note 10.5). This makes us wonder what would be the next set of innovations in store for the future of automobile technology in the new millennium.

In the late 1990s, automobiles used basically the same internal combustion engine technology as was used by automobiles at the turn of the century when they replaced the horse. This shift had a significant impact on the environment. Will the future power systems for automobiles change radically from today's internal combustion engine as it did from the horse? One source of change could come from reducing the environmental pollution due to auto emission. Another source of change could come from reducing mankind's reliance on burning the fossil fuels.

According to president Hiroshi Okuda of Toyota, who gave a talk on this subject of change in November 1997 in Akron, Ohio, the automobile technology was once again on the edge of four major watersheds of change (see Note 10.11). These watershed changes are listed below.

1. Decentralization of automobile production
2. A paradigm shift to system integration of product and process changes
3. Information processing technology transforming the intelligence of the automobile
4. Fast-changing globalized economy

President Okuda cautioned that large and successful companies can easily become the prisoners of their own successes. They then sometimes fail to perceive for many years the watersheds of change facing their technology. Such watersheds of change require bold decisions rather than safe and risk-free incremental improvements.

As one proactive response to such watershed of change, Toyota was changing the car as well as the business model that would adapt quickly to future changes. Since becoming the president of Toyota in August 1995, president Okuda stepped up Toyota's commitment to a greener and safer environment. The Toyota president emphasized that consumers were smart, and they recognized the threat that pollution and global warming presented to them and their children. He defined that Toyota's job was to present customers with a broad selection of clean, green technologies that met their needs. Toyota therefore increased its commitment and efforts for developing affordable products and technologies that safeguarded the global environment (see Note 10.12).

For example, Toyota developed a hybrid car for the December 1997 launch in Japan that used a gasoline engine as well as the electric motor generator. The Toyota hybrid car started on battery power, and when it picked up speed it switched to a combination of gasoline and electric power. A computerized power splitter smoothly directed the power to the engine from the two sources, without the driver noticing the change. The distribution of power was displayed on the dashboard. The battery was charged when its power was low.

According to Toyota, the hybrid car emitted 90% less pollution than ordinary cars in city driving conditions, gave twice the mileage, and thus produced 50% less output of carbon dioxide. The biggest remaining challenge was the cost. The company claimed that the hybrid car was cost-competitive with conventional cars in its class in Japan, where the gasoline costs were three times as much as those in North America. Toyota expected to gain economies of scale as production volume increased. This was expected to help Toyota's hybrid car become competitive around the world.

3. Strides in Semiconductor Technology. Dr. Andy Grove, the Hungarian born pathbreaking CEO of Intel, the maker of microprocessor chips, managed to keep his company as a technology leader for the past many decades. Intel supplied microprocessors for about 75% of all PCs sold. The company often doubled in size roughly every 2 years and achieved a gross profit margin of 58%, the enviable high margin after deducting cost of goods sold from the revenue. Net earnings in 1993 were $2.3 billion on $8.8 billion in sales (see Note 10.13).

How did Andy Grove manage to do all that? By proactive investment of megabucks in product development, along with capital investment to constantly update and keep its semiconductor production technology ahead of the rivals. He invested nearly a third of his company's revenues, $3.5 billion in 1994, on such technology-related efforts. The investment needed to develop the Pentium chip was $5 billion, about five times the capital needed for the 486 chip and an unbelievable 50 times the money spent on the 386 chip. Intel's manufacturing makeover included spending $10 billion from 1991 to 1995. Intel could afford to do so because of the economies of scale generated by the company with proactive expansion of capacity and new product development (see Note 10.14).

In a 1996 best-selling book, *Only the Paranoid Survive*, Dr. Andrew Grove cautioned that many of the old rules of conducting business have changed and do not work anymore. This should make senior executives a lot more nervous about the future of their companies and their technologies (see Note 10.15). He pointed out how the information-processing technologies could completely alter the newspaper and magazine publishing to on-line interactive technology, replace bank tellers with remote automated banking service, and overhaul the health care service industry. These were called 10× changes. Such changes are usually difficult to recognize until they strike a technology-driven enterprise.

Grove urged the middle and the senior managers to pay great attention to the stream of potential nightmares about demanding customers, frustrated employees, and innovative rivals. This, he admitted, required many years of careful practice to eliminate the fear of punishment for those who bring such bad news. Grove wrote that competing in the 1990s was like driving a car in a fog. It was easy as long as there was a car ahead of you, and its tail-lights guided you. But when you became the leading car, you were stuck with your confidence and ability to find your own way ahead. Managing technology and seeking technological leadership seems somewhat the same.

MANAGEMENT OF TECHNOLOGY: A MULTIDISCIPLINE INTERSECTION

The above discussion illustrates that in the years to come, management of technology is not likely to be entrusted to the scientists, engineers, and researchers alone—allowing them to freely do whatever comes to their fancy, with unlimited funding for their quests. Nor will technology be trusted to the snake-oil salesmen, ready to charge us an arm and a leg for a quickly put together concoction, with exaggerated claims of a cure-all and no risks whatsoever. With constraints on corporate resources, coupled with pressures from product liability and perennial budget deficits, technology is likely to be funded in companies very cautiously—perhaps in a far more focused manner than it ever was during the previous romantic eras of technology.

In the mid-1990s the myopic and quarterly bottom-line-driven managers and the get-rich-quick investors were becoming increasingly less inclined to invest enthusiastically in the development of new technology. Their lack of understanding about technology was reflected in their demanding risk-free guarantees from their investments in risk-laden technology. Yet they demanded high levels of returns which is justified only for the highly risky investments. It is hoped that in the future such a trend will be reversed as America's policy makers notice the shrinking gap in technological advancements from America's global competitors. American businesses often invest the least proportion of their funds on capital expenses for technological growth, research, and development compared to the other largest economies of the world in the G7 group of the most industrialized economies.

In the years to come, the management of technology will be best understood as an intersection where the physical and natural sciences meet the humanities and economics, and where the business requirements intersect with the societal expectations.

THE "SYN-CREMENTAL" MANAGEMENT OF TECHNOLOGY

The above discussion can be summarized as the four essential attributes of effective management of technology. These are listed below.

1. Technology involves synthesis.
2. Technology is incremental.

3. Technology is developmental.
4. Technology gets sentimental.

1. Management of Technology Involves Synthesis. Technology is the synthesis of scientific seeds and the market needs put together by investors' greed. It requires the synthesis of scientific ideas in diverse related fields. Management of technology is also a cross-disciplinary and cross-functional activity. It requires integration of scientific research with applied development, designing and engineering, marketing, and social responsibility. All technological growth has taken place as a result of such synthesis across diverse fields. Aspirin was available naturally in plants before it was synthesized by humans in a pharmaceutical company. The precursor to the printing press was available in wine-making. The cotton gin found its origin in the sugar cane mill in pre-Christian Indian civilization. The barbed wire was born out of natural hedges. Technology is the successful development and commercialization of such connections across different disciplines.

2. Management of Technology Is Incremental. Technology sometimes gives an impression of galloping with leaps and bounds. These are called radical innovations. They are radical applications of incremental bits of knowledge known to other fields of knowledge. However, if we look at these innovations carefully, technology is growing at "invisible" incremental paces. Sometimes the steps that are combined together surprise us. The standardization of auto parts in Henry Ford's Model T came from the standard fitting parts of Winchester guns and rifles a few years earlier.

3. Management of Technology Is Developmental. By this we mean that technology needs a proactive developing hand. Technological successes do not happen by accident. Even when useful accidents take place, they need a prepared mind that can support the embryonic technology to commercial success. Charles Goodyear learned from the accidental vulcanization of rubber. That support needs much money and commitment. Technology can cost a lot to get from a concept to the commercial stage. Joe Wilson almost risked his fledgling Haloid Company to pay for the inventor Chester Carlson's yet to be proven photocopying technology. Persistent support, such as that of George Eastman for photographic film and that of 3M for Post-its, is crucial for the development and success of an emerging technology.

4. Technology Is Sentimental. For a technology to be successful, it has to relate to the hearts of the people who are going to use it. User-friendliness of technology is not something critical for computer technology alone. The microwave oven in its current form did not have appeal for potential users for a long time. The potential consumers could not see or believe the heating caused by the microwave technology—that is, not until they saw the potential and visible popping of popcorn by microwave heating. Managers of technology must not overlook the potential of using the emotional side of using a new technology. We have seen Ben & Jerry use that emotional appeal for environmental friendliness to sell ice cream, and we have seen Toyota trying to do the same for selling their new hybrid electric cars with high fuel efficiency and less polluting emissions.

Technology Involves Symbiosis

Technology is sometimes misunderstood because it involves a mosaic of many ideologies. These ideologies come with different intellectual pursuits. In a nutshell, technology is a symbiotic synthesis of the discovery of science coupled with the creativity of art. Technology is then encapsulated in the socio-economic shell of a business organization. This sociotechnical system draws profits from the markets and economy. The visible hand that wields the power of technology in the society and decides how a technology will be put to a commercial use in the market owns the potential returns from the technology. Sometimes this hand belongs to the originator of the technology, but more often it does not.

Science persists in understanding and discovering the underlying laws of nature. Why did the apple fall on Newton's head? Because of the gravitational pull of the earth. Why do different organic molecules of carbon and hydrogen exhibit different properties? Because of their functional groups and electronic exchange. But, the scientists stop at that excitement of discovery. The technologists take those conceptual ideas and match them with the society's unfulfilled wants and needs. They commercialize concepts into useful products, services, and cash.

Technologists take pride in producing. Sometime technologists produce without understanding. Mankind has benefited from many accidental technological creations. The Carothers' polymers which resulted in the synthetic fibers were discovered accidentally. But they were produced very deliberately. Bakelite and nylon were produced as a result of deliberately pursuing such accidents. The Manhattan Project produced the nuclear bomb technology without fully understanding the full implications of its devastating effect on humanity and the environment of planet earth. But it produced a decisive victory in World War II for America, and it led to America's global supremacy for more than half a century.

In many ways a technologist is closer to a sculptor than to a scientist. Leonardo da Vinci, that fountainhead of Renaissance in Europe, was not surprisingly both. The technologists get their excitement from creating and producing something new, rather than from deciphering the details of its genealogical antecedents. This is not free of economic and social risks. But such practical use has been the source of wealth of nations, particularly the wealthiest nation of the world, the United States of America. That excitement to create, manage, and commercialize new technologies should be carefully nurtured and sustained if any nation or organization wants to continue to generate more wealth in the future. This search is ultimately tied to man's freedom and liberty to pursue prosperity and happiness.

A BRIEF REVIEW OF LESSONS LEARNED IN THIS CHAPTER

In this final chapter we had the goal to explore the links between technology, society, and people. We started by considering the perceptions people have about technology. We noted that most people have only a partial view of the elephant of technology. Besides the tangible attributes of technology, we noted certain invisible attributes of technology.

The influence and impact of contextual environment on the success of technology in society was recounted in the examples of the cotton gin, barbed wire, and the Jeep. During the post–World War II industrial state, we reviewed Galbraith's observations about changes in the management of technology. Lester Thurow showed us the tectonic changes in the emerging global economy. These megashifts have redefined the way technology will be managed in the future. We saw a glimpse in this future by how Boeing managed its design-to-build teams that put together the 21st century 777 jet, how Toyota planned to introduce environmentally friendly green cars, and how Intel's Dr. Andy Grove looked for the $10\times$ changes and developed technological leadership by nurturing the complaining customers and the paranoid employees. Finally, we proposed the "syn-cremental" management of technology.

SELF-REVIEW QUESTIONS

1. What are the different perceptions people have about technology?

2. Discuss the product life cycle of the Jeep.

3. How did contextual environment influence the success of different technologies at different times? Compare any two technologies.

4. If America's auto industry could create a Jeep for wartime, why couldn't they offer a fuel-efficient compact car after the two Oil Shocks in the 1970s? Discuss with examples.

5. In an intensely competitive and dynamic market, how will you sustain your technological leadership?

NOTES, REFERENCES, AND ADDITIONAL READINGS

10.1.   Drucker, Peter F. 1970. *Technology, Management and Society,* New York: Harper and Row.

10.2. Gehani, R. Ray. 1994. A note for classroom discussion. Adapted from Basalla, George Basalla. 1988. *The Evolution of Technology.* Cambridge: Cambridge University Press.

10.3. Gehani, R. Ray. 1995. A teaching note prepared for classroom discussions. Adapted from Hartwell, Dickson. 1960, The mighty Jeep. *American Heritage,* December: 38–41. Includes reports from Chicago Daily News and other sources.

10.4. Alex, Taylor, III. 1996. It's the slow lane for automakers. *Fortune,* April 1:59–64.

10.5. Swan, Tony. 1996. Automotive innovation, *Forbes,* November 18. A special advertising section.

10.6. Galbraith, John Kenneth. 1966. *The New Industrial State,* Boston: Houghton Mifflin.

10.7. Thurow, Lester C. 1996. *The Future of Capitalism,* New York: William Morrow.

10.8. Moynihan, Michael. 1996. *The Coming American Renaissance,* New York: Simon & Schuster.

10.9. Labich, Kenneth. 1996. Boeing's new dream machine. *Fortune,* February 19:104.

10.10. Sabbagh, Karl. 1996. *Twenty-First-Century Jet: The Making and Marketing of the Boeing 777,* New York: Scribner.

10.11 . Okuda, Hiroshi, 1997. Management of change—a perspective for the automotive industry. An invited talk presented to the Global Business Institute at the University of Akron on November 12, 1997.

10.12. See Toyota Annual Report. 1997. Toyota's plans for hybrid car system were advertised in a special November 1997 issue of *Times.* Many of president Hiroshi Okuda's comments are taken from this special issue.

10.13. Kirkpatrick, David. 1994. Intel goes for broke. *Fortune,* May 16:62–68.

10.14. Clark, David. 1995. A big bet made Intel what it is today; now, it wagers again. *Wall Street Journal,* June 7:A1, A6.

10.15. Grove, Andrew S. 1996. *Only the Paranoid Survive.* New York, Doubleday/Currency. Also see Labich, Kenneth. 1996. *Fortune,* October 14: 216–218.